T0319387

FREQUENCY-DOMAIN ANALYSIS AND DESIGN OF DISTRIBUTED CONTROL SYSTEMS

FREQUENCY-DOMAIN ANALYSIS AND DESIGN OF DISTRIBUTED CONTROL SYSTEMS

Yu-Ping Tian
Southeast University, China

This edition first published 2012

Registered office
John Wiley & Sons Singapore Pte. Ltd., 1 Fusionopolis Walk, #07-01 Solaris South Tower, Singapore 138628

For details of our global editorial offices, for customer services and for information about how to apply for permission to reuse the copyright material in this book please see our website at www.wiley.com.

A catalogue record for this book is available from the Library of Congress.

ISBN: 9780470828205

Set in 10/12 pt Times by Thomson Digital, Noida, India
Printed and bound in Singapore by Markono Print Media Pte Ltd

I dedicate this book to
Ningning and Ouya for their love, trust and support.

Contents

Preface

This book is devoted to the study of distributed control systems, a very promising research area given the current rapid development of technologies of micro-sensors, micro-motors, sensor networks and communication networks. The main feature of the book is to adopt a frequency-domain approach to cope with analysis and design problems in distributed control systems. Frequency-domain methods utilize frequency response properties of subsystems (agents) and communication channels to characterize the stability and other performance criteria of the entire system. By comparison with time-domain methods, it is usually assumed to be more convenient to use frequency-domain methods for analyzing the robustness of the system against noise and dynamic perturbations. It will be shown in this book that frequency-domain methods are also powerful in scalability analysis of distributed control system.

The book consists of three parts. The first part includes Chapters 1, 2 and 3, and describes common features and mathematical models of distributed control systems; it also introduces basic tools that are useful for further analysis and design of the underlying systems, such as graph theory, frequency-domain stability criteria, scalability analysis based on differential geometric properties of frequency response plots, etc. The second part includes Chapters 4 and 5, and the third part includes Chapters 6 and 7. The second part focuses on the distributed congestion control of communication networks while the third part studies the consensus control of multi-agent systems. The second and third parts can be considered as the application of the general theory introduced in the first part. However, the theory is also developed in the last two parts for two particular types of distributed control systems. Moreover, many interesting and beneficial results are obtained when general theory is applied to concrete systems.

It is quite common to choose the congestion control and the consensus control for the study of distributed control systems. There are at least two reasons. Firstly, both types of control schemes emphasize cooperation in a group of agents although the starting points of the cooperation are somehow different. For the congestion control, the cooperation is realized through distributed real-time optimization of some common performance indexes, i.e., allocation of limited resources in networks. For the consensus control, the basis of the cooperation is to manipulate states (or output) of agents to reach some common value. The two cooperation approaches are widely used in many other practical distributed control systems, such as multi-robot control systems, localization and information fusion sensor networks, etc. Secondly, many theoretical problems such as stability, scalability and delay effect encountered in these two kinds of systems can be treated by using a unified frequency-domain method. Although distributed real-time optimization problems in congestion control usually lead to nonlinear models, the local dynamics around an equilibrium of a congestion control system can be described as a linear distributed feedback control system with a bipartite interconnection topology graph, which essentially has the same structure as a multi-agent system controlled by a consensus protocol. Nevertheless, consensus problems are different from congestion control problems in many aspects. For example, a multi-agent system driven by a consensus protocol has a continuum of equilibriums instead of an isolated equilibrium. This is perhaps why the Laplacian matrix of the interconnection topology graph plays a key role in consensus problems.

The time-delayed feedback control is introduced in this book as one of the basic design methods for distributed control systems. We study not only the effect of time delays, in particular various communication delays, on stability and scalability, but also the potential of time-delayed feedback control for enhancing the stability. This looks somewhat illogical. But in fact it is possible for time delay to have a dual character in feedback control systems, as in many other things in nature.

Some notions and results appearing in this book are new. They are mostly related to bipartite systems, symmetric systems, symmetric communication delays, commensurate self-delays, semi-stability test in the frequency domain, and high-order consensus. However, most of the material is based on the research results of the author with his collaborators and students, notably Jiandong Zhu, Hong-Yong Yang, Cheng-Lin Liu and Ya Zhang, to whom the author would like to express his sincere thanks. The author would also like to acknowledge the continuous support of the National Natural Science Foundation of China in the research topics of this book and in other related areas.

Parts of the book were taught on short courses listed below.

1. "Time-delayed Feedback Control", Pre-conference workshop lecture on Chinese Control Conference, Harbin, 2006.
2. "Internet Congestion Control", Summer school lecture at the School of Mechanics and Engineering Sciences, Peking University, Beijing, 2008.
3. "Coordination Control of Multi-agent Systems", Summer school lecture at School of Mechanics and Engineering Sciences, Peking University, Beijing, 2010.

The author would also like to thank Prof. Zhong-Ping Jiang of New York University, Prof. Lin Huang of Peking University and Prof. Guangrong Chen of City University of Hong Kong for their invitations to present these lectures, which impelled the author to collect related materials scattered in his own and others' research publications.

Yu-Ping Tian
Southeast University, Nanjing

Glossary of Symbols

$:=$	"defined as"
$\triangleq$	"denoted as"
$A \setminus B$	Exclude set B from set A, where $B \subset A$
$A \otimes B$	Kronecker product of matrices A and B
$A \oplus B$	Direct sum of set A and B
$(z)_x^+$	A piece-wise function of $x \geq 0$, which takes value z if $x > 0$, or $\max(z, 0)$ if $x = 0$
$\mathbb{R}, \mathbb{R}^n, \mathbb{R}^{n\times n}$	The set of real numbers, n component real vectors, and n by n real matrices
$\mathbb{R}^+$	The set of nonnegative real numbers
$\mathbb{C}, \mathbb{C}^n, \mathbb{C}^{n\times n}$	The set of complex numbers, n component complex vectors, and n by n complex matrices
C^n	Space of functions which are n-times differentiable
$\mathcal{C}(\omega)$	Curvature of a parametric curve at parameter ω
$\mathrm{Re}(s)$	The real part of complex number s
$\mathrm{Im}(s)$	The imaginary part of complex number s
$\mathbb{C}^-$	The open left half of the complex plane (briefly as LHP), $\{s \in \mathbb{C} : \mathrm{Re}(s) < 0\}$
$\bar{\mathbb{C}}^-$	The closed left half of the complex plane (briefly as closed LHP), $\{s \in \mathbb{C} : \mathrm{Re}(s) \leq 0\}$
$\mathbb{C}^+$	The closed right half of the complex plane (briefly as RHP), $\{s \in \mathbb{C} : \mathrm{Re}(s) \geq 0\}$
$\check{\mathbb{C}}^+$	The open right half of the complex plane (briefly as open RHP), $\{s \in \mathbb{C} : \mathrm{Re}(s) > 0\}$
$\mathbb{D}$	The interior of the unit disc (briefly as IUD), $\{s \in \mathbb{C} : \|s\| < 1\}$

$\bar{\mathbb{D}}$	The closure of $\mathbb{D}$ (briefly as closed IUD), $\{s \in \mathbb{C} : \|s\| \leq 1\}$
$\mathbb{D}^{o}$	The closed outer part of the unit disc (briefly as OUD), $\{s \in \mathbb{C} : \|s\| \geq 1\}$
$\breve{\mathbb{D}}^{o}$	The open outer part of the unit disc (briefly as open OUD), $\{s \in \mathbb{C} : \|s\| > 1\}$
$\mathbb{H}_{\infty}^{n\times n}$	The set of n by n transfer function matrices $G(s)$, with $\\|G\\|_{\infty} < \infty$
$\mathbb{RH}_{\infty}^{n\times n}$	The set of n by n rational transfer function matrices $G(s)$, with $\\|G\\|_{\infty} < \infty$
$\mathrm{Co}(r_1, \cdots, r_n)$	Convex hull of elements $r_1 \cdots, r_n$
$\overline{n_1, n_2}$	The set of integers from n_1 to n_2 satisfying $n_1 \leq n_2$, $\{n_1, \cdots, n_2\}$
$\overline{S}$	The index set of source nodes in a communication network, $\{1, 2, \cdots, S\}$
$\overline{L}$	The index set of link nodes in a communication network, $\{1, 2, \cdots, L\}$
A^{T}	Transpose of matrix (vector) A
A^{*}	Conjugate transpose of matrix (vector) A
j	Unit of imaginary numbers, $\sqrt{-1}$
$\mathbf{1}_n$	$n \times 1$ vector $[1, \cdots, 1]^{\mathrm{T}}$
e_1	$n \times 1$ vector $[1, 0, \cdots, 0]^{\mathrm{T}}$
$\mathrm{diag}\{t_i, i \in \overline{1, n}\}$	Diagonal matrix with diagonal entries t_i, $i \in \overline{1, n}$
$\lambda_i(A)$	i-th eigenvalue of matrix $A \in \mathbb{C}^{n\times n}$.
$\sigma(A)$	Spectrum of matrix $A \in \mathbb{C}^{n\times n}$, $\{\lambda_1(A), \cdots, \lambda_n(A)\}$
$\rho(A)$	Spectral radius of matrix $A \in \mathbb{C}^{n\times n}$, $\max_{i\in\overline{1,n}} \|\lambda_i(A)\|$
$\bar{\sigma}(A)$	Maximum singular value of matrix $A \in \mathbb{C}^{n\times m}$, $\left(\max_{i\in\overline{1,n}}\{\lambda_i(AA^{*})\}\right)^{\frac{1}{2}}$
$\mathrm{rank}(A)$	Rank of matrix A
$\mathrm{span}(A)$	The space spanned by all the columns of matrix A
$[A](:, i)$	The i-th column of matrix A
$[A](:, \overline{i_1, i_2})$	The matrix formed by i_1-th to i_2-th columns of matrix A, where $i_1 \leq i_2$

1

Introduction

Ruling a large country is like cooking a small fish.

—*Lao Dan (580–500* BC*), Tao Te Ching*

This chapter is concerned with some common characteristics of distributed control systems, such as distributive interconnections, local control rules and scalability. Basic notions and results of algebraic graph theory are introduced as a theoretic foundation for modeling interconnections of subsystems (agents) in distributed control systems. Coordination control systems and end-to-end congestion control systems are also introduced as two typical kinds of distributed control systems. The main topics of this book are also highlighted in the introduction of these two kinds of systems.

1.1 Network-Based Distributed Control System

From the "flyball" speed governor of Watt's steam engine to the control of communication systems including the telephone system, cell phones and the Internet, the control mechanism in industrial, social and many other real-world systems has experienced a development process from centralized control to distributed control. In the past decade, this process has been significantly speeded up thanks to rapid advances of communication techniques and their application to control systems.

Recently, workshops, seminars, short courses and even regular courses on distributed control systems emerge in the control society just like bamboo shoots after spring rain. But the following questions, which are often raised in seminars or lectures by post-graduate students, may also puzzle the reader of this book. What kind of control systems can be called distributed control system? Are there any differences between a distributed control system and a so-called large-scale control system or a decentralized control systems? What's the advantage of distributed control over centralized control? etc. We will not attempt to answer all the questions; rather, we will try to grasp some common features of distributed control systems.

Cooperation among agents

Cooperation is perhaps the soul of a distributed control system. Since there is no centralized control unit, the only way through which all the agents in the system can work in harmony

Frequency-Domain Analysis and Design of Distributed Control Systems, First Edition. Yu-Ping Tian.

is to conduct some kind of cooperation. Ignoring cooperation in biology might be one of a few shortcomings by which one could rebuke the great theory of Darwin (1859). Indeed, cooperation has been observed in many biological colonies, such as in birds, fish, bacilli and so on. Through cooperation they increase the probability of discovering food, get rid of prey and other dangers. Breder (1954) proposed a simple mathematic model to characterize the attracting and excluding actions in schools of fish. To describe the flocks and aggregations of schools more effectively, Reynolds (1987) proposed three simple rules: (1) collision avoidance, (2) agreement on velocity and (3) approaching center, which actually implies approaching any neighbor. He successfully simulated the motion of fish by using these rules. These investigations stimulate researchers to develop artificial systems that make decisions and take actions in distributive but cooperative manners. Nowadays, distributed control mechanisms have been widely used in engineering systems, such as aircraft traffic control (Tomlin, Pappas and Sastry 1998), multi-robot control (Rekleitis, Dudek and Milios 2000), coverage control of sensor networks (Qi, Iyengar and Chakrabarty 2001), formation control of unmanned aerial vehicles (UAVs) (Giulietti, Pollini and Innocenti 2000) etc.

Spatially distributed interconnection

In a classic feedback control system, the controller gets the output signal of the plant, compares it with some reference signal and makes decisions on how to act. Such a system also serves as one of basic units (subsystems) in a distributed control system. However, to cooperate with other units in the entire system, it should be equipped with a sensor/communication modular besides the classic decision/action modular (controller), as shown in Figure 1.1. Such a subsystem is sometimes called an *agent*. In a distributed control system, therefore, each subsystem gets not only the information of the output of itself but also the information of some other subsystems via sensor or/and communication networks (Figure 1.2). A distributed control system interconnected with multiple agents is also called a *multi-agent system (MAS)*.

For many distributed control systems such as the Internet, power grids, traffic control systems etc., subsystems (agents) are often distributed across multiple computational units in an immense space and connected through long-distance packet-based communications. In these system packet loss and delay are unavoidable, and hence, computational and communication

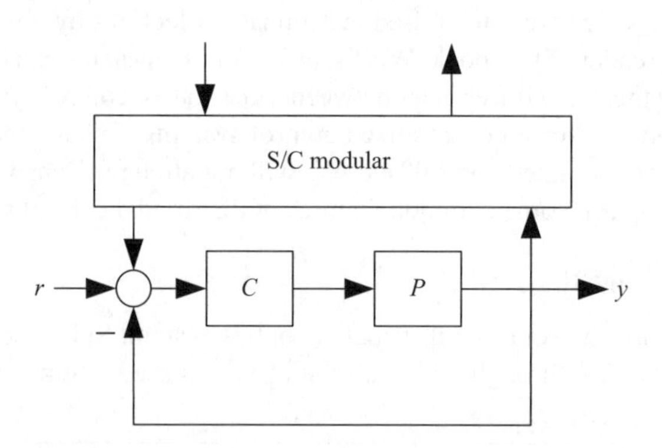

Figure 1.1 An agent in distributed control system.

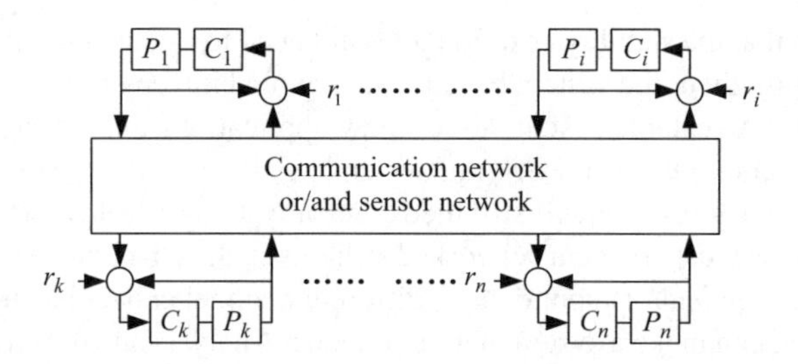

Figure 1.2 A network-based distributed control system.

constraints cannot be ignored. New formalism to ensure stability, performance and robustness is required in the analysis and design of distributed control systems (Murray *et al.* 2003).

Local control rule

A distributed control law should be subject to the *local control rule*, which implies that in most cases there is no kind of centralized supervision or control unit in the system, and each agent makes decisions based only on the information received by its own sensor or from its neighboring agents through communication. The action rules proposed by Reynolds (1987) are typically local control rules. The local control rule is the most important feature of distributed control systems and makes the distributed control distinguished from the decentralized control of large-scale systems, which were extensively studied in the 1970s and 1980s (see, e.g., Sastry (1999) for a survey on large-scale systems). The decentralized control mechanism allows each agent to communicate perfectly with any other agent in the system, ignoring the computational and communication constraints. Moreover, the plant in decentralized control systems is usually a single tightly connected unit and not a rather loosely interconnected group of complete systems, which is the typical case in distributed control systems. Therefore, cooperation among agents in distributed control systems is much more important than in large-scale systems, and hence the coupling between a pair of subsystems can not be dealt with as a disturbance as in some decentralized control designs.

A remarkable advantage of the local control rule over the non-local control rules is its higher fault-tolerance capability. This is extremely important for many large-scale engineering systems which must operate continuously even when some individual units fail. Therefore, building a very reliable distributed control system with fault-tolerance ability from unreliable parts is a very promising research direction (Murray *et al.* 2003), although the topic is beyond the scope of this book.

Scalability

Scalability is perhaps another very important reason why distributed control systems prefer the local control rule. In this book the scalability of a distributed control system implies that the controller of the system and its maintenance utilize only local information around each agent and rarely depends on the scale of the system. In other words, by scalability we mean that not only the control law but also most important properties of the system rely on the local information. For example, in checking the stability of the entire distributed control system the scalability requires that the stability criterion does not need to use information about the global interconnection topology of the system because such global information is usually unavailable to individual agents. A scalable system allows new applications to be designed and deployed without requiring changes to the underlying system.

The scalability of heterogeneous distributed control systems has drawn much more attention of researchers because most practical networked systems such as the Internet are heterogeneous. Diverse communication and/or input delays, different channel capacities, non-identical agent dynamics and so on, can make a system heterogeneous. Analysis and design of heterogeneous systems are much more difficult in comparison with homogeneous systems. One reason for that is the analysis and/or design of a large but homogenous systems can usually be treated as a task for a small system through diagonalization of the original system. But in most cases such a simplification method is not applicable to heterogeneous systems.

1.2 Graph Theory and Interconnection Topology

1.2.1 Basic Definitions

Graph theory plays a crucial role in describing the interconnection topology of distributed control systems. In this section we only present basic definitions about graph theory. For systematic study of graph theory we refer the reader to, for example, Biggs (1994); Bollobás (1998); Diestel (1997); Godsil and Royle (2001).

Digraph

A *directed graph* (in short, *digraph*) of order n is a pair $G = (V, E)$, where V is a set with n elements called *vertices* (or *nodes*) and E is a set of ordered pairs of vertices called *edges*. In other words, $E \subseteq V \times V$. We denote by $V(G)$ and $E(G)$ the vertex set and edge set of graph G, respectively, and denote by $\overline{1, n} := \{1, \cdots, n\}$ a finite set for vertex index. For two vertices $v_i, v_j \in V$, i.e., $i, j \in \overline{1, n}$, $i \neq j$, the ordered pair (v_i, v_j) represents an edge from v_i to v_j and is also simply denoted by e_{ij}.

A digraph $G(V', E')$ is said to be a *subgraph* of a digraph (V, E) if $V' \subset V$ and $E' \subset E$. In particular, a digraph (V', E') is said to be a *spanning subgraph* of a digraph (V, E) if it is a subgraph and $V' = V$. The digraph (V', E') is the subgraph of (V, E) induced by $V' \subset V$ if E' contains all edges in E between two vertices in V'.

Path and connectivity

A *path* in a digraph is an ordered sequence of vertices such that any ordered pair of vertices appearing consecutively in the sequence is an edge of the digraph. A path is simple if no vertices appear more than once in it, except possibly for the initial and final vertices. The *length* of a path is defined as the number of consecutive edges in the path. For a simple path, the path length is less than the number of vertices contained in the path by unity.

A vertex v_i in digraph G is said to be *reachable* from another vertex v_j if there is a path in G from v_i to v_j. A vertex in the digraph is said to be *globally reachable* if it is reachable from every other vertex in the digraph. A digraph is *strongly connected* if every vertex is globally reachable. In Figure 1.3, v_1, v_2, v_4, v_5 are globally reachable vertices. But the digraph is not strongly connected because v_3 is unreachable from the other vertices.

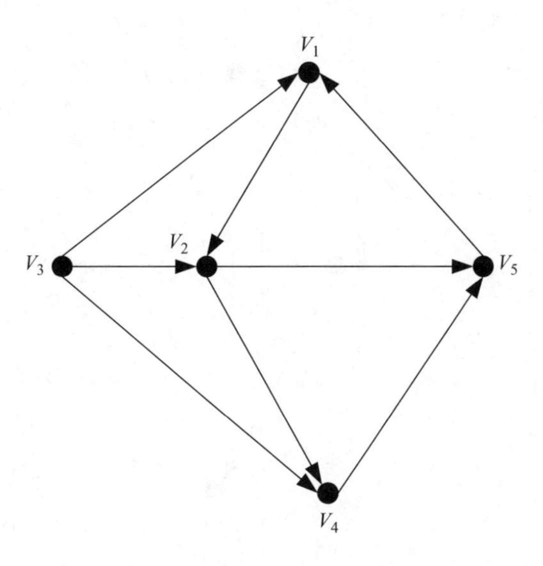

Figure 1.3 A digraph.

Cycle and tree

A *cycle* is a simple path that starts and ends at the same vertex. A cycle containing only one vertex is called a *self-cycle* (or *self-loop*). The *length* of a cycle is defined as the number of edges contained in the cycle. A cycle is *odd* (*even*) if it's length is odd (even). If a vertex in a cycle is globally reachable, then any other vertex in the cycle is also globally reachable. In Figure 1.3, the path (v_1, v_2, v_5, v_1) is a cycle. The path $\{v_2, v_4, v_5, v_2\}$ and the path $\{v_1, v_2, v_4, v_5, v_1\}$ are also cycles. This digraph has no self-cycle. A digraph with self-cycle is shown in Figure 1.4.

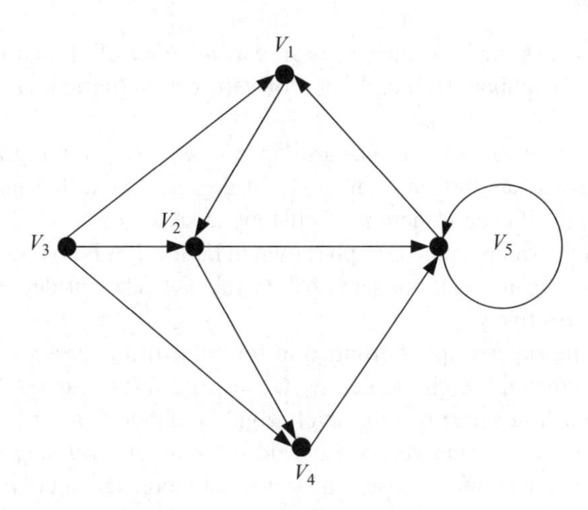

Figure 1.4 A digraph with a self-cycle.

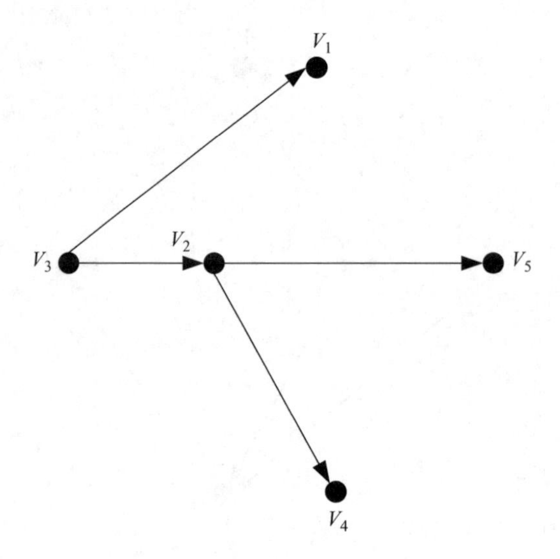

Figure 1.5 A directed tree.

A digraph is *acyclic* if it contains no cycles. An acyclic digraph is called a *directed tree* if it satisfies the following property: there exists a vertex, called the *root*, such that any other vertex of the digraph can be reached by one and only one path starting at the root. A *directed spanning tree* of a digraph is a spanning subgraph that is a directed tree.

The digraph shown in Figure 1.5 is a directed tree. Obviously, it is a directed spanning tree of both the digraph in Figure 1.3 and the digraph in Figure 1.4.

Neighbor and degree

If (v_i, v_j) is an edge of digraph G, then v_j is an *out-neighbor* of v_i, and v_i is an *in-neighbor* of v_j. The set of out-neighbors (in-neighbors) of vertex v_i in digraph $G(V, E)$ is denoted by $N_i^{\text{out}}(G)$ $(N_i^{\text{in}}(G))$.

The *out-degree* (*in-degree*) of v_i is the cardinality of N_i^{out} (N_i^{in}). A digraph is *topologically balanced* if each vertex has the same in- and out-degrees. Note that neither N_i^{out} nor N_i^{in} contains the vertex i itself even if there is a self-loop at vertex i.

Let us consider the example of a digraph shown in Figure 1.3. For vertex 1, its out-neighbor set is $N_1^{\text{out}} = \{v_2\}$, and its in-neighbor set is $N_1^{\text{in}} = \{v_3, v_5\}$. The out-degree and the in-degree of v_1 are 1 and 2, respectively.

In this book, if the superscript is dropped in formulae or the prefix is omitted in texts, it will be referred to as the out-neighbor, i.e., $N_i(G) = \{v_j \in V : (v_i, v_j) \in E\}$.

Sometimes we may use the term multi-level neighbor. If there is an m-length path in digraph G from vertex i to vertex j, then vertex j is said to be an *m-level neighbor* of vertex i. Of course, a neighbor in conventional sense is a 1-level neighbor. The set of m-level neighbors of agent i in digraph G is denoted by $N_i^m(G)$. For example, in the digraph shown in Figure 1.3, $N_1^2(G) = \{v_4, v_5\}$.

Digraph and information flow

In a distributed control system, each agent can be considered as a vertex in a digraph, and the information flow between two agents can be regarded as a directed path between the vertices in the digraph. Thus, the interconnection topology of a distributed control system can be described by a digraph. However, differing from the classic signal-flow graph (Chen 1984), in this book and in many other references on the distributed control system, the direction of an edge in the digraph does not mean the direction of an information flow. Let us consider the digraph shown in Figure 1.3 for an instance. Denote by $x_i \in \mathbb{R}$, $i \in \overline{1,5}$, the state of agent i associated with vertex i. The existence of edge e_{ij} implies that agent i gets the state information x_j from agent j. For example, agent 1 gets information from agent 2.

1.2.2 Graph Operations

We shall construct new graphs from old ones by graph operations.

For two digraphs $G_1 = (V_1, E_1)$ and $G_2 = (V_2, E_2)$, the *intersection* and *union* of G_1 and G_2 are defined by

$$G_1 \cap G_2 = (V_1 \cap V_2, E_1 \cap E_2),$$
$$G_1 \cup G_2 = (V_1 \cup V_2, E_1 \cup E_2).$$

For a digraph $G = (V, E)$, the *reverse digraph* of G is a pair $\text{rev}(G) = (V, \text{rev}(E))$, where $\text{rev}(E)$ consists of all edges in E with reversed directions.

If $W \subset V(G)$, then $G - W = G[V \backslash W]$ is the subgraph of G obtained by deleting the vertices in W and all edges incident with them. Obviously, $G - W$ is the subgraph of G induced by $V \backslash W$. Similarly, if $E' \subset E(G)$, then $G - E' = (V(G), E(G) \backslash E')$. If W (or E') contains a single vertex w (or a single edge xy, respectively), the notion is simplified to $G - w$ (or $G - xy$, respectively). Similarly, if x and y are non-adjacent vertices of G, then $G + xy$ is obtained from G by joining x to y.

Undirected graph

An *undirected graph* (in short, *graph*) G consists of a vertex set V and a set E of unordered pairs of vertices. If each edge of the graph G is given a particular orientation, then we get an oriented graph of G, denoted by $G^{\rightarrow}$, which is a digraph. Denote by $G^{\leftarrow}$ the reverse of $G^{\rightarrow}$. Then, $G = G^{\rightarrow} \cup G^{\leftarrow}$.

For an undirected graph G, the in-neighbor set of any vertex is always equal to the out-neighbor set of the same vertex. Therefore, in the undirected case we simply use the terminations neighbor, neighbor set and degree.

For an undirected graph, if it contains a globally reachable node, then any other vertex is also globally reachable. In that case we simply say that the undirected graph is *connected*.

For an undirected graph, it is said to be a *tree* if it is connected and acyclic.

Theorem 1.1 *$G(V, E)$ is a tree if and only if G is connected and $|E| = |V| - 1$. Alternatively, $G(V, E)$ is a tree if and only if G is acyclic and $|E| = |V| - 1$.*

Theorem 1.2 *A graph is connected if and only if it contains a spanning tree.*

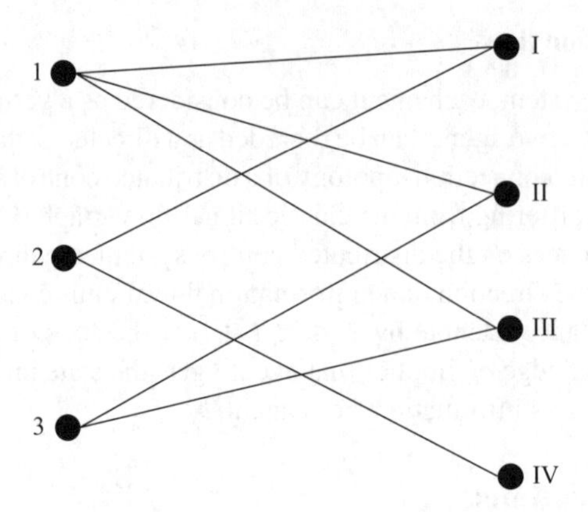

Figure 1.6 A bipartite graph.

Bipartite graph

A graph G is a *bipartite graph* with vertex classes V_1 and V_2 if $V(G)$ is a direct sum of V_1 and V_2, i.e., $V = V_1 \oplus V_2$, which implies that $V = V_1 \cup V_2$ and $V_1 \cap V_2 = \emptyset$, and every edge joins a vertex of V_1 to a vertex of V_2. It is also said that G has bipartition (V_1, V_2). Figure 1.6 shows an example of a bipartite graph. For this graph, the vertex set V is a direct sum of $V_1 = \{v_1, v_2, v_3\}$ and $V_2 = \{v_{\mathrm{I}}, v_{\mathrm{II}}, v_{\mathrm{III}}, v_{\mathrm{IV}}\}$. Each vertex in V_1 has neighbors only in V_2, and vice versa.

Theorem 1.3 *A graph is bipartite if and only if it does not contain an odd cycle.*

For the bipartite graph G with vertex classes V_1 and V_2, define

$$E'_{12} = \bigcup_{v_l \in V_2} \{(v_i, v_j) : v_i, v_j \in N_l\}, \tag{1.1}$$

$$E'_{21} = \bigcup_{v_r \in V_1} \{(v_i, v_j) : v_i, v_j \in N_r\}. \tag{1.2}$$

Note that in both (1.1) and (1.2), v_i and v_j can be the same vertex. In this case, (v_i, v_i) is a self-loop. Let

$$G_1 = G - V_2 + E'_{12} \tag{1.3}$$

and

$$G_2 = G - V_1 + E'_{21}. \tag{1.4}$$

G_1 is a new graph obtained from G by deleting all the vertices in V_2 and adding the new edges in E'_{12}. By definition (1.1), E'_{12} is the set of new edges, each of which adds a self-loop to one vertex or joins two vertices in V_1 that are neighbors of a vertex in V_2. Similarly, G_2 is a new graph obtained from G by deleting all the vertices in V_1 and adding the new edges in E'_{21}.

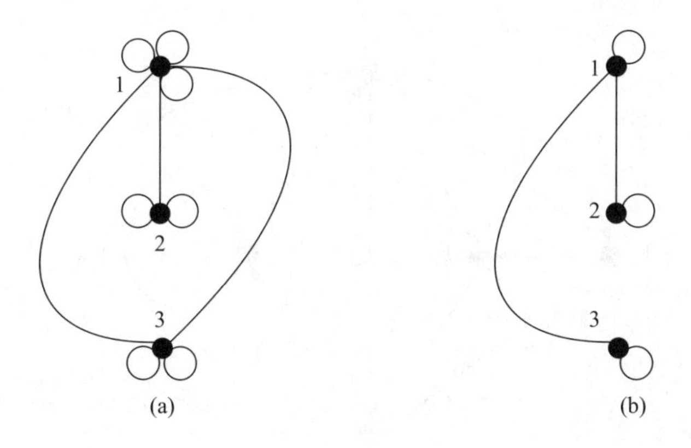

Figure 1.7 Graph operation and equivalent graph for V_1.

E'_{21} is the set of new edges, each of which adds a self-loop to one vertex or joins two vertices in V_2 that are neighbors of a vertex in V_1.

G_1 (G_2) is said to be the *equivalent graph* for V_1 (V_2) deduced from G. Obviously, neither G_1 nor G_2 is a subgraph of G. But the interconnection topology between any two vertices in one vertex class remains unchanged for G and the equivalent graph G_1 (or G_2).

The graph operation defined by (1.3) for the bipartite graph shown in Figure 1.6 is illustrated by Figure 1.7 (a). Since only one edge is defined between any pair of vertices in graph, the equivalent graph for V_1 is given by Figure 1.7 (b). Similarly, the graph operation defined by (1.4) for the same bipartite graph is illustrated by Figure 1.8 (a). The equivalent graph for V_2 is given by Figure 1.8 (b).

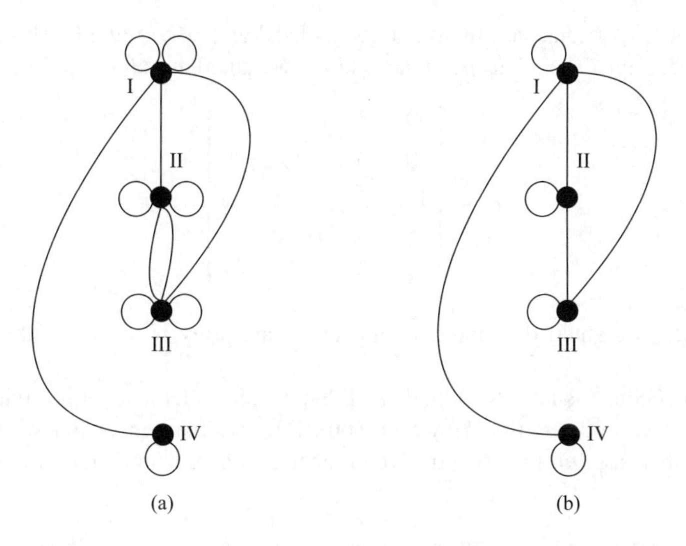

Figure 1.8 Graph operation and equivalent graph for V_2.

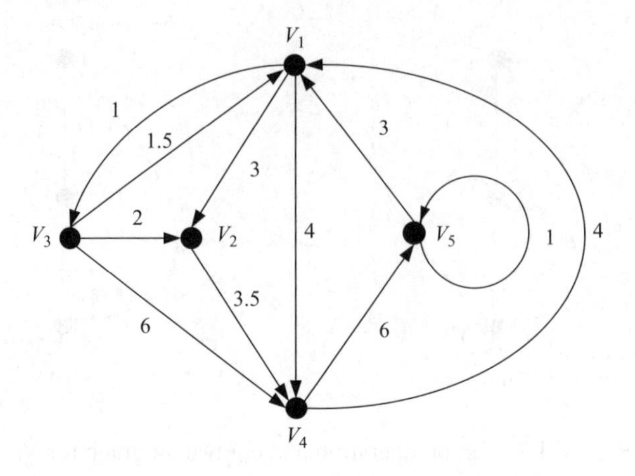

Figure 1.9 A weighted digraph.

1.2.3 Algebraic Graph Theory

Algebraic graph theory studies matrices associated with digraphs.

Weighted digraph and adjacency matrix

A *weighted digraph* of digraph $G(V, E)$ is a triplet $G = (V, E, A)$, where $A = [a_{ij}] \in \mathbb{R}^{n\times n}$ is an *adjacency matrix* satisfying

$$a_{ij} = \begin{cases} > 0, & \text{if } (v_i, v_j) \in E, \\ 0, & \text{otherwise.} \end{cases}$$

We denote by $A(G)$ the adjacency matrix of a weighted digraph G. Figure 1.9 shows an example of a weighted digraph. The adjacency matrix of the weighted digraph is

$$A = \begin{bmatrix} 0 & 3 & 1 & 4 & 0 \\ 0 & 0 & 0 & 3.5 & 0 \\ 1.5 & 2 & 0 & 6 & 0 \\ 1 & 0 & 0 & 0 & 6 \\ 3 & 0 & 0 & 0 & 1 \end{bmatrix}.$$

Obviously, a weighted digraph is undirected if and only if $a_{ij} = a_{ji}$, that is, $A(G)$ is symmetric. [1]

If digraph G contains no self-loop, then all the diagonal elements of its adjacency matrix A are zero, i.e., $a_{ii} = 0$, $i \in \overline{1, n}$. However, sometimes we may encounter with graphs with self-loops. In this case, a_{ii} may be positive numbers. To avoid confusion in terminology, by

[1] If $A \in \mathbb{C}^{n\times n}$, under the symmetry, we mean the conjugate symmetry, i.e., $A^* = A$, where A^* is the conjugate transpose of A.

adjacency matrix we still mean the matrix A with zero entries in the diagonal. Actually, $A(G)$ is associated with the digraph that is obtained by cutting off all the self-loops in G. We denote by $\bar{A}(G)$ the matrix that characterizes the existence of all edges including self-loops in G, i.e.,

$$\bar{A} = \mathrm{diag}\{a_{ii},\ i \in \overline{1, n}\} + A, \tag{1.5}$$

and refer to $\bar{A}$ as the generalized adjacency matrix of G.

For a weighted digraph $G(V, E, \bar{A})$ of order n, we use $\bar{A}_{0,1} = \{d_{ij}\}^{n\times n}$ to denote its *un-weighted adjacency matrix*, where

$$d_{ij} = \begin{cases} 1, & \text{if } a_{ij} > 0, \\ 0, & \text{if } a_{ij} = 0. \end{cases}$$

Let $G(V, E, \bar{A})$ be a weighted digraph of $G(V, E)$ of order n. Given a matrix $\bar{A}' = \{a'_{ij}\}^{n\times n}$ with $a'_{ij} \in \mathbb{R}^+$, if

$$(\bar{A}')_{0,1} = \bar{A}_{0,1}, \tag{1.6}$$

then, by the definition of adjacency matrix, $G(V, E, \bar{A}')$ is also a weighted digraph of $G(V, E)$.

Naturally, a digraph $G = (V, E)$ can be considered as a weighted digraph with $\{0, 1\}$-weights, i.e.,

$$a_{ij} = \begin{cases} 1, & \text{if } (v_i, v_j) \in E, \\ 0, & \text{otherwise}. \end{cases}$$

The weighted out-degree matrix of digraph G, denoted by $D^{\mathrm{out}}(G)$, is the diagonal matrix with the weighted out-degree of each node along its diagonal, i.e.,

$$D^{\mathrm{out}}(G) = \mathrm{diag}\{d^{\mathrm{out}}(v_i), i \in \overline{1, n}\}. \tag{1.7}$$

The weighted in-degree matrix of digraph G, denoted by $D^{\mathrm{in}}(G)$, is the diagonal matrix with the weighted in-degree of each node along its diagonal, i.e.,

$$D^{\mathrm{in}}(G) = \mathrm{diag}\{d^{\mathrm{in}}(v_i), i \in \overline{1, n}\}. \tag{1.8}$$

While the adjacency matrix characterizes the location of edges among vertices in a digraph, the following result shows that powers of the adjacency matrix characterize the relationship between directed paths and vertices in the digraph.

Lemma 1.4 *Let $G(V, E, \bar{A})$ be a weighted digraph of order n possibly with self-loops. $\bar{A}_{0,1}$ is its un-weighted adjacency matrix. Then, for all $i, j, k \in \overline{1, n}$,*

(1) the (i, j) entry of $\bar{A}_{0,1}^k$ equals the number of directed paths of length k (including paths with self-loops) from vertex i to vertex j; and

(2) the (i, j) entry of $\bar{A}^k$ is positive if and only if there exists a directed path of length k (including paths with self-loops) from vertex i to vertex j.

Gain, measurement matrix and state-transfer matrix

Let $x_i(k) \in \mathbb{R}$, $i \in \overline{1, n}$, be the state of agent i associated with vertex i, where k represents the time. By our stipulation of the physical meaning of the edge direction in a digraph, the existence of e_{ij} implies that agent i gets the state information x_j from agent j. Then, the weight a_{ij} associated with e_{ij} can be regarded as the gain for the information flow. So, for a system with $G(V, E, A)$ as its interconnection topology digraph, the existence of e_{ij} implies that the agent i gets the amplified state information $a_{ij}x_j(k)$ from agent j. Let the measurement $y_i(k)$ of agent i be equal to the sum of all the amplified states at the present time received by agent i, i.e.,

$$y_i(k) = a_{ii}x_i(k) + \sum_{j \in N_i} a_{ij}x_j(k). \tag{1.9}$$

Such a measurement will be also referred to as aggregated measurement. Sometimes, each agent can get only some relative measurement which can be expressed as

$$y_i(k) = a_{ii}x_i(k) - \sum_{j \in N_i} a_{ij}x_j(k). \tag{1.10}$$

Denote by $x(k) = [x_1(k), \cdots, x_n(k)]^{\mathrm{T}}$ the state vector and $y(k) = [y_1(k), \cdots, y_n(k)]^{\mathrm{T}}$ the measurement vector. Then, the matrix form of the aggregated measurement (1.9) is

$$y(k) = \bar{A}x(k). \tag{1.11}$$

Equation (1.11) provides an interpretation of the generalized adjacency matrix $\bar{A}$ from a viewpoint of system theory: it can be considered as a measurement matrix.

Suppose the state formation of each agent is updated by the following local law:

$$x(k+1) = Ky(k), \tag{1.12}$$

where $K = \mathrm{diag}\{\kappa_i \in \mathbb{R}, i \in \overline{1, n}\}$. Then, we have

$$x(k+1) = K\bar{A}x(k). \tag{1.13}$$

So, with the updating law (1.12), $K\bar{A}$ is a state-transfer matrix of a closed-loop system, as shown in Figure 1.10 .

Weighted bipartite graph

Let $G(V, E, A)$ be a weighted bipartite graph with vertex classes $V_1 = \{v_1, \cdots, v_{n_1}\}$ and $V_2 = \{v_{n_1+1}, \cdots, v_n\}$. Then, by the definition of bipartite graph, the adjacency matrix

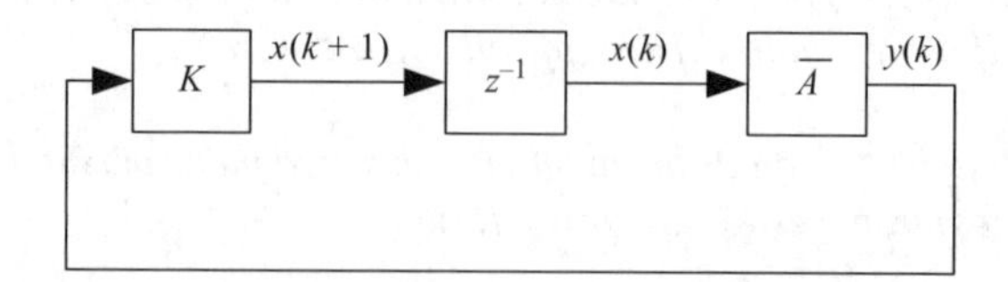

Figure 1.10 A closed-loop system.

$A(G)$ can be partitioned as

$$A = \begin{bmatrix} 0 & A_{12} \\ A_{12}^{\mathrm{T}} & 0 \end{bmatrix}, \tag{1.14}$$

where $A_{12} \in \mathbb{R}^{n_1 \times (n-n_1)}$. Hence,

$$A^2 = \begin{bmatrix} A_{12}A_{12}^{\mathrm{T}} & 0 \\ 0 & A_{12}^{\mathrm{T}}A_{12} \end{bmatrix}. \tag{1.15}$$

By Lemma 1.4 we know that any vertex in V_1 has two-level neighbors in V_1, and any vertex in V_2 has two-level neighbors in V_2. Furthermore, the existence of paths of length 2 (including self-loops) in G between any pair of vertices in V_1 (or V_2) is characterized by the sign of entries of $A_{12}A_{12}^{\mathrm{T}}$ (or $A_{12}^{\mathrm{T}}A_{12}$). So, by the definition of equivalent graph, $A_{12}A_{12}^{\mathrm{T}}$ (or $A_{12}^{\mathrm{T}}A_{12}$) can be defined as a weighted adjacency matrix associated with the equivalent graph of the bipartite graph G for V_1 (or V_2). Summarizing the discussion we have the following proposition.

Proposition 1.5 *Let $G_1(V_1, E_1)$ and $G_2(V_2, E_2)$ be the equivalent graphs of the bipartite graph $G(A)$ for V_1 and V_2, respectively, where A is given by (1.14). Then,*

$$\bar{A}_1 = A_{12}A_{12}^{\mathrm{T}}, \tag{1.16}$$

$$\bar{A}_2 = A_{12}^{\mathrm{T}}A_{12}, \tag{1.17}$$

are weighted adjacency matrices associated with G_1 and G_2, respectively.

By the definition of un-weighted adjacency matrix,

$$A_{0,1} = \begin{bmatrix} 0 & (A_{12})_{0,1} \\ (A_{12}^{\mathrm{T}})_{0,1} & 0 \end{bmatrix}. \tag{1.18}$$

So, we have

$$A_{0,1}^2 = \begin{bmatrix} (A_{12})_{0,1}(A_{12}^{\mathrm{T}})_{0,1} & 0 \\ 0 & (A_{12}^{\mathrm{T}})_{0,1}(A_{12})_{0,1} \end{bmatrix}.$$

So, by Lemma 1.4, the number of paths of length 2 (including self-loops) in G from vertex i to vertex j ($i, j \in \overline{1, n_1}$) in V_1 equals the value of the (i, j) entry of $(A_{12})_{0,1}(A_{12}^{\mathrm{T}})_{0,1}$; and the number of paths of length 2 (including self-loops) in G from vertex i to vertex j ($i, j \in \overline{n - n_1, n}$) in V_2 equals the value of the (i, j) entry of $(A_{12}^{\mathrm{T}})_{0,1}(A_{12})_{0,1}$). However, the signs of entries of $(A_{12})_{0,1}(A_{12}^{\mathrm{T}})_{0,1}$ (or $(A_{12}^{\mathrm{T}})_{0,1}(A_{12})_{0,1}$) also characterize the existence of paths of length 2 (including self-loops) in G between any pair of vertices in V_1 (or V_2) (see Exercise 1.7). So, by the definition of equivalent graph, the following proposition is also true.

Proposition 1.6 *Let $G_1(V_1, E_1)$ and $G_2(V_2, E_2)$ be the equivalent graphs of the bipartite graph $G(A)$ for V_1 and V_2, respectively, where A is given by (1.14). Denote*

$$\bar{A}_I = (A_{12})_{0,1}(A_{12}^{\mathrm{T}})_{0,1}, \tag{1.19}$$

$$\bar{A}_{II} = (A_{12}^{\mathrm{T}})_{0,1}(A_{12})_{0,1}. \tag{1.20}$$

Then, $G_1(V_1, E_1, \bar{A}_I)$ and $G_2(V_2, E_2, \bar{A}_{II})$ are weighted equivalent graphs of the bipartite graph G for V_1 and V_2, respectively.

Remark. $\bar{A}_I$ (or $\bar{A}_{II}$) is not the un-weighted matrix of $\bar{A}_1$ (or $\bar{A}_2$), i.e.,

$$(A_{12})_{0,1}(A_{12}^{\mathrm{T}})_{0,1} \neq (A_{12}A_{12}^{\mathrm{T}})_{0,1},$$
$$(A_{12}^{\mathrm{T}})_{0,1}(A_{12})_{0,1} \neq (A_{12}^{\mathrm{T}}A_{12})_{0,1}.$$

Exercise 1.7 *Show*

$$[(A_{12})_{0,1}(A_{12}^{\mathrm{T}})_{0,1}]_{0,1} = (A_{12}A_{12}^{\mathrm{T}})_{0,1}, \tag{1.21}$$
$$[(A_{12}^{\mathrm{T}})_{0,1}(A_{12})_{0,1}]_{0,1} = (A_{12}^{\mathrm{T}}A_{12})_{0,1}. \tag{1.22}$$

Let $x_1(k) \in \mathbb{R}^{n_1}$, $x_2(k) \in \mathbb{R}^{n-n_1}$ be the state vectors associated with V_1 and V_2, respectively. Denote $x(k) = [x_1^{\mathrm{T}}(k), x_2^{\mathrm{T}}(k)]^{\mathrm{T}}$. Then, under the measurement rule (1.11) and the state updated law (1.12), we have

$$\begin{bmatrix} x_1(k+1) \\ x_2(k+1) \end{bmatrix} = K \begin{bmatrix} 0 & A_{12} \\ A_{12}^{\mathrm{T}} & 0 \end{bmatrix} \begin{bmatrix} x_1(k) \\ x_2(k) \end{bmatrix}.$$

Partition K as $K = \mathrm{diag}\{K_1, K_2\}$, where $K_1 = \mathrm{diag}\{\kappa_i, i \in \overline{1, n_1}\}$, $K_2 = \mathrm{diag}\{\kappa_i, i \in \overline{(n_1+1), n}\}$. Hence,

$$\begin{cases} x_1(k+1) = K_1 A_{12} x_2(k), \\ x_2(k+1) = K_2 A_{12}^{\mathrm{T}} x_1(k). \end{cases} \tag{1.23}$$

which yields

$$x_1(k+1) = K_1 A_{12} x_2(k) = K_1 A_{12} K_2 A_{12}^{\mathrm{T}} x_1(k-1). \tag{1.24}$$

If $K_2 > 0$, we can define a new weight matrix for the bipartite graph G as

$$A' = \begin{bmatrix} 0 & A_{12}K_2^{\frac{1}{2}} \\ K_2^{\frac{1}{2}}A_{12}^{\mathrm{T}} & \end{bmatrix}.$$

Then, by Proposition 1.5, $A_{12}K_2A_{12}^{\mathrm{T}}$ is a weighted adjacency matrix of the equivalent graph for vertex class V_1 in G. Equation (1.24) links $x(k+1)$ to $x(k-1)$ instead of $x(k)$ because any vertex in V_1 has no (one-level) neighbors in V_1. So, $K_1A_{12}K_2A_{12}^{\mathrm{T}}$ is also the state-transfer matrix for the states associated with the vertices in V_1.

Laplacian matrix

Now, we can generalize the notions of in- and out-degree to weighted digraphs. In a weighted digraph $G(V, E, A)$ with $V = \{v_1, \cdots, v_n\}$, the weighted out-degree and the weighted in-degree of vertex v_i are defined by, respectively,

$$d^{\text{out}}(v_i) = \sum_{j=1}^{n} a_{ij}, \tag{1.25}$$

and

$$d^{\text{in}}(v_i) = \sum_{j=1}^{n} a_{ji}. \tag{1.26}$$

The weighted digraph G is *weight-balanced* if $d^{\text{out}}(v_i) = d^{\text{in}}(v_i)$ for all $v_i \in V$.

Definition 1.8 *The Laplacian matrix of the weighted digraph $G(V, E, A)$ is defined as*

$$L(G) = D^{\text{out}}(G) - A(G). \tag{1.27}$$

If we let

$$a_{ii} = \sum_{j=1}^{n} a_{ij},$$

then, the relative measurement (1.10) can be rewritten as

$$y_i(k) = \sum_{j \in N_i} l_{ij} x_j(k),$$

or in the matrix form

$$y(k) = Lx(k).$$

So, the Laplacian matrix can be interpreted as a kind of *relative-measurement matrix*.

The following theorems give some useful properties of Laplacian matrices.

Theorem 1.9 *Let G be a weighted digraph of order n. The following statements hold:*

(1) $L(G)\mathbf{1}_n = \mathbf{0}_n$, that is, 0 is an eigenvalue of $L(G)$ with eigenvector $\mathbf{1}_n$;

(2) for any eigenvalue $\lambda_i, i = 1 \cdots, n$, of $L(G)$, either $\lambda_i = 0$ or $\text{Re}\lambda_i > 0$ (thus, if G is undirected, then $L(G)$ is positively semi-definite);

(3) digraph G contains a globally reachable vertex (or say, undirected graph G is connected) if and only if rank$(L(G))=n-1$.

Theorem 1.10 *Digraph G is weight-balanced if and only if one of the following statements holds:*

(1) $\mathbf{1}_n^{\text{T}} L(G) = \mathbf{0}_n^{\text{T}}$;

(2) $L(G) + L(G)^{\text{T}}$ is positively semi-definite.

1.3 Distributed Control Systems

In this section we introduce two typical kinds of distributed control systems, which will be studied more carefully in the remaining chapters of this book.

1.3.1 End-to-End Congestion Control Systems

One effective way for agents in a distributed control system to cooperate with each other is to optimize some performance index which concerns all of the agents in the system, such as allocation of limited resources in a communication network (Kelly 1997; Kelly, Maulloo and Tan 1998), coverage of as much as possible an area by a sensing network (Cortés *et al.* 2004), etc. Congestion control of the Internet is a typical system designed by using a distributed real-time optimization approach, which will be presented in Chapter 4. Here we just introduce some basic models with some necessary notions of end-to-end congestion control systems.

Basic model

Let us consider an end-to-end congestion control system for a network with a set of L link nodes shared by a set of S source nodes. Denote by $\overline{L} = \{1, \cdots, L\}$ the set of all link nodes and by $\overline{S} = \{1, \cdots, S\}$ the set of source nodes. Link node $l \in \overline{L}$ is used by a set of sources denoted by $S_l \subset \overline{S}$, and source node $i \in \overline{S}$ sends packets to a set of links denoted by $L_i \subset \overline{L}$. The sets L_i define an $L \times S$ routing matrix $R = \{R_{li}\}$, where

$$R_{li} = \begin{cases} 1, & \text{if } l \in L_i \\ 0, & \text{otherwise.} \end{cases}$$

From a viewpoint of graph theory, each node in $\overline{S}$ has neighbors only in $\overline{L}$ and each node in $\overline{L}$ has neighbors only in $\overline{S}$. So, the topology of such a system can be described by a bipartite graph $G(V, E)$, where $V = \overline{L} \oplus \overline{S}$. A weighted adjacency matrix of the bipartite graph can be given by

$$A = \begin{bmatrix} 0 & K_1 R K_2 \\ K_2 R^{\mathrm{T}} K_1 & 0 \end{bmatrix}, \tag{1.28}$$

where $K_1 \in \mathbb{R}^{L \times L}$, $K_2 \in \mathbb{R}^{S \times S}$ are two positively definite diagonal matrices. Source $i \in \overline{S}$ transmits packets to all the links used by source i at a rate $x_i(t)$; in other words, all the links in L_i receive packets at rate $x_i(t)$.

The primal algorithm for the congestion-avoidance rate control is given by (Kelly, Maulloo and Tan 1998)

$$\dot{x}_i(t) = \kappa_i(w_i - x_i(t) \sum_{l \in L_i} q_l(t)), \quad i \in \overline{S} \tag{1.29}$$

$$q_l(t) = p_l(\sum_{r \in S_l} x_r(t)), \tag{1.30}$$

where the control gain k_i is a positive constant, and w_i is some desired value of the rate of marked packets received back at sources i. The first equation describes the time evolution of the transmission rate $x_i(t)$ of source i. The second equation describes the generation of congestion signal $q_l(t)$ at link l, by means of a congestion indication function $p_l(y)$, which is assumed to be monotonically increasing, nonnegative, and not identically zero (a candidate of such a function can be taken as $p(y) = 1 - \mathrm{e}^y(1-y)$, which is positive and strictly increasing for all $y > 0$).

Suppose that matrix R is of full row rank. Then, one can get the unique equilibrium point of system (1.29)–(1.30) as $x^\star = [x_1^\star, \cdots, x_S^\star]^{\mathrm{T}}$, where $x_i^\star = w_i / \sum_{l\in L_i} p_l(\sum_{r\in S_l} x_r^\star)$. Define

$$y_i(t) = (x_i(t) - x_i^\star)/\sqrt{k_i x_i^\star}, \quad i \in \overline{S}.$$

The dynamic property of the system around the equilibrium is determined by the following linearized model:

$$\dot{y}_i(t) = -\kappa_i w_i x_i^{\star -1} y_i(t) - \sqrt{\kappa_i x_i^\star} \sum_{l\in\overline{L}} R_{li} p_l' \sum_{r\in\overline{S}} R_{lr} \sqrt{\kappa_r x_r^\star} y_r(t) \tag{1.31}$$

where p_l' is the derivative of p_l evaluated at $y = \sum_{i\in S_l} x_i^\star$, i.e. $p_l' = p_l'(\sum_{i\in S_l} x_i^\star)$. Since $p_l(y)$ is assumed to be monotonically increasing, $p_l' > 0$, $\forall l \in \overline{L}$. By defining

$$K_1 = \mathrm{diag}\{\sqrt{p_l'}, \quad i \in \overline{L}\},$$
$$K_2 = \mathrm{diag}\{\sqrt{\kappa_r x_r^\star}, \quad r \in \overline{S}\},$$

and

$$\bar{A}_S = K_2 R^T K_1^2 R K_2,$$

it is easy to verify that the interconnection between y_l and y_r can be described by a weighted equivalent graph $G_S(\overline{S}, E_S, \bar{A}_S)$ for vertex class $\overline{S}$ in the bipartite graph $G(V, E, A)$, where $V = \overline{L} \oplus \overline{S}$ and A is given by (1.28). Denote

$$y(t) = [y_1(t), \cdots, y_N(t)]^{\mathrm{T}},$$
$$K = \mathrm{diag}\{\kappa_i w_i x_i^{\star -1}, \quad i \in \overline{S}\}.$$

Then, the matrix form of (1.31) is given by

$$\dot{y}(t) = -Ky(t) - \bar{A}_S y(t). \tag{1.32}$$

Model with delays

When propagation delays are considered, the congestion control algorithms can be modeled by the following delayed differential equations (Johari and Tan 2001)

$$\dot{x}_i(t) = \kappa_i(w_i - x_i(t - D_i) \sum_{l\in L_i} q_l(t - d_{li}^{\leftarrow})), \quad i \in \overline{S}, \tag{1.33}$$

$$q_l(t) = p_l(\sum_{r\in S_l} x_r(t - d_{lr}^{\rightarrow})), \tag{1.34}$$

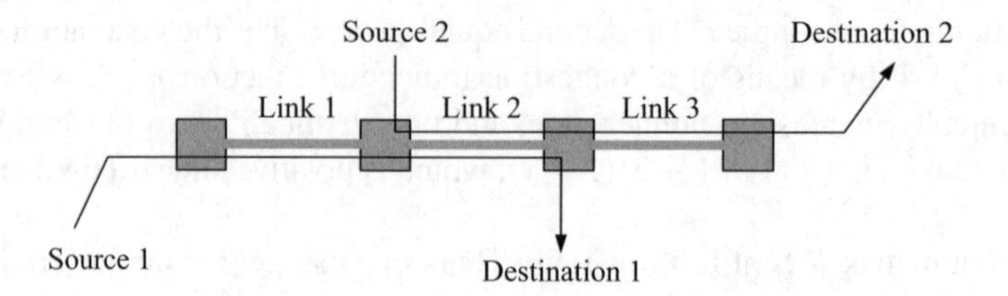

Figure 1.11 Communication network.

where $d_{li}^{\rightarrow}$ denotes the forward delay of delivering packets from source i to link l, $d_{li}^{\leftarrow}$ denotes the backward delay of sending the feedback signal from link l to source i, and D_i is the round-trip delay associated with source i. In the current Internet, it is assumed that propagation delay is much more significant than queueing delay. Therefore, ignoring the queueing delay, the following relationship holds:

$$d_{li}^{\rightarrow} + d_{li}^{\leftarrow} = D_i, \quad \forall l \in L_i \subseteq \overline{L}.$$

The linearized model around the equilibrium is given by

$$\dot{y}_i(t) = -\kappa_i w_i x_i^{\star -1} y_i(t - D_i) - \sqrt{\kappa_i x_i^\star} \sum_{l \in \overline{L}} R_{li} p_l' \sum_{r \in \overline{S}} R_{lr} \sqrt{\kappa_r x_r^\star} y_r(t - d_{lr}^{\rightarrow} - d_{li}^{\leftarrow}). \quad (1.35)$$

Delay effects on stability

Let us use the congestion control system to check the influence of delays. Consider a simple network shown in Figure 1.11. There are two sources transmitting packages and three links through which packages are sent to destinations. Source 1 uses link 1 and link 2, source 2 uses link 2 and link 3. The interconnection topology of sources and links can be represented by a bipartite graph shown in Figure 1.12. According to (1.33) the congestion control algorithm is given by

$$\dot{x}_i(t) = \kappa_i(w_i - x_i(t - D_i) \sum_{l \in L_i} q_l(t - d_{li}^{\leftarrow})), \quad i = 1, 2,$$

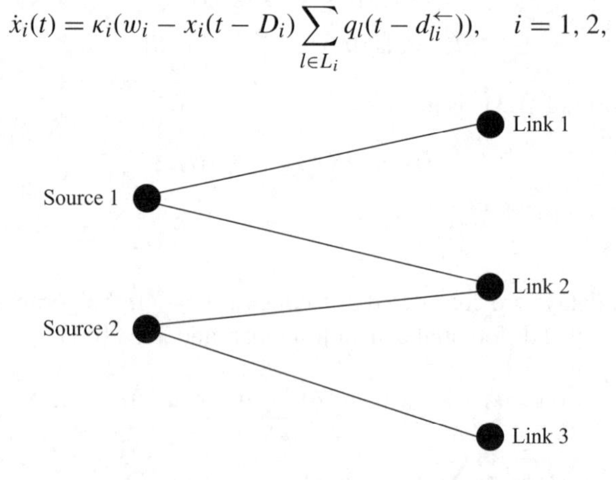

Figure 1.12 Interconnection graph.

where

$$
\begin{aligned}
L_1 &= \{1, 2\},\\
L_2 &= \{2, 3\},\\
q_1(t) &= p_1(x_1(t - d_{11}^{\rightarrow})),\\
q_2(t) &= p_2(x_1(t - d_{21}^{\rightarrow}) + x_2(t - d_{22}^{\rightarrow})),\\
q_3(t) &= p_3(x_2(t - d_{32}^{\rightarrow}))),\\
p_1(y) &= p_2(y) = p_3(y) = 1 - (1 - y)e^y.
\end{aligned}
$$

The parameters in the simulation are selected as follows: $\kappa_1 = w_1 = 0.05$, $\kappa_2 = w_2 = 0.07$, $D_1 = 22(\mathrm{s})$, $D_2 = 25(\mathrm{s})$, $d_{21}^{\rightarrow} + d_{22}^{\leftarrow} = d_{22}^{\rightarrow} + d_{21}^{\leftarrow} = \frac{1}{2}(D_1 + D_2) = 23.5(\mathrm{s})$. Under this set of parameters, the system is asymptotically stable. Suppose that the transmission process of the network is influenced by some external factors so that D_2 changes from 25(s) to 50(s) when $t \geq 1000(\mathrm{s})$, and suppose that the increments of both $d_{22}^{\leftarrow}$ and $d_{22}^{\rightarrow}$ are equal to 12.5(s). In this case, the equilibrium point becomes unstable and periodic oscillation takes place. The simulation of this process is shown in Figure 1.13.

The above example shows that the size of delay constant plays a significant role in the stability analysis. The problem of the stability of time-delayed feedback control systems has been extensively studied in recent years (see, e.g., Gu, Kharitonov and Chen (2003) and Niculescu (2001)). In this book we will pay more attention to one specific aspect of the problem: under what condition can one get a scalable stability criterion for a time-delayed distributed control

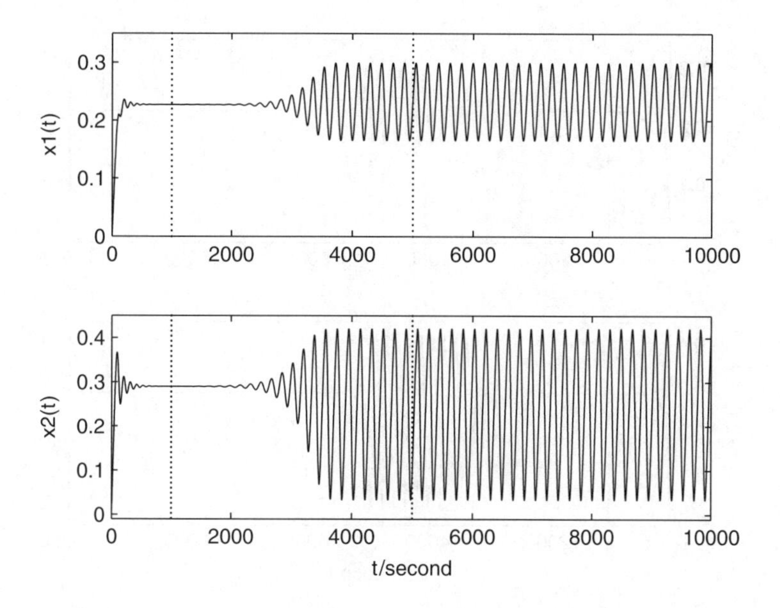

Figure 1.13 Periodic oscillations due to communication delays.

system? Here, under the scalability we mean the criterion checks only dynamics and parameters locally around each individual agent. The stability and scalability of distributed congestion control systems will be studied in Chapter 5.

Stabilization by time-delayed feedback control

On the one hand, propagation delay may destroy the stability of the congestion system as we have shown by numerical experiments. On the other hand, an appropriately introduced time-delayed feedback controller may serve as a stabilizer. Let us introduce the following time-delayed feedback control (TDFC)

$$u_d(t) = \frac{h}{\tau}(x(t) - x(t-\tau))$$

into the equation of $x_2(t)$. Then, the closed-loop system is modified as

$$\begin{aligned}\dot{x}_1 &= k_1(w_1 - x_1(t - D_1)(p(x_1(t - D_1)) + p(x_1(t - D_1) + x_2(t - \overleftarrow{d_{21}} - \overrightarrow{d_{22}})))) \\ \dot{x}_2 &= k_2(w_2 - x_2(t - D_2)(p(x_2(t - D_2)) + p(x_2(t - D_2) + x_1(t - \overleftarrow{d_{22}} - \overrightarrow{d_{21}})))) \\ &\quad - \frac{h}{\tau}(x_2(t) - x_2(t-\tau)).\end{aligned}$$

Now, let us conduct numerical experiments again for the foregoing oscillating system. In the simulation, the gain and the delay time are selected as $h = 0.8$ and $\tau = 4$(s), respectively. For the system, the TDFC is introduced when $t \geq 5000$(s). Figure 1.14 shows that the closed-loop system becomes stable again at a steady transmission rate: $x^{\star} = (0.2271, 0.2901)$. Figure 1.15 shows that the control force returns to zero when the equilibrium is stabilized by the TDFC.

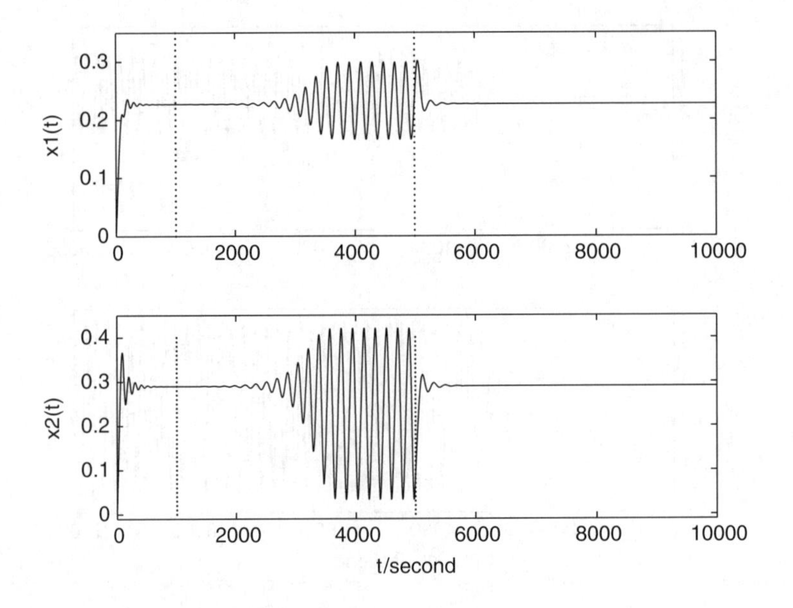

Figure 1.14 Stabilization of oscillations by TDFC.

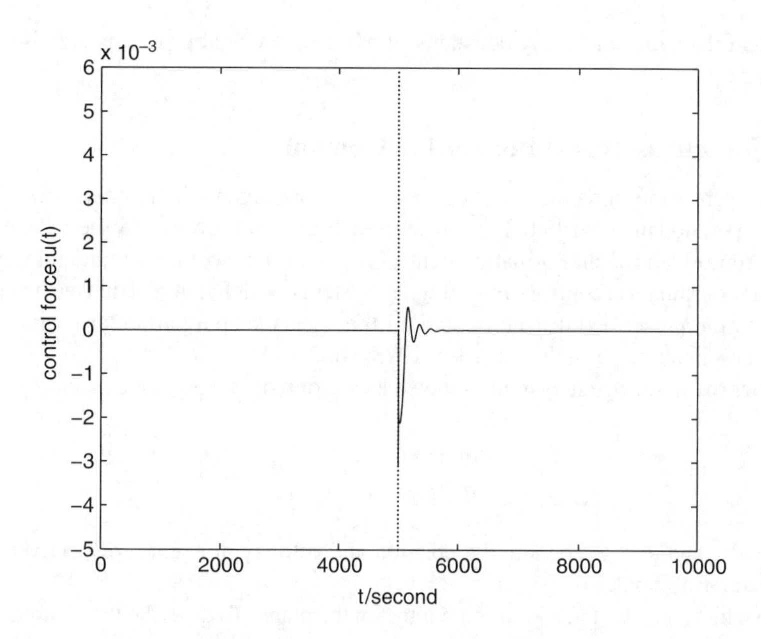

Figure 1.15 Control signal of TDFC.

It is natural to guess that the delay constant of the TDFC can not be arbitrarily large. In the simulation, the algorithm becomes unstable again when τ is greater than a certain value. The simulation shown in Figure 1.16 demonstrates that oscillations of the sending rates of the sources come about when $\tau \geq 84$(s). So, an interesting problem is why the TDFC works and

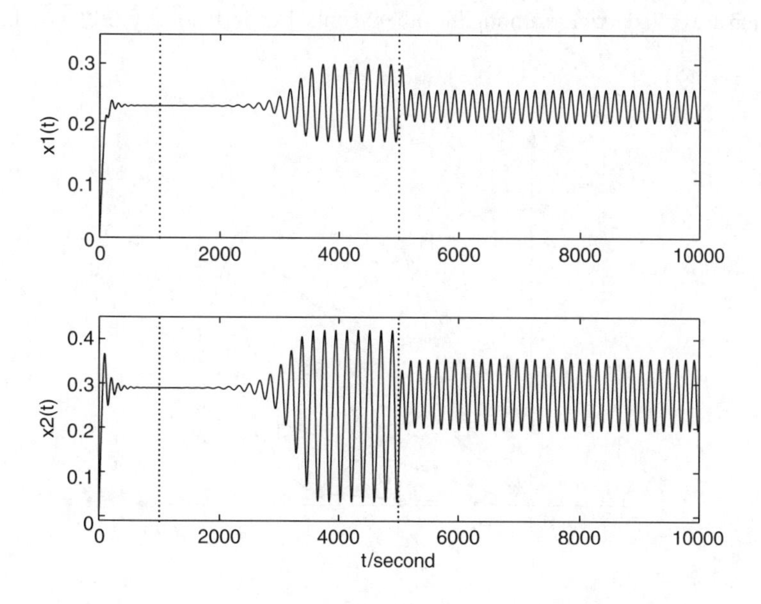

Figure 1.16 Oscillations due to large delay in TDFC.

how to select the gains and delay constants of a TDFC. Related problems are also discussed in Chapter 5.

1.3.2 Consensus-Based Formation Control

Trying to reach some agreement (or consensus) among agents is another effective way to conduct cooperation in a distributed control system. Here we briefly review the consensus-based formation control approach. Formation control is one of the most important and fundamental issues in the coordination control of multi-agent systems, which requires that each agent moves according to the prescribed trajectory, and all the agents keep a particular spatial formation pattern at the same time (Balch and Arkin 1998; Hu 2001).

Consider the multi-agent system modeled as a group of Newtonian particles

$$\begin{aligned} \dot{p}_i &= q_i, \\ \dot{q}_i &= u_i, \quad i \in \overline{1, n}, \end{aligned} \tag{1.36}$$

where $p_i \in \mathbb{R}^2$ and $q_i \in \mathbb{R}^2$ denote the position and velocity of agent i, respectively; $u_i \in \mathbb{R}^2$ denotes its control input.

Here we just consider the formation control in the plane. To describe the desired geometric pattern of the multi-agent system, let us introduce vector $c_i \in \mathbb{R}^2$ to specify the position of agent i in the pattern. Then, the desired formation can be described by a set of vectors, $F = \{c_i \in \mathbb{R}^2, i \in \overline{1, n}\}$. Figure 1.17 shows a pentagon formation with vertex-position vectors c_i, $i \in \overline{1, 5}$. Note that for any $c_0 \in \mathbb{R}^2$, $F' = \{(c_i + c_0) \in \mathbb{R}^2, i \in \overline{1, n}\}$ describes the same formation as F. The objective of formation control includes that the agents asymptotically converge to the prescribed geometric pattern F and each agent's velocity asymptotically approaches to a desired constant $v_0 \in \mathbb{R}^2$. Suppose the desired velocity v_0 is known by only one or a few agents which are called *leaders* among the other agents. Let us denote by $L \subset \overline{1, n}$ the index set of leaders.

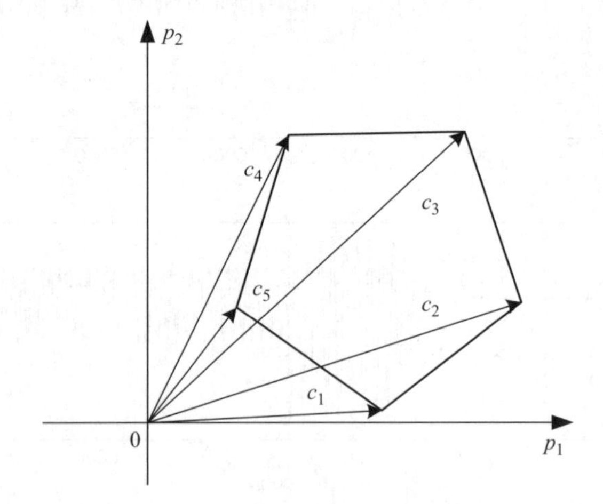

Figure 1.17 A pentagon formation.

For the system (1.36), a formation control algorithm can be given by

$$u_i(t) = \begin{cases} u_{i1} + u_{i2}, & \text{if } i \in L \\ u_{i2}, & \text{otherwise} \end{cases} \tag{1.37}$$

where

$$u_{i1} = -\beta(q_i(t) - v(t)) \tag{1.38}$$

is the velocity tracking part for leader agents, and

$$u_{i2} = -\gamma_1 \sum_{j \in N_i} a_{ij}(q_i(t) - q_j(t)) - \gamma_0 \sum_{j \in N_i} a_{ij}((p_i(t) - c_i) - (p_j(t) - c_j)) \tag{1.39}$$

is the coordination control part. In the control law, $\gamma_i > 0, i = 0, 1$, and $\beta > 0$ are control parameters, $a_{ij} > 0$ is a weight used by agent i for the information obtained from its neighboring agent j, and N_i is the set of neighbors of agent i. Obviously, the interconnection between the agents in such a system can be described by a digraph $G = (V, E, A)$: each agent can be considered as a vertex in V; the information flow between two agents can be considered as a directed edge in E, i.e., if agent i uses position and velocity information of agent j, then there is an edge from vertex i to vertex j; and the weight a_{ij} can be regarded as the entry of the adjacency matrix A. The formation control algorithm (1.37) is indeed a local control law because each agent uses only the information of itself and its neighbors.

For such a distributed control system, of course, the first problem we are interested in is stabilizability and stability, i.e., under what conditions do there exist available control parameters such that the closed-loop system is stable? and how can control parameters be selected in order to ensure the stability? The second problem we are interested in is: can the desired formation indeed be achieved when the system is asymptotically stabilized? In the first problem, the scalability of the stabilizability and/or the stability criteria is the profile to which we will pay more attention due to the possible huge scale of distributed control systems. While the problem of stabilizability and stability is of concern in all feedback control systems, the second problem raised here is a more distinct feature of distributed control systems. Indeed, even if the system is stable, does it imply the conclusion $u_{i1} \to 0, \forall i \in L$ and $u_{i2} \to 0, \forall i \in \overline{1, n}$? Even if $u_{i1} \to 0, \forall i \in L$ and $u_{i2} \to 0, \forall i \in \overline{1, n}$, does it imply that all the agents track the desired velocity v_0 and that the relative displacement of each pair of agents satisfies the formation requirement F? By taking a first look at the questions, we can see that the condition of

$$u_{i2} \to 0, \quad \forall i \in \overline{1, n}, \tag{1.40}$$

and the implication

$$\begin{gathered} u_{i2} \to 0, \quad \forall i \in \overline{1, n} \\ \Longrightarrow \\ q_i(t) = q_j(t), \text{ and } p_i(t) - p_j(t) = c_i - c_j, \quad \forall i, j \in \overline{1, n},\ i \neq j \end{gathered} \tag{1.41}$$

are critical for getting the answer to all the questions.

To consider the problem formulated by (1.40) and (1.41) one can let $c_i = 0, \forall i \in \overline{0,n}$. In this case, the formation problem reduces to the so-called *rendezvous* problem (i.e., position agreement). Actually, the basic algorithm for the rendezvous problem is

$$u_i(t) = -\gamma_1 \sum_{j \in N_i} a_{ij}(q_i(t) - q_j(t)) - \gamma_0 \sum_{j \in N_i} a_{ij}(p_i(t) - p_j(t)), \tag{1.42}$$

This algorithm is also called the *consensus* (or agreement) algorithm. If the multi-agent system (1.36) is stable under this control, then the states (velocities and positions) of all the agents in the system approach to the same value.

From the above discussion, we see that the stability of the consensus algorithm is at the center of the problem. The problem, in particular, includes the analysis and design of the available values of control parameters γ_i, β and even adjacent weights a_{ij} in the sense of stability. We should note the difference between the stability of the consensus problem and the stability of conventional control systems. In most conventional (linear) feedback control systems, the origin is the only equilibrium and the stability implies that all the eigenvalues of the system matrix of the closed-loop system have positive real parts. But, it is easy to see that the system matrix of the closed-loop system of (1.36) with (1.42) is singular, and thus the system has a continuum of equilibrium points. And therefore, the stability of the consensus algorithm does not imply that all the eigenvalues of the closed-loop have positive real parts. The stability of consensus control systems and other related problems are discussed in Chapter 6.

If the agents are distributed in a huge range of space and connected with the help of a communication network, delay is inevitable in information transmission. At current time t agent i can get only the delayed information $p_{ij}(t - \tau_{ij})$ and $q_{ij}(t - \tau_{ij})$ from its neighboring agent j, where $\tau_{ij} > 0$ is the communication delay of information flow from agent j to agent i. In this case, if the value of the delay constant is available for agent i, then the consensus algorithm (1.42) can be modified as follows:

$$\begin{aligned} u_i(t) = -\gamma_1 &\sum_{j \in N_i} a_{ij}(q_i(t - \tau_{ij}) - q_j(t - \tau_{ij})) \\ &-\gamma_0 \sum_{j \in N_i} a_{ij}(p_i(t - \tau_{ij}) - p_j(t - \tau_{ij})). \end{aligned} \tag{1.43}$$

Note that when diverse communication delays are involved in the communication network, the stability and scalability analysis of such consensus algorithms will become much more complicated and challenging. Related problems are discussed in Chapter 7.

In the algorithm (1.43) the communication delay τ_{ij} is assumed to be available for agent i and is used as a "self-delay". In practice, however, only approximate estimations of communication delays are available. If the self-delay in the (1.43) is replaced by the estimation of the communication delay, then we have to use the following consensus algorithm:

$$\begin{aligned} u_i(t) = -\gamma_1 &\sum_{j \in N_i} a_{ij}(q_i(t - \tau'_{ij}) - q_j(t - \tau_{ij})) \\ &-\gamma_0 \sum_{j \in N_i} a_{ij}(p_i(t - \tau'_{ij}) - p_j(t - \tau_{ij})). \end{aligned} \tag{1.44}$$

In this case, a new question is raised naturally: does there exists any consensus solution of the closed-loop system (1.36) under feedback control (1.44) when $\tau'_{ij} \neq \tau_{ij}$? This question will be also answered in Chapter 7 for this and more generalized systems.

In closing, we would like to note that although the consensus control and the congestion control are designed from quite different principles of cooperation, they share a lot of similarities in system structure, and hence the stability and scalability of the two kinds of systems can be dealt with by using some common tools which are introduced in this chapter and the following two chapters as well.

1.4 Notes and References

1.4.1 Graph Theory and Distributed Control Systems

Graph theory is a fundamental tool in distributed control and distributed computation. The basic definitions of graph theory introduced in this book are quite standard; see, for example, Biggs (1994), Bollobás (1998), Diestel (1997), Godsil and Royle (2001). For powers of the adjacency matrix the reader is referred to literature on algebraic graph theory (Biggs 1994; Bullo, Cortés and Martínez 2009; Godsil and Royle 2001). Laplacian matrices have numerous remarkable properties and some of them play a key role in the analysis of consensus problems. The reader is referred to Mohar (1991) and Merris (1994) for elegant surveys of Laplacian matrices. Theorem 1.9 and Theorem 1.10, characterizing the properties of the Laplacian matrix, contain some recent results. A proof of statement (2) of Theorem 1.9 was given in Olfati-Saber and Murray (2004). Statement (3) of Theorem 1.9 was proved by Lin, Francis and Maggiore (2005). An equivalent version of statement (3) was proved in Ren and Beard (2005). Statement (1) of Theorem 1.10 was proved by Olfati-Saber and Murray (2004), and statement (2) of Theorem 1.10 was proved by Moreau (2005).

Two typical kinds of distributed control systems are introduced in this chapter. The first is the end-to-end congestion control and the second is the consensus-based formation control.

The first congestion avoidance control algorithm for the Internet was proposed by Jacobson (1988). Motivated by the theory of economics, Kelly (1997) and Kelly, Maulloo and Tan (1998) introduced the concept of price into the congestion control, and formulated the rate control in communication networks as a problem of optimization of allocation resources (utilization of channel capacities) in networks. For a mathematic treatment of congestion control problems the reader is referred to, for example, Low and Lapsley (1999); Low, Paganini and Doyle (2002); Srikant (2004), but this is the first book to use bipartite graphs to describe the architecture of congestion control systems. It will be shown in the remaining chapters of this book that we benefit from such a treatment in characterizing the symmetry of congestion control systems and relating them to multi-agent systems discussed in consensus problems.

Consensus problems are closely related to many other problems in coordination control, such as formation control (Balch and Arkin 1998; Fax and Murray 2004; Hu 2001; Wang and Hadaegh 1994) and flocking (Olfati-Saber and Murray 2006; Toner and Tu 1998). For a survey on the relationship between consensus problems and coordination control strategies the reader is referred to Olfati-Saber, Fax and Murray (2007).

1.4.2 Delay in Control and Control by Delay

The fact that delay in feedback may cause oscillation and instability was pointed out as early as the 1930s (see, e.g., Wiener (1948)). The delay effects on the stability of feedback control systems have been extensively studied with tools from both classic Lyapunov stability theory and contemporary robust control theory (Gu, Kharitonov and Chen 2003; Niculescu 2001).

Johari and Tan (2001) studied the stability of the primal algorithm of the end-to-end congestion control system with propagation delays. They gave a stability criterion for systems with identical round-trip delay and also proposed a conjecture for systems with diverse round-trip delays. The conjecture was proved in its original version and generalized to a less conservative form by Tian and Yang (2004) via a frequency-domain method. Tian (2005) further showed that such a frequency-domain method was powerful in studying the stability of second-order congestion control systems with diverse propagation delays.

The study on the consensus problem with communication delays started as early as the 1980s in the context of distributed computation (Bertsekas and Tsitsiklis 2007; Tsitsiklis, Bertsekas and Athans 1986). In some other communities such as physics, this problem has also been extensively studied as synchronization of coupled oscillators (see, e.g., Yeung and Strogatz 1999). Recently, these kinds of consensus protocols have been studied for fixed or even switched graphs by using different analysis methods, such as the contraction analysis method (Wang and Slotine 2006), the passivity-based method (Chopra and Spong 2006), the method based on delayed and hierarchical graphs (Cao, Morse and Anderson 2006), among others. Tian and Liu (2008, 2009) noticed the symmetry of multi-agent systems with diverse input delays, and used the frequency-domain method developed initially for congestion control to study the stability of consensus control systems.

The idea of using a time-delayed feedback controller (TDFC) to stabilize unstable periodic orbits of chaotic systems was first proposed by Pyragas (1992). Kokame *et al.* (2001) illustrated the TDFC as a kind of differential control and applied it to stabilize uncertain steady states. Liu and Tian (2008) noticed the fact that the equilibrium point of the Internet is usually not available for each source node or link node and thus proposed a stabilization scheme for the congestion control system of the Internet by introducing the TDFC into the primal algorithm and proved that the system with any round-trip delay can be locally stabilized by the modified algorithm.

Many other applications of the TDFC method have also been reported in the literature, such as stabilization of coherent models of lasers (Bleich and Socolar 1996; Naumenko *et al.* 1998) and magneto-elastic systems (Hilkihara, Touno and Kawagoshi 1997), control of cardiac conduction model (Brandt, Shih and Chen 1997), control of stick-slip friction oscillations (Elmer 1998), traffic models (Konishi, Kokame and Hirata 1999) and PWM-controlled buck convertors (Battle, Fossas and Olivar 1999), just to name a few. The reader is referred to Tian, Zhu and Chen (2005) for a recent survey of time-delayed feedback control.

References

Balch T and Arkin RC (1998). Behavior-based formation control for multirobot terms. *IEEE Transanctions on Robotics and Automation*, 14, 926–939.

Battle C, Fossas E and Olivar G (1999). Stabilization of periodic orbits of the buck convertor by time-delayed feedback. *International Journal of Circuits Theory and Application*, 27(6), 617–631.

Bertsekas DP and Tsitsiklis JN (2007). Comments on "Coordination of groups of mobile autonomous agents using nearest neighbor rules". *IEEE Transactions on Automatic Control*, 52, 968–969.

Biggs N (1994). *Algebraic Graph Theory*, second edition. Cambridge University Press, Cambridge.

Bleich ME and Socolar JES (1996). Controlling spatiotemporal dynamics with time-delay feedback. *Physics Review E*, 54(1), R17–R20.

Bollobás B (1998). *Modern Graph Theory*. Graduate Text in Mathematics, 184, Sringer, New York.

Brandt MEH, Shih T and Chen G (1997). Linear time-delayed feedback control of a pathological rhythm in a cardiac conduction model. *Physics Review E*, 56(2), R1334–R1337.

Breder CM (1954). Equations descriptive of fish schools and other animal aggregations. *Ecology*, 35(3), 361–370.

Bullo F, Cortés J and Martínez (2009). *Distributed Control of Robotic Networks*. Princeton University Press. available online at http://coordinationbook.info

Cao M, Morse AS and Anderson BDO (2006). Reaching an agreement using delayed information. *IEEE Conference on Decision and Control*, San Diego, CA, USA, 3375–3380.

Chen CT (1984). *Linear System Theory and Design*. CBS College Publishing, New Yoyk.

Chopra N and Spong MW (2006). Passivity-based control of multi-agent systems. *Advances in Robot Control: From Everyday Physics to Muman-like Movements*, S Kawamura and M Svinin, Editors, pp.107–134, Spinger-Verlag, Berlin.

Cortés J, Martinez S, Karatas T, and Bullo F (2004). Coverage control for mobile sensing networks. *IEEE Transactions on Robotics and Automation*, 20(2), 243–255.

Darwin R (1859). *The Origin of Species*. Chinese Translation by Qian Xun, Chong Qin Publishing Company, Chong Qin, China.

Diestel R (1997). *Graph Theory*. Springer-Verlag, New York.

Elmer FJ (1998). Controlling friction. *Physics Review E*, 57(5), R4903–R4906.

Fax JA and Murray RM (2004). Information flow and cooperative control of vehicle formations. *IEEE Transactions on Automatic Control*, 49, 1465–1476.

Godsil C and Royle G (2001). *Algebraic Graph Theory*. Graduate Text in Mathematics, 207, Sringer, New York.

Gu K, Kharitonov VL and Chen J (2003). *Stability of Time-delayed Systems*. Birkhäuser, Boston.

Giulietti F, Pollini L and Innocenti M (2000). Autonomous formation flight. *IEEE Control Systems Magazine*, 20(12), 34–44.

Hilkihara T, Touno M and Kawagoshi T (1997). Experiment stabilization of unstable periodic orbit in magneto-elastic chaos by delayed feedback control. *International Journal of Bifurcation and Chaos*, 7(12), 2837–2846.

Hu X (2001). Formation control with virtual leaders and reduced communications. *IEEE Transactions on Robotics and Automation*, 17, 947–951.

Jacobson V (1988). Congestion avoidance and control. *Proceedings of ACM SIGCOMM' 88*, Stanford, CA, 314–329.

Johari R and Tan D (2001). End to end congestion control for the internet: delays and stability. *IEEE/ACM Transactions on Networking*, 9, 818–832.

Kelly FP Charging and rate control for elastic traffic. *European Transactions on Telecommunication*, 8, 33–37.

Kelly FP, Maulloo A and Tan D (1998). Rate control for communication networks: shadow prices proportional fairness and stability. *J. Oper. Res. Soc.*, 49, 237–252.

Kokame H, Hirata K, Konishi K and Mori T (2001). Difference feedback can stabilize uncertain steady states. *IEEE Transactions on Automatic Control*, 46, 1908–1913.

Konishi K, Kokame H and Hirata K (1999). Coupled map car-following model and its delayed-feedback control. *Physics Review E*, 60(4), 4000–4007.

Lin Z, Francis B and Maggiore M (2005). Necessary and sufficient graphical conditions for formation control of unicycles. *IEEE Transactions on Automatic Control*, 50, 121–127.

Liu C-L and Tian Y-P (2008). Eliminating oscillations in the Internet by time-delayed feedback control. *Chaos, Solitons and Fractals*, 35, 878–887.

Low SH and Lapsley DE (1999). Optimization flow control, I: basic algorithm and convergence. *IEEE/ACM Transactions on Networking*, 7, 861–874.

Low SH, Paganini F and Doyle JC (2002). Internet congestion control. *IEEE Control Systems Magazine*, 22, 28–43.

Lunch NA (1997). *Distributed Algorithms*. Morgan Kaufmann, San Francisco.

Merris R (1994). Laplacian matrices of a graph: a survey. *Linear Algebra and its Applications*, 197, 143–176.

Mohar B (1991). The Laplacian spectrum of graphs. *Graph Theory, Combinatorics, and Applications*, Alavi Y, Chartrand G, Oellermann OR, and Schwenk AJ, editors, 2, 871–898, John Wiley.

Moreau L (2005). Stability of multiagent systems with time-dependent communication links. *IEEE Transactions on Automatic Control*, 50, 169–182.

Murray RM, Åström KJ, Boyd SP, Brockett RW and Stein G (2003). Future directions in control in an information-rich world. *IEEE Control Systems Magazine*, 23(April), 20–33.

Naumenko AV, Loiko NA, Turovets SI, Spencer PS and Shore KA (1998). Chaos control in external cavity laser diodes using electronic impulsive delayed feedback. *International Journal of Bifurcation and Chaos*, 8(9), 1791–1799.

Niculescu S-I (2001). *Delay effects on stability: a robust control approach*. Lecture Notes in Control and Information Sciences, 269, Springer, Berlin.

Olfati-Saber R and Murray RM (2004). Consensus problems in networks of agents with switching topology and time-delays. *IEEE Transactions on Automatic Control*, 49, 1520–1533.

Olfati-Saber R and Murray RM (2006). Flocking for multi-agent dynamic systems: algorithms and theory. *IEEE Transactions on Automatic Control*, 51, 401–420.

Olfati-Saber R, Fax JA and Murray RM (2007). Consensus and cooperation in networked multi-agent systems. *Proceedings of the IEEE*, 95(1), 215–223.

Pyragas K (1992). Continuous control of chaos by self-controlling feedback. *Physics Letters A*, 170, 421–428.

Qi HR, Iyengar SS and Chakrabarty K (2001). Distributed sensor networks – a review of recent research. *Journal of the Franklin Institute*, 338(6), 655–668.

Rekleitis IM, Dudek G and Milios EE (2000). Multi-robot collaboration for robust exploration. *Proceeding of the IEEE International Conference on Robotics and Automation*, 4, 3164–3168.

Ren W and Beard RW (2005). Consensus seeking in multi-agent systems under dynamically changing interaction topologies. *IEEE Transactions on Automatic Control*, 50, 655–661.

Reynolds CW (1987). Flocks, herds, and schools: a distributed behavioral model. *Computer Graphics*, 21(4), 25–34.

Sastry SS (1999). *Nonlinear Systems: Analysis, Stability, and Control*. Interdisciplinary applied mathematics, 10, Springer, New York.

Srikant R (2004). *The Mathematics of Internet Congestion Control*. Birkhäuser, Boston.

Tian Y-P and Yang H-Y (2004). Stability of the Internet congestion control with diverse delays. *Automatica*, 40, 1533–1541.

Tian Y-P (2005). Stability analysis and design of the second-order congestion control for networks with heterogeneous delays. *IEEE/ACM Transactions on Networking*, 13, 1082–1093.

Tian Y-P, Zhu J and Chen G (2005). A survey on delayed feedback control of chaos. *Journal of Control Theory and Applications*, 3, 311–319.

Tian Y-P and Liu C-L (2008). Consensus of multi-agent systems with diverse input and communication delays. *IEEE Transactions on Automatic Control*, 53, 2122–2128.

Tian Y-P and Liu C-L (2009). Robust consensus of multi-agent systems with diverse input delays and asymmetric interconnection perturbations. *Automatica*, 45, 1347–1353.

Tomlin C, Pappas GJ and Sastry S (1998). Conflict resolution for air traffic management: a study in multiagent hybrid systems. *IEEE Transactions on Automatic Control*, 43(4), 509–521.

Toner J and Tu Y (1998). Flocks, herds, and schools: A quantitative theory of flocking. *Physical Review W*, 58, 4828–4858.

Tsitsiklis JN, Bertsekas DP and Athans M (1986). Distributed asynchronous deterministic and stochastic gradient optimisation algorithms. *IEEE Transactions on Automatic Control*, 31, 803–812.

Ushio T (1996). Limitation of delayed feedback control in nonlinear discrete-time systems. *IEEE Transactions on Circuits and Systems*, 43, 815–816.

Wang PKC and Hadaegh FY (1994). Coordination and control of multiple microspacecraft moving in formation. *J. Astronaut. Sci.*, 44, 315–355.

Wang W and Slotine JJE (2006). Contraction analysis of time-delayed communications and group cooperation. *IEEE Transactions on Automatic Control*, 51, 712–717.

Wiener N (1948). *Cybernetics: Or Control and Communication in the animal and the machine*. Wiley, New York.

Yeung MKS and Strogatz SH (1999). Time delay in the Kuramoto model of coupled oscillators. *Physical Review Letter*, 82, 648–651.

2

Symmetry, Stability and Scalability

People today see not the moon the ancients saw, and yet this moon shone once upon the ancients.

—*Li Bai (701–762), 'Inquiring of the moon, with wine in hand'*

This chapter introduces general models of distributed control systems based on graph theory and linear system theory. The symmetry property of distributed control systems in the frequency domain is defined and investigated. Classic results of the frequency-domain test of the stability of multivariable feedback control systems are reviewed with the emphasis on their extension to the semi-stability and the scalability for distributed control systems.

2.1 System Model

2.1.1 Graph-Based Model of Distributed Control Systems

Consider a distributed control system with n subsystems, each of which is also called an agent in this book. Let the model of the dynamics of the ith agent be given by the following state-space model

$$\begin{cases} \dot{x}_i(t) = A_i x_i(t) + B_i u(t - T_i) \\ y_i(t) = C_i x_i(t) + D_i u(t - T_i) \end{cases} \tag{2.1}$$

where $x_i(t) \in \mathbb{R}^{n_i}$, $u_i(t) \in \mathbb{R}$ and $y_i(t) \in \mathbb{R}$ denote the state, input and output, respectively, of the ith agent, T_i is the input delay. Note that in distributed control systems input delays are usually caused by the processing time for the packets arriving at each agent and/or the communication time between actuators and controllers.

The agents are interconnected according a fixed topology which can be described by a digraph $G = (V, E, \bar{A})$ of order n. Denote by N_i the set of neighbors of agent i. By (1.9), the aggregated measurement of agent i is based on the information from its neighbors and itself:

$$z_i(t) = a_{ii} y_i(t - \tau_{ii}) + \sum_{j \in N_i} a_{ij} y_j(t - \tau_{ij}), \tag{2.2}$$

Frequency-Domain Analysis and Design of Distributed Control Systems, First Edition. Yu-Ping Tian.

where τ_{ij} is the communication delay from agent j to agent i which mainly consists of propagation delay and queueing delay, τ_{ii} is called self-delay which is often introduced to match the communication delays, a_{ij} is the weight for the output channel from agent j to agent i and a_{ii} is the weight for the self-loop of agent i. To incorporate the aggregated measurement (1.9) and the relative measurement (1.10) in a unified form, throughout this chapter, we assume that a_{ij} in (2.2) can be either positive or negative.

Suppose that agent i is manipulated by the following local controller

$$\begin{cases} \dot{\hat{x}}_i(t) = A_i^k \hat{x}_i(t) + B_i^k (r_i(t) - z_i(t)) \\ u_i(t) = C_i^k \hat{x}_i(t) + D_i^k (r_i(t) - z_i(t)) \end{cases} \tag{2.3}$$

where $\hat{x}_i(t) \in \mathbb{R}^{m_i}$ is the state of the local controller for agent i, $A_i^k \in \mathbb{R}^{m_i \times m_i}$, $B_i^k \in \mathbb{R}$, $C_i^k \in \mathbb{R}$, $D_i^k \in \mathbb{R}$ are parameters of the controller and $r_i(t)$ is some reference signal for agent i.

Denote by $G_i(s)$ the transfer function of agent i, i.e.,

$$G_i(s) = [C_i(sI - A_i)^{-1} B_i + D_i] \mathrm{e}^{-T_i s}, \tag{2.4}$$

and by $\kappa_i(s)$ the transfer function of the ith controller, i.e.,

$$\kappa_i(s) = C_i^k (sI - A_i^k)^{-1} B_i^k + D_i^k. \tag{2.5}$$

The feedback control system for agent i is shown in Figure 2.1. Let $\hat{y}_i(s)$ and $\hat{u}_i(s)$ be the Laplace transformation of the output and input, respectively, of the i-agent. Denote

$$Y(s) = [\hat{y}_1(s), \cdots, \hat{y}_n(s)]^{\mathrm{T}}, \tag{2.6}$$

$$U(s) = [\hat{u}_1(s), \cdots, \hat{u}_n(s)]^{\mathrm{T}}, \tag{2.7}$$

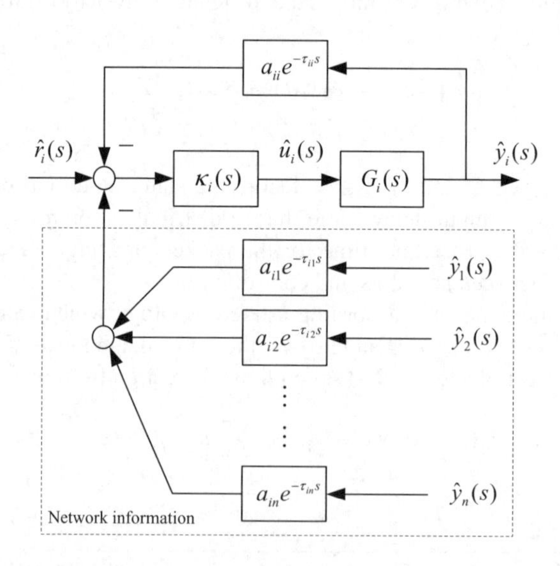

Figure 2.1 Feedback control system of agent i.

$$G(s) = \operatorname{diag}\left\{G_i(s),\ i \in \overline{i,n}\right\}, \tag{2.8}$$

$$K(s) = \operatorname{diag}\{\kappa_i(s),\ i \in \overline{i,n}\}, \tag{2.9}$$

$$\bar{A}(s) = \{a_{ij}\mathrm{e}^{-\tau_{ij}s}\}, \tag{2.10}$$

$$\Lambda(s) = \operatorname{diag}\{a_{ii}\mathrm{e}^{-\tau_{ii}s},\ i \in \overline{1,n}\}, \tag{2.11}$$

and

$$A(s) = \bar{A}(s) - \Lambda(s). \tag{2.12}$$

$A(s)$ and $\bar{A}(s)$ can be considered as the weighted adjacency matrix and generalized weighted adjacency matrix, respectively, of topology digraph G. As a matter of fact, in our model the matrix $\bar{A}(s)$ can be further extended to the following form

$$\bar{A}(s) = \{\alpha_{ij}(s)\mathrm{e}^{-\tau_{ij}s}\}, \tag{2.13}$$

where the transfer function $\alpha_{ij}(s) \in \mathbb{RH}_\infty$ describes the possible dynamics of the communication channel from agent j to agent i.

With notations introduced above, the closed-loop system composed of (2.1), (2.2) and (2.3) can be sketched in Figure 2.2. The transfer function matrix from $R(s)$ to $Y(s)$ is given by

$$W(s) = (I + G(s)K(s)\bar{A}(s))^{-1}G(s)K(s). \tag{2.14}$$

The system shown in Figure 2.2 is an interconnection of two parts. The first part $G(s)K(s)$ is a diagonal matrix, which represents the agent dynamics. The second part $A(s) + \Lambda(s)$ can be considered as a generalized complex adjacency matrix of the topology graph of the network; it reflects both the topology and the channel dynamics of the network used by the multi-agent system.

Definition 2.1 *The dynamics of the agents in the distributed control system shown in Figure 2.1 are said to be homogeneous if $\kappa_i(s)G_i(s) = \kappa_j(s)G_j(s)$ for all $i, j \in \overline{1,n}$; the network through which the agents are interconnected is said to be homogeneous if $\tau_{ij} = \tau_0$ and $\alpha_{ij}(s) = \alpha_0(s)$, $\forall i, j \in \overline{1,n}$, where $\tau_0 \in \mathbb{R}^+$ and $\alpha_0(s) \in \mathbb{RH}_\infty$. The distributed control system is said to be homogeneous if all the agents and the network are all homogeneous. Otherwise, the agents (network) are said to be heterogeneous if they are not homogeneous. The system is said to be heterogeneous if it contains heterogeneous agents or heterogeneous network.*

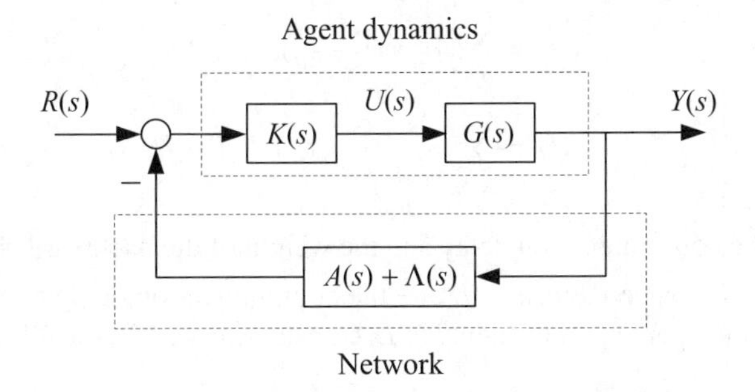

Figure 2.2 Diagram of distributed control system.

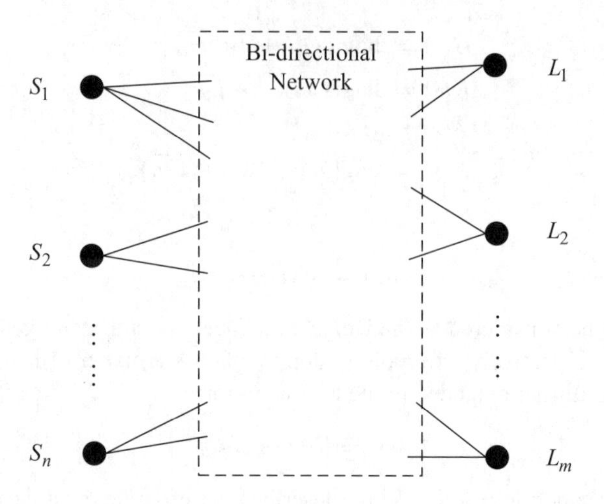

Figure 2.3 Bipartite graph for distributed control systems.

2.1.2 Bipartite Distributed Control Systems

There are many distributed control systems in which information exchanges are conducted through forward and backward channels. The interconnection of agents in such systems can be described by the weighted bipartite digraph shown in Figure 2.3. We call these systems *bipartite systems*. Denote by (S, L) and A the vertex bipartition and the weighted adjacency matrix of the graph, respectively, where

$$S = \{S_1, \cdots, S_n\}, \tag{2.15}$$

$$L = \{L_1, \cdots, L_m\}, \tag{2.16}$$

$$A = \begin{bmatrix} 0 & A_{12} \\ A_{12}^{\mathrm{T}} & 0 \end{bmatrix}. \tag{2.17}$$

Let the agents associated with vertex class S be described by transfer functions $G_i(s), i \in \overline{1, n}$; and the agents associated with L be described by transfer functions $P_j(s),\ j \in \overline{1, m}$. Agent i in S receives output information from its neighbors in L, and agent j in L receives output information from its neighbors in S, i.e.,

$$z_i(t) = \sum_{l \in N_i} a_{il}^b y_l(t - \tau_{li}^b), \quad i \in S, \tag{2.18}$$

$$z_l(t) = \sum_{i \in N_l} a_{li}^f y_i(t - \tau_{li}^f), \quad l \in L, \tag{2.19}$$

where τ_{li}^b, a_{il}^b are the communication delay and the weight of the backward channel from agent l in L to agent i in S, respectively; τ_{li}^f, a_{li}^f are the communication delay and the weight of the forward channel from agent i in S to agent l in L. Considering (2.17), the following relationship should hold

$$a_{il}^b = a_{li}^f, \quad \forall i \in S, \forall l \in L. \tag{2.20}$$

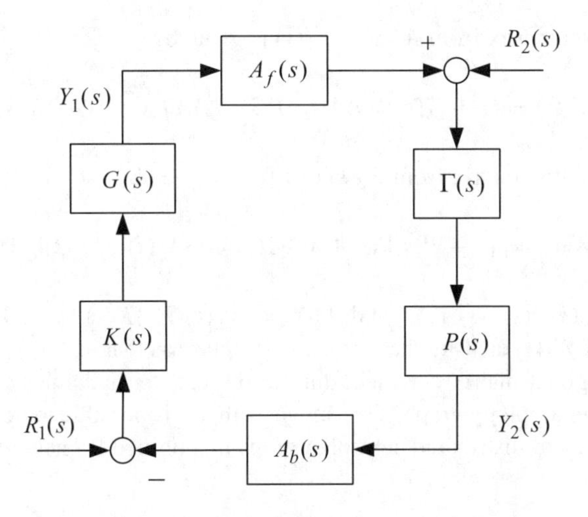

Figure 2.4 Bipartite distributed control system.

Agent i in S is manipulated by dynamic controller $\kappa_i(s)$, $i = 1, \cdots, n$, and agent l in L is manipulated by dynamic controller $\gamma_l(s)$, $l \in 1, \cdots, m$. Denote by

$$R_1(s) = [\hat{r}_1^1(s), \cdots, \hat{r}_n^1(s)]^{\mathrm{T}}$$

and

$$Y_1(s) = [\hat{y}_1^1(s), \cdots, \hat{y}_n^1(s)]^{\mathrm{T}}$$

the reference input vector and output vector of all the agents in S, respectively; by

$$R_2(s) = [\hat{r}_1^2(s), \cdots, \hat{r}_m^2(s)]^{\mathrm{T}}$$

and

$$Y_2(s) = [\hat{y}_1^2(s), \cdots, \hat{y}_m^2(s)]^{\mathrm{T}}$$

the reference input vector and output vector of all the agents in L, respectively. Then, the bipartite system with forward and backward channels can be shown by Figure 2.4, where

$$\begin{aligned}
G(s) &= \mathrm{diag}\left\{G_i(s),\ i \in \overline{1,n}\right\}, \\
K(s) &= \mathrm{diag}\{\kappa_i(s),\ i \in \overline{1,n}\}, \\
P(s) &= \mathrm{diag}\left\{P_l(s),\ l \in \overline{1,m}\right\}, \\
\Gamma(s) &= \mathrm{diag}\{\gamma_l(s),\ l \in \overline{1,m}\}, \\
\bar{A}_b(s) &= \{a_{il}^b \mathrm{e}^{-\tau_{li}^b s},\ i \in \overline{1,n},\ l \in \overline{1,m}\}, \\
\bar{A}_f(s) &= \{a_{li}^f \mathrm{e}^{-\tau_{li}^f s},\ l \in \overline{1,m},\ i \in \overline{1,n}\}.
\end{aligned}$$

The transfer function matrix from $R_1(s)$ to $Y_1(s)$ is given by

$$W_1(s) = (I + G(s)K(s)A_b(s)P(s)\Gamma(s)A_f(s))^{-1}G(s)K(s), \tag{2.21}$$

and the transfer function matrix from $R_2(s)$ to $Y_2(s)$ is given by

$$W_2(s) = (I + P(s)\Gamma(s)A_f(s)G(s)K(s)A_b(s))^{-1}P(s)\Gamma(s). \tag{2.22}$$

Denote $\bar{A}_1(s) = A_b(s)P(s)\Gamma(s)A_f(s)$ and $\bar{A}_2(s) = A_f(s)G(s)K(s)A_b(s)$. Then, the open-loop transfer function of $W_1(s)$ or $W_2(s)$ has the same structure as that of $W(s)$ given by (2.14), i.e., it consists of a diagonal matrix corresponding to the agents' dynamics and a square matrix corresponding to the topology graph. This suggests that it is possible to deal with the stability problem of two kinds of distributed control systems in a unified framework.

2.2 Symmetry in the Frequency Domain

2.2.1 Symmetric Systems

In practice, many multi-agent systems are based on networks with undirected topology graph. In this case, if there is no delay and all the channels have constant dynamics, i.e., $\tau_{ij} = 0$ $\alpha_{ij}(s) = 1$, $\forall i, j \in \overline{1, n}$, then the system shown in Figure 2.2 is said to be *symmetric* because the adjacency matrix of the topology graph is symmetric. Can such a conception of symmetry be extended to the case when there are non-zero delays or non-constant dynamics in network channels? In this book, we define the symmetry of networked-based distributed control systems as follows.

Definition 2.2 *Let the open-loop transfer function of a distributed control system be given by $\bar{G}(s)\bar{A}(s)$, where $\bar{G}(s)$ is a diagonal matrix and $\bar{A}(s)$ is a square matrix. The system is said to be symmetric if $\bar{A}(s)$ can be decomposed as*

$$\bar{A} = \Theta(s)\hat{A}(s), \tag{2.23}$$

where $\Theta(s)$ is a diagonal matrix and $\hat{A}(s)$ satisfies the following condition:

$$\hat{A}(s) = \hat{A}^{\mathrm{T}}(-s). \tag{2.24}$$

A starting point for this definition is that a complex generalized adjacency matrix of an undirected graph should be a Hermitian matrix as we defined in Chapter 1. Indeed, letting $s = \mathrm{j}\omega$, then from (2.24) we have $\hat{A}(\mathrm{j}\omega) = \hat{A}^*(\mathrm{j}\omega)$. So, if we absorb the diagonal matrix $\Theta(s)$ in the agents' dynamics which also form a diagonal matrix, then $\hat{A}(\mathrm{j}\omega)$ can be regarded as a generalized adjacency matrix of the undirected topology graph in the frequency domain.

Example 2.3 *Symmetry in the frequency domain.*

Consider the simplest undirected graph of two nodes, shown by Figure 2.5. Suppose the communication delays from node 1 to node 2 and from node 2 to node 1 are $\tau_{21} > 0$ and $\tau_{12} > 0$,

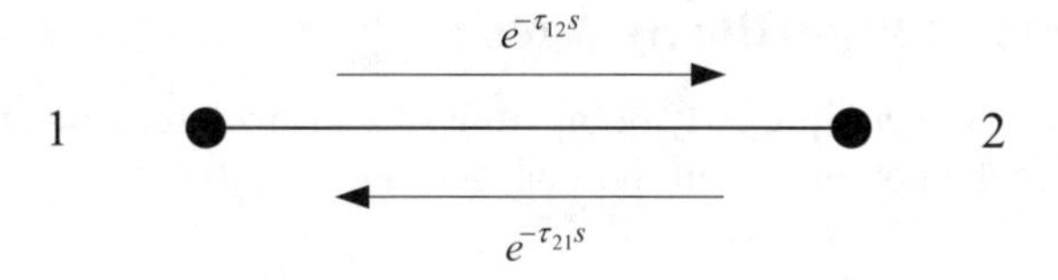

Figure 2.5 Symmetry in the frequency domain.

respectively. It is natural to set $\mathrm{e}^{-\tau_{12}s}$ *and* $\mathrm{e}^{-\tau_{21}s}$ *as the weights in the Laplace domain. But, the matrix*

$$A(\mathrm{j}\omega) = \begin{bmatrix} 0 & \mathrm{e}^{-\tau_{12}s} \\ \mathrm{e}^{-\tau_{21}s} & 0 \end{bmatrix}$$

does not satisfy conjugate symmetry even in the case $\tau_{12} = \tau_{21} = \tau$. *In the frequency domain,* $A(\mathrm{j}\omega)$ *is not a Hermitian matrix, either. Now, let us scale the matrix* $A(s)$ *by multiplying*

$$\begin{bmatrix} 1 & 0 \\ 0 & \mathrm{e}^{-Ts} \end{bmatrix} \begin{bmatrix} 1 & 0 \\ 0 & \mathrm{e}^{Ts} \end{bmatrix},$$

where

$$T = \tau_{12} + \tau_{21}, \tag{2.25}$$

which is called the round-trip time between node 1 and node 2. Then, we find that

$$\begin{aligned} A(s) &= \begin{bmatrix} 1 & 0 \\ 0 & \mathrm{e}^{-Ts} \end{bmatrix} \begin{bmatrix} 0 & \mathrm{e}^{-\tau_{12}s} \\ \mathrm{e}^{\tau_{12}s} & 0 \end{bmatrix} \\ &\triangleq \Theta(s)\hat{A}(s), \end{aligned}$$

where $\Theta = \begin{bmatrix} 1 & 0 \\ 0 & \mathrm{e}^{-Ts} \end{bmatrix}$ *is a diagonal matrix,* $\hat{A}(s) = \begin{bmatrix} 0 & \mathrm{e}^{-\tau_{12}s} \\ \mathrm{e}^{\tau_{12}s} & 0 \end{bmatrix}$ *satisfies* $\hat{A}(s) = \hat{A}^{\mathrm{T}}(-s)$, *and hence,* $\hat{A}(\mathrm{j}\omega)$ *is a Hermitian matrix. So, by Definition 2.2, the system based on the graph shown in Figure 2.5 is symmetric in the frequency domain even if* $\tau_{12} \neq \tau_{21}$.

Absorbing $\Theta(s)$ into the diagonal transfer function matrix of the agents implies that the agent associated with node 2 obtains an additional time delay T. So, this example shows that sometimes the frequency-domain symmetry of a communication topology can be achieved by introducing a lag in the time domain for agents!

Now, let us consider the general distributed control system composed of (2.1), (2.2) and (2.3) or equivalently the system shown in Figure 2.1. Assume the interconnection topology graph of the system is undirected. Then, to match the symmetry requirement of Definition 2.2, there should exist $T_i \in \mathbb{R}$, $i \in \overline{1, n}$, such that

$$\tau_{ij} + \tau_{ji} = T_i + T_j, \quad \forall i, j \in \overline{1, n}. \tag{2.26}$$

It is said that the system has symmetric communication delays if the requirement (2.26) is satisfied. Note that the symmetric communication delay does not imply $\tau_{ij} = \tau_{ji}$.

2.2.2 Symmetry of Bipartite Systems

Before we study the symmetry property of bipartite systems with forward and backward channels, let us introduce the notion of semi-homogeneousness as follows.

Definition 2.4 *Consider a bipartite system which is based on the bipartite graph $G(S \oplus L)$. It is said to be semi-homogeneous by vertex L if for each agent i in S, the following conditions are satisfied:*

(1) all the neighbors of agent i have homogeneous dynamics, i.e.,

$$P_l(s)\Gamma_l(s) = \gamma_l \alpha_i(s), \quad \forall l \in N_i \tag{2.27}$$

where $\gamma_l \in \mathbb{R}$ is a constant and $\alpha_i(s)$ is a scalar transfer function that is independent of l;

(2) the round-trip delay from agent i to its any neighbor in L is a constant, i.e.,

$$\tau_{li}^f + \tau_{li}^b = T_i, \quad \forall l \in N_i. \tag{2.28}$$

Similarly, one can also define the semi-homogeneousness of the system by vertex class S.

Remark. Obviously, if all the agents in vertex class L have homogeneous dynamics, (2.27) holds.

The next theorem shows that the technique used in Example 2.3 can be applied to general semi-homogeneous bipartite systems.

Theorem 2.5 *If the system based on the bipartite graph $G(S \oplus L)$ is semi-homogeneous by L or by S, then it is symmetric in the frequency domain.*

Proof. We just consider the case that the system is semi-homogeneous by L. The proof for the other case is the same.

Denote $\Gamma_0 = \mathrm{diag}\{\gamma_1, \cdots, \gamma_m\}$ and $\Xi(s) = \mathrm{diag}\{\alpha_1(s), \cdots, \alpha_n(s)\}$. Then, under assumption (2.27) we have

$$A_b(s)P(s)\Gamma(s) = \Xi(s)A_b(s)\Gamma_0.$$

Denote $T(s) = \mathrm{diag}\{\mathrm{e}^{-T_1 s}, \cdots, \mathrm{e}^{-T_n s}\}$. Then, by using (2.20) and (2.28), one can see that

$$A_b(s) = T(s)T(-s)A_b(s) = T(s)A_f^{\mathrm{T}}(-s).$$

So, under the assumption of the semi-homogeneousness of the system by L, we get

$$\begin{aligned} W_1(s) &= (I + G(s)K(s)A_b(s)P(s)\Gamma(s)A_f(s))^{-1}G(s)K(s) \\ &= (I + G(s)K(s)\Xi(s)A_b(s)\Gamma_0 A_f(s))^{-1}G(s)K(s) \\ &= (I + G(s)K(s)\Xi(s)T(s)A_f^{\mathrm{T}}(-s)\Gamma_0 A_f(s))^{-1}G(s)K(s). \end{aligned}$$

So, we get the open-loop transfer function of the system as

$$\begin{aligned} L(s) &= G(s)K(s)\Xi(s)T(s)A_f^{\mathrm{T}}(-s)\Gamma_0 A_f(s) \\ &\triangleq \hat{G}(s)\hat{A}(s), \end{aligned}$$

where $\hat{G}(s) = G(s)K(s)\Xi(s)T(s)$ is a diagonal matrix, $\hat{A}(s) = A_f^{\mathrm{T}}(-s)\Gamma_0 A_f(s)$ satisfies $\hat{A}(s) = \hat{A}^{\mathrm{T}}(-s)$. Then, by Definition 2.2, the system is symmetric in the frequency domain. □

2.3 Stability of Multivariable Systems

2.3.1 Poles and Stability

Let the state-space representation of the closed-loop system (2.14) be

$$\dot{x}(t) = A_0 x + \sum_{i=1}^{n_d} A_i x(t - \tau_i) + Br(t) \tag{2.29}$$

$$y(t) = Cx(t) + Dr(t) \tag{2.30}$$

where $x(t) \in \mathbb{R}^n$ is the state-vector of the system, $\{A_0, B, C, D\}$ is the minimal state-space realization of the system with all delays being zero, τ_i, $i \in \overline{1, n_d}$, are all possible time-delays in the closed-loop system, A_i, $i \in \overline{1, n_d}$, are the matrices corresponding to time-delayed states. Then, the *poles* of the system are the roots of the characteristic equation

$$\phi(s) := \det(sI - A_0 - \sum_{i=1}^{n_d} \mathrm{e}^{-s\tau_i}) = 0, \tag{2.31}$$

where is $\phi(s)$ is the characteristic (or pole) quasi-polynomial of the system. We also call $\phi(s)$ characteristic (or pole) quasi-polynomial corresponding to the transfer function $W(s)$ given by (2.14). By the theory of the time-delayed system (Gu, Kharitonov and Chen 2003), the stability of system (2.29) is fully determined by the characteristic quasi-polynomial $\phi(s)$. Namely, the system is stable if and only if $\phi(s)$ has no pole in the closed right half of the complex plane.

In this book we call the open left half of the complex plane simply the LHP and denote it as $\mathbb{C}^-$, and call the closed right half of the complex plane simply as RHP and denote it as $\mathbb{C}^+$. For simplicity of statement we also call the closed left half of the complex plane simply the closed LHP and denote it as $\bar{\mathbb{C}}^-$, and call the open right half of the complex plane simply the open RHP and denote it as $\check{\mathbb{C}}^+$.

Definition 2.6 *The time-delayed system (2.29) is said to be stable if*

$$\phi(s) \neq 0, \quad \forall s \in \mathbb{C}^+. \tag{2.32}$$

By this definition there is no difference between "stable" and "asymptotically stable", which are both used in this book. The following notion of semi-stability is also used in this book.

Definition 2.7 *The time-delayed system (2.29) is said to be semi-stable if*

$$\phi(s) \neq 0, \quad \forall s \in \check{\mathbb{C}}^{+}. \tag{2.33}$$

In particular, it is said to be steady semi-stable if

$$\phi(s) \neq 0, \quad \forall s \in \mathbb{C}^{+} \backslash 0. \tag{2.34}$$

By this definition, a linear time-invariant system is semi-stable if all its poles are inside the closed LHP which includes the imaginary axis, and steady semi-stable if all its poles are inside the LHP or at the origin of the complex plane. Obviously, if each root on the imaginary axis is simple, the semi-stability reduces to the so-called marginal stability.

The following theorem from MacFarlane and Karcanias (1976) allows us to determine the characteristic quasi-polynomial of the system directly from the transfer function matrix.

Theorem 2.8 *The pole polynomial $\phi(s)$ corresponding to transfer function $W(s)$ is the least common denominator of all non-identically-zero minors of all orders of $W(s)$.*

Let us illustrate the theorem by the following example, which shows how to determine the poles of the system directly from the transfer function matrix.

Example 2.9 *Consider the transfer function matrix*

$$W(s) = \frac{1}{2s(s+2)} \begin{bmatrix} s-1 & 2 \\ 3 & s-2 \end{bmatrix}. \tag{2.35}$$

The minors of order 1 are $\frac{s-1}{s(s+2)}$, $\frac{3}{s(s+2)}$, $\frac{3}{s(s+2)}$ and $\frac{s-2}{s(s+2)}$. The minor of order 2 is the determinant

$$\begin{aligned} \det W(s) &= \frac{(s-1)(s-2)-6}{s^2(s+2)^2} \\ &= \frac{(s+1)(s-4)}{s^2(s+2)^2}. \end{aligned}$$

The least common denominator of all the minors is

$$\phi(s) = s^2(s+2)^2.$$

Therefore, the system has four poles: double at $s = 0$ and double at $s = -2$.

For distributed control systems time-delay is inevitable. For a time-delayed system $W(s) = G(s)e^{-\tau s}$, where $G(s)$ is a rational function matrix and $\tau > 0$ is the delay constant, one can use Theorem 2.8 to handle the rational part $G(s)$ and calculate its "regular" poles. Then, by using a definition of the exponential function we have

$$e^{-\tau s} = \lim_{n \to \infty} \frac{1}{\left(1 + \frac{\tau}{n} s\right)^n}. \tag{2.36}$$

So, if a transfer function has a factor $e^{-\tau s}$, equation (2.36) implies that it also has an infinite number of LHP poles at $s = -\frac{n}{\tau}$.

2.3.2 Zeros and Pole-Zero Cancelation

In general, zeros are the values of complex variable s at which the transfer function matrix $W(s)$ loses rank. This is the basis of the following definition of zeros for multivariable systems (or multiple-input multiple-output systems, or briefly MIMO systems) (MacFarlane and Karcanias 1976).

Definition 2.10 *z_i is a zero of $W(s)$ if the rank of $W(z_i)$ is less than the normal rank of $W(s)$. The zero polynomial is defined as $z(s) = \prod_{i=1}^{n_z}(s - z_i)$ where n_z is the number of finite zeros of $W(s)$.*

Note that the normal rank of $W(s)$ is the rank of $W(s)$ at all values of s except at a finite number of points (which are the zeros). The zeros defined above are sometimes called *transmission zeros*.

If the system is described by a state-space model as follows:

$$\begin{bmatrix} 0 \\ y \end{bmatrix} = P(s) \begin{bmatrix} x \\ u \end{bmatrix}, \tag{2.37}$$

$$P(s) = \begin{bmatrix} sI - A & -B \\ C & D \end{bmatrix}, \tag{2.38}$$

then the transmission zeros of $W(s)$ are the values $s = z$ for which the matrix $P(s)$ loses rank.

Based on this definition of zeros we can introduce the notion of minimum phase systems.

Definition 2.11 *(Chen 1984; Davison 1983) The system defined by (2.37) and (2.38) is said to be minimum phase if all its transmission zeros lie inside the LHP.*

The following theorem from MacFarlane and Karcanias (1976) is often used for calculating the zeros of a transfer function matrix.

Theorem 2.12 *The zero polynomial corresponding to a minimal realization of the system $W(s)$ is the greatest common divisor of all the numerators of all order-r minors of $W(s)$, where r is the nominal rank of $W(s)$, provided that these minors have been adjusted in such a way that the pole polynomial $\phi(s)$ is their denominator.*

For single-input single-output (SISO) systems, if the transfer function has a pole and a zero at the same place on the complex plane, it is said that there is a pole-zero cancelation. For multivariable systems the situation becomes much more complicated. The following example shows a case of pole-zero cancelation.

Example 2.13 *Consider the transfer function matrix*

$$W(s) = \frac{1}{s(s+2)} \begin{bmatrix} s-1 & 2 \\ 6 & s-2 \end{bmatrix}. \tag{2.39}$$

The minors of order 1 are $\frac{s-1}{s(s+2)}, \frac{2}{s(s+2)}, \frac{6}{s(s+2)}$ *and* $\frac{s-2}{s(s+2)}$. *The minor of order 2 is the determinant*

$$\begin{aligned} \det W(s) &= \frac{(s-1)(s-2)-12}{s^2(s+2)^2} \\ &= \frac{s-5}{s^2(s+2)}. \end{aligned}$$

The least common denominator of all the minors is

$$\phi(s) = s^2(s+2).$$

So, by Theorem 2.8, the system has three poles: two at $s = 0$ *and one at* $s = -2$. *And by Theorem 2.12, the zero polynomial of the system is* $z(s) = s - 5$. *Note that there is a pole-zero cancelation at* $s = -2$ *when calculating the determinant of* $W(s)$.

The next example shows that even if a multivariable system has a pole and a zero at the same place, it does not mean that there must be a pole-zero cancelation.

Example 2.14 *Consider the transfer function matrix*

$$W(s) = \begin{bmatrix} \frac{2s}{s+2} & 0 \\ 0 & \frac{s+2}{s} \end{bmatrix}. \tag{2.40}$$

By Theorem 2.8, $W(s)$ *has poles at* $s = 0$ *and* $s = -2$. *We also notice that* $W(s)$ *loses rank at* $s = 0$ *and* $s = -2$. *So,* $s = 0$ *and* $s = -2$ *are also zeros of* $W(s)$, *but there is no pole-zero cancelation.*

If W(s) loses rank for any value of the complex variable s, by Definition 2.10 or Theorem 2.12, it does not mean that $W(s)$ has (transmission) zeros at any place in the complex plane. However, such a "non-existing" zero may also cause a pole-zero cancelation. Sometimes, we call such a zero an *invariant zero*.

Example 2.15 *Let*

$$W_1(s) = \begin{bmatrix} \frac{1}{s} & -\frac{1}{s} \\ -\frac{1}{s} & \frac{1}{s} \end{bmatrix}, \quad W_2(s) = \begin{bmatrix} \frac{1}{s} & -\frac{2}{s} \\ -\frac{2}{s} & \frac{1}{s} \end{bmatrix}.$$

By Definition 2.10, both $W_1(s)$ *and* $W_2(s)$ *have no zero. Now, let us check the poles of* $W_1(s)$ *and* $W_2(s)$. *By Theorem 2.8,* $W_1(s)$ *has only a single pole at* $s = 0$ *but* $W_2(s)$ *has double poles at* $s = 0$. *Where is another "pole"* $s = 0$ *for* $W_1(s)$? *It may be assumed that the pole is canceled by an invariant zero at* $s = 0$ *because* $\det W(s) = 0$ *for any value of* s *including zero, although such a zero is not counted by Definition 2.10.*

Now, we consider the poles of the cascaded system $W_1(s)W_2(s)$. At the first look, one may guess that the cascaded system has triple poles at $s = 0$. But, since

$$W_1(s)W_2(s) = \begin{bmatrix} \frac{3}{s^2} & -\frac{3}{s^2} \\ -\frac{3}{s^2} & \frac{3}{s^2} \end{bmatrix}$$

is singular for any complex variable s, by Theorem 2.8, we know that $W_1(s)W_2(s)$ only has double poles at $s = 0$. The third pole at $s = 0$ is canceled again by the invariant zero.

Invariant zeros are actually caused by the singularity of system matrices. Note that in consensus problems, which will be discussed in Chapter 6 and Chapter 7, open-loop transfer function matrices are usually singular for any complex variable since the Laplacian matrix is singular.

2.4 Frequency-Domain Criteria of Stability

The return difference equation of the closed-loop system shown in Figure 2.2 is

$$I + G(s)K(s)\bar{A}(s) = 0. \tag{2.41}$$

According to linear system theory (Chen 1984; Desoer and Vidyasagar 1975), it holds that

$$\det(I + G(s)K(s)\bar{A}(s)) = \det(I + G(\infty)K(\infty)\bar{A}(\infty))\frac{\phi_c(s)}{\phi_o(s)}, \tag{2.42}$$

where $\phi_c(s)$ is the characteristic polynomial (or quasi-polynomial for time-delayed systems) of the closed-loop system, and $\phi_o(s)$ is the characteristic polynomial (or quasi-polynomial for time-delayed systems) of the open-loop system. $\hat{G}(s) := G(s)K(s)\bar{A}(s)$ is referred to as the open-loop transfer function of the system. For the purpose of the well-posedness of the system we always assume that

$$\det(I + G(\infty)K(\infty)\bar{A}(\infty)) \neq 0.$$

Equation (2.42) shows that the solutions of the following equation

$$\det(I + \hat{G}(s)) = 0 \tag{2.43}$$

may be a proper subset of the poles of the closed-loop system since cancelations may occur on the right side of (2.42). Nevertheless, if there is no cancelation between RHP pole(s) of the closed-loop system and RHP pole(s) of the open-loop system, the system is stable if and only if all the solutions of the equation (2.43) have strictly negative real parts.

Exercise 2.16 *Let the transfer function matrix $W(s)$ be given by (2.14). Show that if*

$$\det(I + G(\infty)K(\infty)\bar{A}(\infty)) = 0,$$

then (1) $W(s) \to \infty$ as $s \to \infty$, (2) the closed-loop system does not have a state representation in the standard form (A, B, C, D).

2.4.1 Loop Transformation and Multiplier

For a given closed-loop system, one can make some loop transformation as shown by Figure 2.6 for stability analysis. Obviously, we have

$$\begin{aligned}&\det(I + G_1(s)K(s))\det(I + (I + G_1(s)K(s))^{-1}G_1(s)(G_2(s) - K(s)))\\&= \det(I + G_1(s)G_2(s)),\end{aligned}$$

which yields the conclusion that

$$\det(I + G_1(s)G_2(s)) \neq 0, \quad \forall s \in \mathbb{S}$$

if and only if

$$\det(I + (I + G_1(s)K(s))^{-1}G_1(s)(G_2(s) - K(s)) \neq 0, \quad \forall s \in \mathbb{S}$$

and

$$\det(I + G_1(s)K(s)) \neq 0, \quad \forall s \in \mathbb{S},$$

where $\mathbb{S}$ denotes a region in the complex plane, which can be $\mathbb{C}^+$, $\check{\mathbb{C}}^+$, or $\mathbb{C}^+\backslash 0$ depending on the stability requirement. Therefore, if the feedback system composed of $G_1(s)$ and $K(s)$ is (semi-)stable, the system shown by Figure 2.6(a) has the same (semi-)stability property as the system shown in Figure 2.6(b).

One can also introduce some multiplier into the loop as shown by Figure 2.7 and Figure 2.8.

Loop transformation is a very useful tool in the analysis of the stability of feedback systems. When we apply some stability criteria such as the Nyquist criterion, ignorance of pole cancelations between the open-loop transfer function and the closed-loop transfer function may result in incorrect results as we will show later in this chapter. Proper loop transformation can remove such cancelations because it changes open-loop transfer functions. Moreover, since some stability criteria such as spectral radius theorem, small-gain theorem and passivity

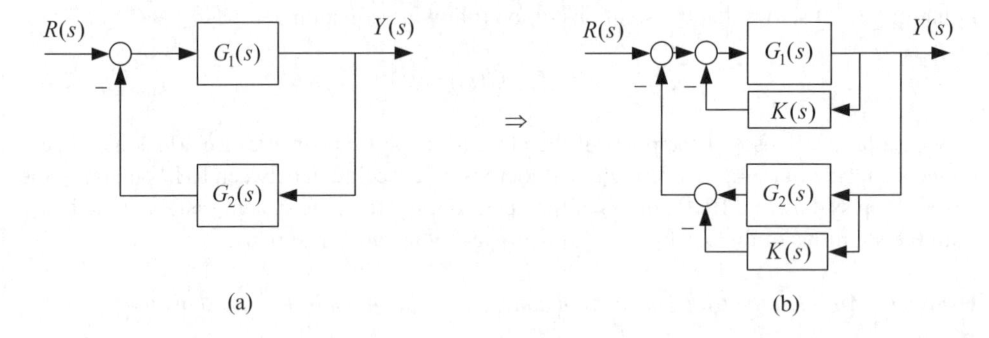

Figure 2.6 (a) Original feedback system; (b) System with loop transformation.

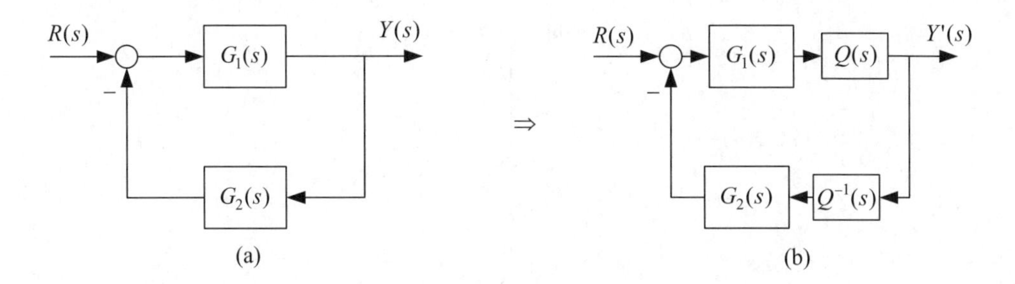

Figure 2.7 (a) Original feedback system; (b) System with multiplier.

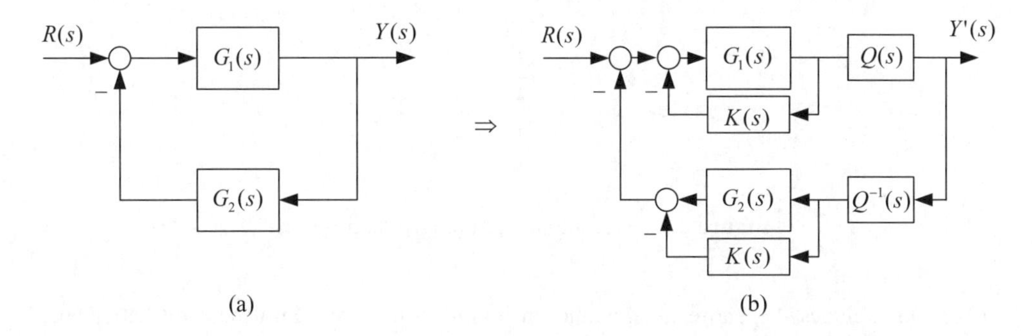

Figure 2.8 (a) Original feedback system; (b) System with mixed loop transformation.

theorem (positive realness theorem) provide only sufficient conditions of stability and hence are conservative, loop transformations and multipliers change open-loop transfer functions and are helpful in reducing the conservatism of these stability criteria.

2.4.2 Multivariable Nyquist Stability Criterion

The stability test for closed-loop systems can be performed by use of frequency-domain techniques based on the multivariable (or generalized) Nyquist stability criterion. The following theorem is a modified version of the generalized Nyquist stability criterion which has been widely used in references (see, e.g., Vidyasagar 1985).

Theorem 2.17 *Consider a system with an open-loop transfer function $\hat{G}(s)$ which has η RHP poles. Then, the closed-loop system has no RHP poles if and only if the Nyquist diagram of* $\det(I + \hat{G}(s))$

(1) does not pass through the origin (i.e., $\det(I + \hat{G}(j\omega)) \neq 0$, *for all* $\omega \in \mathbb{R}$*);*

(2) encircles the origin η times anticlockwise.

Remark 1. By the Nyquist diagram or the Nyquist plot of $\det(I + \hat{G}(s))$ we mean the image of $\det(I + \hat{G}(s))$ as s goes clockwise around the Nyquist D-contour. The Nyquist

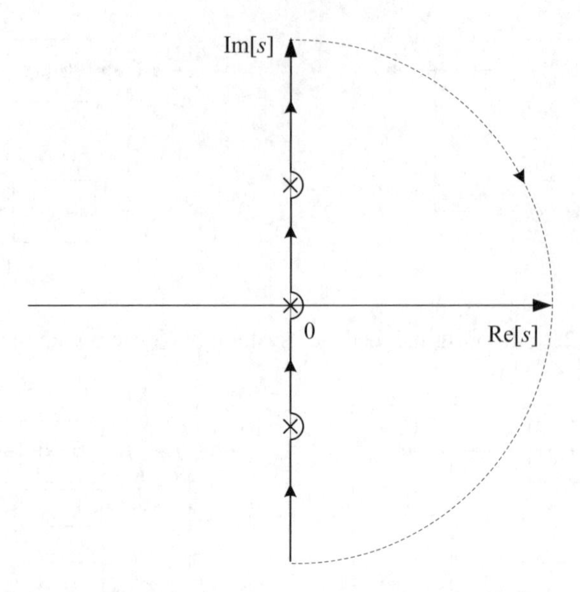

Figure 2.9 D-contour excluding jω-axis poles.

D-contour includes the entire jω-axis and an infinite semi circle into the right-half plane. If $\hat{G}(s)$ has jω-axis poles, one can use the D-contour illustrated in Figure 2.9 to avoid these poles by making sufficiently small semi circles into the RHP around them. By using such a D-contour the jω-axis poles of $\hat{G}(s)$ are obviously excluded from the number of RHP poles of the open-loop system, which are sometimes briefly referred to as open-loop RPH poles (note that the "open-loop RHP pole" is not equal to the "open RHP pole"!). Since this D-contour goes around jω-axis poles of $\hat{G}(s)$ in the *anti-clockwise* direction, when drawing the Nyquist plot of $\det(I + \hat{G}(\mathrm{j}\omega))$ one should add a semi circle with an infinite radius in the *clockwise* direction to connect the undefined part of the plot corresponding to each jω-axis pole. In this way, the jω-axis poles of the open-loop system are not regarded as "RHP poles". So, the term "RHP poles" in the theorem actually should be understood as "open RHP poles". If there is no cancelation of any of the jω-axis poles between the open-loop transfer function and the closed-loop transfer function, conditions (1) and (2) imply stability. Otherwise, only semi-stability is ensured.

Remark 2. The theorem itself does not tell if there is any RHP pole cancelation between the closed-loop system and the open-loop system. In Chapter 7 we will discuss how to justify if there exist multiple pole cancelations between the closed-loop system and the open-loop system at $s = 0$.

Remark 3. Since loop-transformation changes the open-loop transfer function but does not change the closed-loop transfer function, one can also use some loop transformation to avoid pole-zero cancelations of open-loop transfer functions and get the correct conclusion on the stability of the closed-loop system.

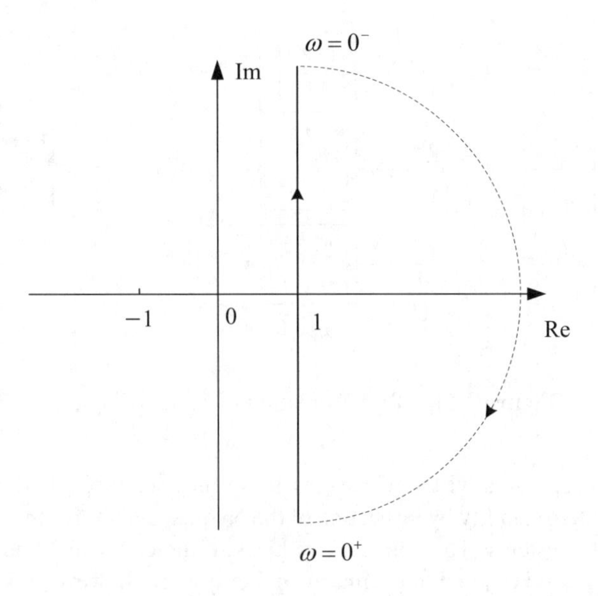

Figure 2.10 Nyquist plots without $s = 0$ as an open-loop RHP pole.

We use the following example to illustrate Theorem 2.17 and the associated remarks as well.

Example 2.18 *Let* $\hat{G}(s) = G(s)A$*, where* $G(s) = \mathrm{diag}\{\frac{1}{s}, \frac{1}{s}\}$*, and* $A = \begin{bmatrix} 1 & -1 \\ -1 & 1 \end{bmatrix}$*. By Theorem 2.8 we know that* $\hat{G}(s)$ *has a single pole at* $s = 0$*. Actually, as we have shown in Example 2.15, there is another open-loop pole at* $s = 0$ *which is canceled because* $\det A = 0$*. Using the modified Nyquist D-contour given in Figure 2.9, the poles at* $s = 0$ *are not regarded as the RHP poles of* $\hat{G}(s)$*. So we have* $\eta = 0$*. Because*

$$\det(I + \hat{G}(s)) = 1 + \frac{2}{s},$$

we can draw the Nyquist plot of $\det(I + K\hat{G}(\mathrm{j}\omega))$ *as shown in Figure 2.10. In the plot we see that it neither passes through nor encircles the origin. Therefore, one may conclude that the closed-loop system has no RHP poles. But this does not lead to the conclusion that the system is stable. Instead, only the steady semi-stability can be derived, i.e., all the poles of the closed-loop system are inside the LHP or at* $s = 0$*.*

We can investigate the stability of this system by making a loop transformation. The system diagram is sketched in Figure 2.11(a). It can be transformed into the form given by Figure 2.11(c). From Figure 2.11(c) we have the transfer function matrix of the open-loop system as $\bar{G}(s) = \mathrm{diag}\{\frac{1}{s+1}, \frac{1}{s+1}\} \begin{bmatrix} 0 & -1 \\ -1 & 0 \end{bmatrix}$*, and* $\det(I + \bar{G}(s)) = 1 - \frac{1}{(s+1)^2}$*. Therefore, we have* $\eta = 0$*. The Nyquist plot of* $\det(I + \bar{G}(\mathrm{j}\omega))$ *is given in Figure 2.12. It passes through the origin, and thus, the system is just marginally stable. The conclusion is the same as that obtained from the original diagram.*

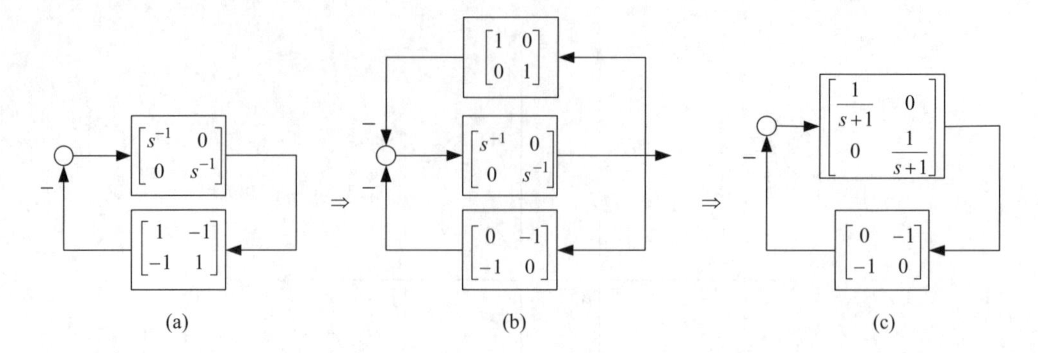

Figure 2.11 Transformation of system diagram.

As is well known, a great virtue of the classic Nyquist stability criterion is that it checks the closed-loop system stability by inspection of the Nyquist diagram of the open-loop transfer function. For SISO systems, Theorem 2.17 reduces to the classical Nyquist criterion since $\det(1 + k\hat{G}(j\omega)) = 1 + k\hat{G}(j\omega)$. For the multivariable systems, however, this theorem requires a plot of $\det(I + K\hat{G}(j\omega))$. Desoer and Wang (1980) proved a version of the generalized Nyquist stability criterion which checks the stability of the closed-loop system by inspection of the eigenloci of the open-loop system.

The eigenloci are defined as the set of all the eigenvalues of the frequency response of the open-loop transfer function matrix, $\lambda_i(K\hat{G}(j\omega))$, $i \in \overline{1, n}$, as $j\omega$ goes clockwise around the Nyquist D-contour. Note that, unlike the SISO case, in general $\lambda_i(K\hat{G}(s))$ is not a rational function of s, and hence the conjugate property does not hold, i.e.,

$$\lambda_i(s^*) \neq \lambda_i^*(s).$$

This leads to the fact that a single eigenlocus $\lambda_i(K\hat{G}(j\omega))$ may not form a closed path when ω varies from $-\infty$ to ∞. Fortunately, when $\lambda_i(s)$ is irrational on s, there always exists another

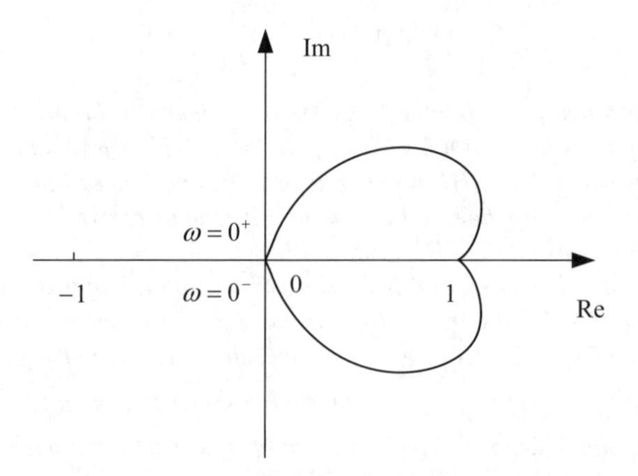

Figure 2.12 Nyquist plot of the transformed system.

eigenvalue $\lambda_{i+1}(s)$ which is conjugate with $\lambda_i(s)$, i.e.,

$$\lambda_i(s^*) = \lambda_{i+1}^*(s), \quad \lambda_{i+1}(s^*) = \lambda_i^*(s).$$

Hence, the part of the eigenlocus $\lambda_i(K\hat{G}(j\omega))$, $\omega \in (0, \infty)$ is conjugate with the part of the eigenlocus $\lambda_{i+1}(K\hat{G}(j\omega))$, $\omega \in (-\infty, 0)$, and the part of the eigenlocus $\lambda_i(K\hat{G}(j\omega))$, $\omega \in (-\infty, 0)$ is conjugate with the part of the eigenlocus $\lambda_{i+1}(K\hat{G}(j\omega))$, $\omega \in (0, \infty)$. Therefore, when ω varies from $-\infty$ to ∞, the eigenloci of $\lambda_i(K\hat{G}(j\omega))$ and $\lambda_{i+1}(K\hat{G}(j\omega))$ form two closed paths. In such a way, we can count the times in which the eigenloci $\lambda_i(K\hat{G}(j\omega))$, $i \in \overline{1, n}$, enclose the point $(-1, j0)$.

Let N be the sum of the times in which the eigenloci $\lambda_i(K\hat{G}(j\omega))$, $i \in \overline{1, n}$, enclose the point $(-1, j0)$ anticlockwise as $j\omega$ goes clockwise around the Nyquist D-contour. Then, the generalized Nyquist stability criterion can be stated as follows.

Theorem 2.19 *Consider a system with an open-loop transfer function $\hat{G}(s)$ which has η open RHP poles. Then, the closed-loop system has no open RHP poles if and only if*

(1) $\lambda_i(\hat{G}(j\omega)) \neq -1$, *for all* $\omega \in \mathbb{R}$;

(2) $N = \eta$.

Furthermore, if there is no cancelation of $j\omega$*-axis poles between the open-loop system and the closed-loop system, the closed-loop system is stable.*

Remark. The assumption of no cancelation of $j\omega$-axis poles between the open-loop system and the closed-loop system can always be satisfied by introducing a proper loop transformation.

For many network-based distributed control systems, open-loop systems composed of agents have no open RHP poles. The following theorem, which is actually a corollary of Theorem 2.19, is very useful for such systems.

Theorem 2.20 *Suppose that the open-loop transfer function $\hat{G}(s)$ has no open RHP poles. Then, the closed-loop system has no open RHP poles if and only if the eigenloci $\lambda_i(\hat{G}(j\omega))$, $i \in \overline{1, n}$, do not enclose the point $(-1, j0)$ as $j\omega$ goes clockwise around the Nyquist D-contour. Furthermore, if there is no cancelation of $j\omega$-axis poles between the open-loop system and the closed-loop system, the closed-loop system is stable.*

Theorem 2.17, Theorem 2.19 and Theorem 2.20 can be easily extended to discrete-time systems after replacing the D-contour by the unit circle in the complex plane. Note that if the z-transfer function matrix $\hat{G}(z)$ of the open-loop system has a pole just on the unit circle, the unit circle should be modified by making a sufficiently small semi circle into the region outside the unit circle around the pole (see Figure 2.13).

For convenience of statement, we denote by $\mathbb{D}$ the interior of the unit disc (briefly as IUD), i.e., $\mathbb{D} = \{z \in \mathbb{C} : |z| < 1\}$. The closure of $\mathbb{D}$ is denoted by $\bar{\mathbb{D}} = \{z \in \mathbb{C} : |z| \leq 1\}$ and briefly referred to as the closed IUD. We also denote by $\mathbb{D}^o$ the closed outer part of the unit disc (briefly as OUD), i.e., $\mathbb{D}^o = \{z \in \mathbb{C} : |z| \geq 1\}$ and denote by $\breve{\mathbb{D}}^o$ the open outer part of the unit disc (briefly as the open OUD), i.e., $\breve{\mathbb{D}}^o = \{z \in \mathbb{C} : |z| > 1\}$.

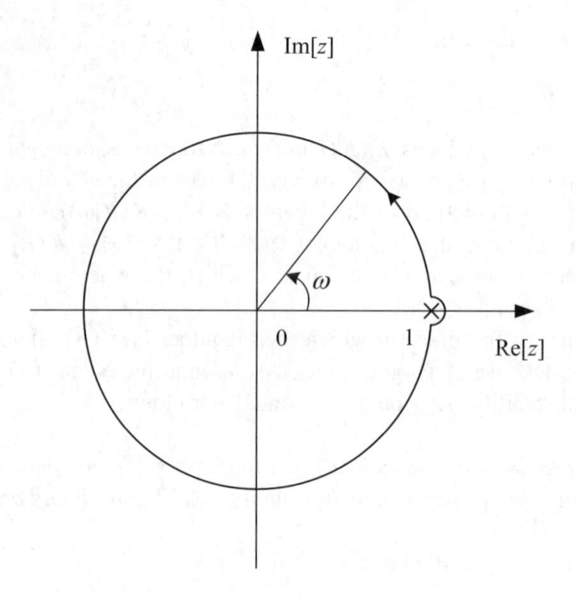

Figure 2.13 Modified unit circle.

Definition 2.21 *Let $\phi(z)$ be the pole (quasi-)polynomial of a discrete-time. The system is said to be stable if*

$$\phi(z) \neq 0, \quad \forall z \in \mathbb{D}^{\mathrm{o}}. \tag{2.44}$$

The system is said to be semi-stable if all its poles are in

$$\phi(z) \neq 0, \quad \forall z \in \breve{\mathbb{D}}^{\mathrm{o}}. \tag{2.45}$$

In particular, it is said to be steady semi-stable if

$$\phi(z) \neq 0, \quad \forall z \in \mathbb{D}^{\mathrm{o}} \backslash (1, -\mathrm{j}0). \tag{2.46}$$

Then, the discrete-time analog of Theorem 2.20 can be stated as follows. The discrete-time analogs of the other versions of the generalized Nyquist stability criterion are similar.

Theorem 2.22 *Suppose that the open-loop transfer function $\hat{G}(z)$ of a discrete-time system has no open OUD poles. Then, the closed-loop system has no open OUD poles if and only if the eigenloci $\lambda_i(\hat{G}(\mathrm{e}^{\mathrm{j}\omega}))$, $i \in \overline{1, n}$, do not enclose the point $(-1, \mathrm{j}0)$ as ω varies from $-\pi$ to π. Furthermore, if there is no cancelation of unit-circle poles between the open-loop system and the closed-loop system, the closed-loop system is stable.*

2.4.3 Spectral Radius Theorem and Small-Gain Theorem

Using the generalized Nyquist stability criterion, Theorem 2.17, it can be proved that the stability of a closed-loop system with a stable open-loop transfer function can be verified by testing the spectral radius of the open-loop transfer function at each frequency, which is defined as the maximum eigenvalue magnitude, i.e.,

$$\rho(\hat{G}(\mathrm{j}\omega)) = \max_i |\lambda_i(\hat{G}(\mathrm{j}\omega))|. \tag{2.47}$$

Theorem 2.23 *Consider a system with a stable open-loop transfer function $\hat{G}(s)$. Then, the closed-loop system is stable if*

$$\rho(\hat{G}(\mathrm{j}\omega)) < 1, \ \forall \omega \in \mathbb{R}. \tag{2.48}$$

This theorem which is referred to as the *spectral radius theorem* was proved in Skogestad and Postlethwaite (1996). With some modification of the proof we can prove the following *extended spectral radius theorem* for the test of steady semi-stability and marginal stability.

Theorem 2.24 *Consider a system with a stable open-loop transfer function $\hat{G}(s)$. Then, the closed-loop system is steady semi-stable, if*

$$\rho(\hat{G}(\mathrm{j}\omega)) < 1, \ \forall \omega \in \mathbb{R}, \omega \neq 0. \tag{2.49}$$

Moreover, if $\rho(\hat{G}(0)) = 1$, and $\det(I + \hat{G}(0)) = 0$, then the closed-loop system is marginally stable with $s = 0$ as a simple pole.

Proof. Firstly, we note that the closed-loop system has no pure imaginary poles because $\rho(\hat{G}(\mathrm{j}\omega)) < 1, \forall \omega \in \mathbb{R}, \omega \neq 0$.

Secondly, we prove that the closed-loop system has no open RHP poles. We prove this by contradiction. Suppose that it has at least one open RHP pole. By the generalized Nyquist theorem (Theorem 2.17), $\det(I + \hat{G}(\mathrm{j}\omega))$ must enclose the origin, which implies that there exists an $\omega_0 \in \mathbb{R}$ such that $\det(I + \hat{G}(\mathrm{j}\omega_0)) < 0$. Then, there exists a gain $\epsilon \in (0, 1)$ such that

$$\det(I + \epsilon\hat{G}(\mathrm{j}\omega_0)) = 0 \tag{2.50}$$

because $\det(I + \epsilon\hat{G}(\mathrm{j}\omega_0)) = 1$ for $\epsilon = 0$. Then, the following derivation leads a contradiction:

$$\begin{aligned}
(2.50) &\Leftrightarrow \prod_i \lambda_i(I + \epsilon\hat{G}(\mathrm{j}\omega_0)) = 0 && (2.51)\\
&\Leftrightarrow 1 + \epsilon\lambda_i(\hat{G}(\mathrm{j}\omega_0)) = 0 \ \text{ for some } i && (2.52)\\
&\Leftrightarrow \lambda_i(\hat{G}(\mathrm{j}\omega_0)) = -\frac{1}{\epsilon} \ \text{ for some } i && (2.53)\\
&\Rightarrow |\lambda_i(\hat{G}(\mathrm{j}\omega_0))| > 1 \ \text{ for some } i && (2.54)\\
&\Leftrightarrow \rho(\hat{G}(\mathrm{j}\omega_0)) > 1. && (2.55)
\end{aligned}$$

Based on the above discussion we conclude that the closed-loop system is steady semi-stable.

Thirdly, $\det(I + \hat{G}(0)) = 0$ implies that the closed-loop system has at least one pole at $s = 0$. Now, we show that any possible pole of the closed-loop system at $s = 0$ must be simple. Indeed, let us consider poles of a perturbed system with characteristic equation

$$\det(I + \epsilon\hat{G}(s)) = 0, \tag{2.56}$$

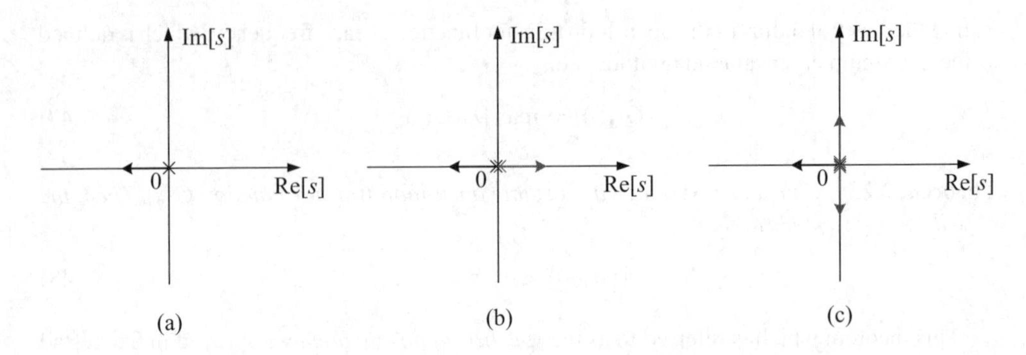

Figure 2.14 Root locus near $s = 0$: (a) simple pole, (b) double poles, (c) triple poles.

where $\epsilon \in (0, 1)$. Obviously, $\rho(\epsilon\hat{G}(j\omega)) < 1$, for all $\omega \in \mathbb{R}$ including $\omega = 0$ because $\rho(\hat{G}(0)) = 1$. So, by Theorem 2.23 the perturbed system is stable. By the rule of root locus (Nise 2000), root loci for perturbed systems with simple pole, double poles and triple poles, respectively, at $s = 0$ are shown in Figure 2.14. It is clear that only the case of simple pole can be stable. □

Note that the open-loop system $\hat{G}(s)$ in the above theorem may be obtained based on some loop transformation.

Example 2.25 *Let us consider the system given in Example 2.18 again. After the loop transformation shown by Figure 2.11 we have*

$$\hat{G}(s) = \mathrm{diag}\left\{\frac{1}{s+1}, \frac{1}{s+1}\right\}\begin{bmatrix} 0 & -1 \\ -1 & 0 \end{bmatrix}.$$

It is easy to check that

$$\rho(\hat{G}(j\omega)) < 1, \ \forall\omega \in \mathbb{R}, \omega \neq 0;$$
$$\rho(\hat{G}(j\omega)) = 1, \ \text{for } \omega = 0.$$

and $\det(I + \bar{G}(0)) = 0$. *So, the closed-loop system has no RHP poles except for a simple pole at $s = 0$, and hence, is marginally stable.*

If we choose a matrix norm satisfying $\|AB\| \leq \|A\| \cdot \|B\|$, we have $\rho(\hat{G}) \leq \|\hat{G}\|$. Then, some *small-gain theorems* can be directly obtained from Theorem 2.23 and Theorem 2.24 as follows.

Theorem 2.26 *Consider a system with a stable open-loop transfer function $\hat{G}(s)$. Then, the closed-loop system is stable if*

$$\|\hat{G}(j\omega)\| < 1, \ \forall\omega \in \mathbb{R}. \tag{2.57}$$

Theorem 2.27 *Consider a system with a stable open-loop transfer function $\hat{G}(s)$. Then, the closed-loop system has all poles in the LHP except for a simple pole at $s = 0$ and hence is* *marginally stable, if*

$$\|\hat{G}(j\omega)\| < 1, \ \forall \omega \in \mathbb{R}, \omega \neq 0; \tag{2.58}$$

$$\|\hat{G}(j\omega)\| = 1, \ \text{for } \omega = 0. \tag{2.59}$$

2.4.4 Positive Realness Theorem

According to the generalized Nyquist stability criterion, Theorem 2.17, one can also analyze the stability of the closed-loop system by testing the positive realness of the eigenvalues of the transfer function matrix.

Theorem 2.28 *Consider a system with a stable open-loop transfer function $\hat{G}(s)$. Then, the closed-loop system is stable if*

$$1 + \text{Re}(\lambda_i(\hat{G}(j\omega))) > 0, \ \forall \omega \in \mathbb{R}, \forall i \in \overline{1, n}. \tag{2.60}$$

With slight modification of the proof of Theorem 2.24 one can also obtain the following theorem for the test of marginal stability.

Theorem 2.29 *Consider a system with a stable open-loop transfer function $\hat{G}(s)$. Then, the closed-loop system has all poles in the LHP except for a simple pole at $s = 0$ and hence is marginally stable, if*

$$1 + \text{Re}(\lambda_i(\hat{G}(j\omega))) > 0, \ \forall \omega \in \mathbb{R}, \omega \neq 0, \forall i \in \overline{1, n}; \tag{2.61}$$

$$1 + \text{Re}(\lambda_i(\hat{G}(j\omega))) = 0, \ \text{for } \omega = 0, i \in \overline{1, n}. \tag{2.62}$$

2.5 Scalable Stability Criteria

2.5.1 Estimation of Spectrum of Complex Matrices

To apply Theorem 2.17 or Theorem 2.19 it is necessary to get the eigenloci $\lambda_i(K\hat{G}(s))$. Generally, there is no analytical formula for calculating eigenvalues of matrices. The following lemmas can be used for the estimation of the spectrum of $A \in \mathbb{C}^{n \times n}$ denoted by $\sigma(A) := \{\lambda_1(A), \cdots, \lambda_n(A)\}$.

The first lemma is usually called Gershgorin's disc lemma (Horn and Johnson 1985).

Lemma 2.30 *For any $A = (a_{ij}) \in \mathbb{C}^{n \times n}$ it holds*

$$\sigma(A) \subset \bigcup_{i=1}^{n} D_i, \tag{2.63}$$

where

$$D_i = \{z \in \mathbb{C} : |z - a_{ii}| \leq \sum_{j=1, j \neq i}^{n} |a_{ij}|\}. \tag{2.64}$$

Remark. The eigenvalues of A are also in the union of the discs:

$$|z - a_{jj}| \leq \sum_{i=1, i \neq j}^{n} |a_{ij}|, \; j \in \overline{1, n}. \tag{2.65}$$

From (2.64) it follows that $|z| \leq \sum_{j=1}^{n} |a_{ij}|$, which implies that any complex number in the disk D_i has the absolute value less than or equal to the sum of the absolute values of the entries of the same row. So, the following corollary is straightforward.

Corollary 2.31 *For any $A = (a_{ij}) \in \mathbb{C}^{n \times n}$ it holds*

$$\rho(A) \leq \max_{i \in \overline{1,n}} \left\{ \sum_{j=1}^{n} |a_{ij}| \right\}. \tag{2.66}$$

Define

$$H = \text{diag}\{h_i, i \in \overline{1, n}\}, \tag{2.67}$$

where $h_i > 0, i \in \overline{1, n}$. We have $\rho(A) = \rho(HAH^{-1})$ for $A \in \mathbb{C}^{n \times n}$. Based on this fact we get the following corollary of Gershgorin's disc lemma for reducing the conservatism of Corollary 2.31.

Corollary 2.32 *For any $A = (a_{ij}) \in \mathbb{C}^{n \times n}$ it holds*

$$\rho(A) \leq \min_{h_1, \cdots, h_n > 0} \max_{i \in \overline{1,n}} \left\{ \sum_{j=1}^{n} \frac{h_i}{h_j} |a_{ij}| \right\}. \tag{2.68}$$

Let $\xi \in \mathbb{C}^{n \times 1}$ and $\xi^* \xi = 1$. Then, for any eigenvalue λ_i of A we have $\lambda_i = \lambda_i \xi^* \xi = \xi^* A \xi$. So, by defining the field of values of A as follows (Horn and Johnson 1991):

$$F(A) = \{\xi^* A \xi : \; \xi^* \xi = 1\}, \tag{2.69}$$

the following lemma is obvious.

Lemma 2.33 *For any $A = (a_{ij}) \in \mathbb{C}^{n \times n}$, it holds*

$$\sigma(A) \subset F(A). \tag{2.70}$$

If the complex matrix A can be written as a product of a diagonal matrix and a Hermitian matrix, then its spectrum can be further estimated by Vinnicombe's lemma (Vinnicombe 2000), which can be derived from Lemma 2.33.

Lemma 2.34 *Let $Q \in \mathbb{C}^{n \times n}$, $Q = Q^* \geq 0$ and $T = \text{diag}(t_i, t_i \in \mathbb{C}, i \in \overline{1, n})$. Then,*

$$\sigma(TQ) \subset \rho(Q)\text{Co}(0 \cup \{t_i, i \in \overline{1, n}\}), \tag{2.71}$$

where $\rho(\cdot)$ denotes the matrix spectral radius, and Co$(\cdot)$ *the convex hull.*

Proof. Denote by $N(A)$ the null space of the complex matrix A. Then,

$$\begin{aligned}\sigma(Q^{1/2}TQ^{1/2}) &\subset N(Q^{1/2}TQ^{1/2})\\ &= \{v^*Q^{1/2}TQ^{1/2}v : v \in \mathbb{C}^n, \|v\| = 1\}\\ &\subset \rho(Q)\{w^*Tw : w \in \mathbb{C}^n, \|w\| \le 1\}\\ &= \rho(Q)\left\{\sum_{i=1}^{n} |w_i|^2 t_i : w_i \in \mathbb{C}, \sum_{i=1}^{n} |w_i| \le 1\right\}\\ &= \rho(Q)\mathrm{Co}(0 \cup \{t_i, i \in \overline{1, n}\}).\end{aligned}$$

So, we have

$$\{0\} \cup \sigma(Q^{1/2}TQ^{1/2}) = \{0\} \cup \sigma(TQ) \subset \rho(Q)\mathrm{Co}(0 \cup \{t_i, i \in \overline{1, n}\}),$$

which implies that $\sigma(TQ) \subset \rho(Q)\mathrm{Co}(0 \cup \{t_i, i \in \overline{1, n}\})$. The lemma is thus proved. □

Remark. The lemma was proved by Vinnicombe (2000) for the case of $Q = Q^* > 0$. The current form of the lemma and the proof are given by Lestas and Vinnicombe (2005).

The following example illustrates that in most cases Vinnicombe's lemma is much less conservative than Gershgorin's disc lemma if A is a product of a diagonal matrix and a Hermitian matrix.

Example 2.35 *Suppose $t_i = t_0 e^{j\theta_i}$, $t_0 \in \mathbb{R}^+$, $i \in \overline{1, n}$, and $Q \in \mathbb{C}^{n\times n}$ is a Hermitian matrix with zero as diagonal entries, i.e., $q_{ii} = 0$. Then, by Gershgorin's disc lemma we have*

$$\begin{aligned}\sigma(TQ) &\subset \bigcup_{i=1}^{n} D_i\\ &= \bigcup_{i=1}^{n}\left\{z \in \mathbb{C} : |z - t_i q_{ii}| \le \sum_{j=1, j\neq i}^{n} |t_i q_{ij}|\right\}\\ &= \left\{z \in \mathbb{C} : |z| \le t_0 \max_{i\in\overline{1,n}}\left\{\sum_{j=1}^{n} |q_{ij}|\right\}\right\}.\end{aligned}$$

Now, by using Vinnicombe's lemma and Corollary 2.31 of Gershgorin's disc lemma we get

$$\begin{aligned}\sigma(TQ) &\subset \rho(Q)\mathrm{Co}(0 \cup \{t_i, i \in \overline{1, n}\})\\ &\subseteq \max_{i\in\overline{1,n}}\left\{\sum_{j=1}^{n} |q_{ij}|\right\}\mathrm{Co}(0 \cup \{t_i, i \in \overline{1, n}\})\\ &\triangleq \mathrm{Co}.\end{aligned}$$

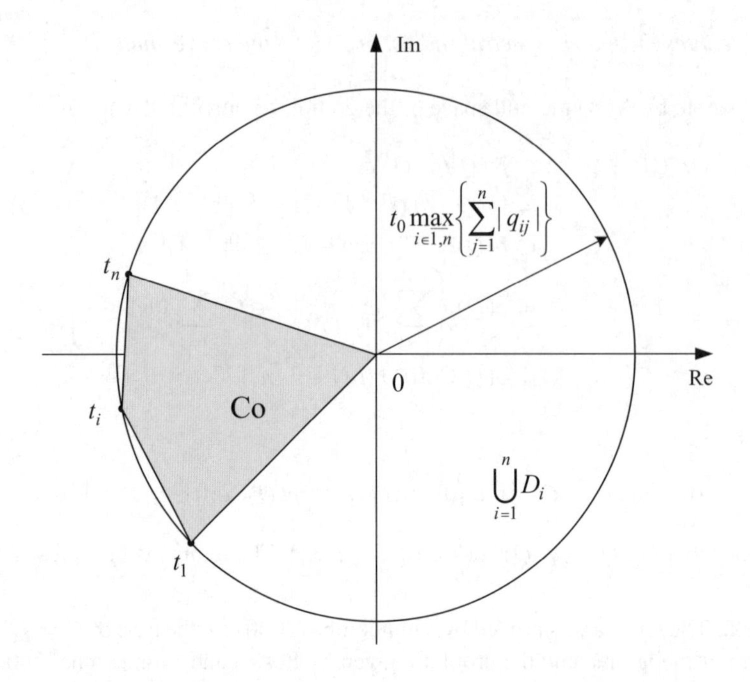

Figure 2.15 Comparison of Gershgorin's lemma and Vinnicombe's lemma.

$\bigcup_{i=1}^{n} D_i$ *is a disc centered at the origin with radius as* $t_0 \max_{i \in \overline{1,n}} \left\{ \sum_{j=1}^{n} |q_{ij}| \right\}$. Co *is a convex polygon contained in the disc as shown by Figure 2.15. Obviously, the higher the diversity degree of* θ_i *is, the closer the two regions are.*

2.5.2 Scalable Stability Criteria for Asymmetric Systems

Now, we consider the stability of the distributed control system described by (2.1), (2.2) and (2.3), which is shown by Figure 2.2. For convenience, let us re-sketched the diagram of the system as shown by Figure 2.16, where $\Lambda(s)$ is defined by (2.11) and $A(s)$ is defined by (2.12).

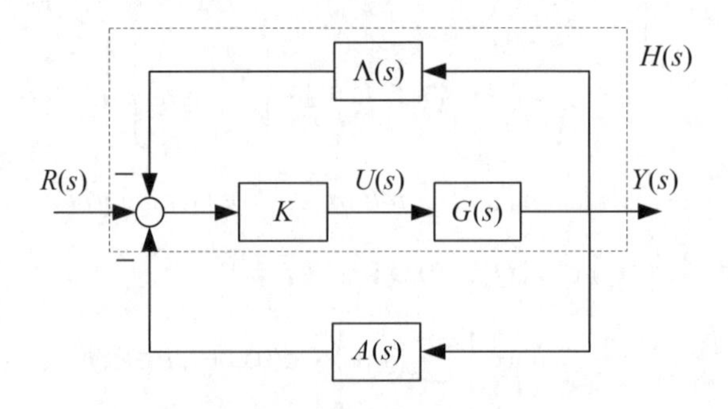

Figure 2.16 Distributed control system.

The following theorem is a direct consequence of the spectral radius theorem and Gershgorin's disc lemma.

Theorem 2.36 *Suppose $G_i(s)$, $i \in \overline{1,n}$, have no open RHP poles. The closed-loop system shown by Figure 2.16 has no open RHP poles, and hence is semi-stable if*

$$\left|\frac{\kappa_i G_i(\mathrm{j}\omega)}{\mathrm{e}^{\mathrm{j}\omega\tau_{ii}} + \kappa_i a_{ii} G_i(\mathrm{j}\omega)}\right| < \left(\sum_{j=1, j\neq i}^{n} |a_{ij}|\right)^{-1}, \quad \forall\omega \in [0, +\infty), \quad \forall i \in \overline{1,n}. \tag{2.72}$$

Furthermore, if there is no cancelation of $\mathrm{j}\omega$-axis poles between the open-loop system and the closed-loop system, the closed-loop system is stable.

Proof. Denote by $H(s)$ the upper part of the system as shown by Figure 2.16, i.e.,

$$\begin{aligned} H(s) &= (I + KG(s)\Lambda(s))^{-1}KG(s) = \mathrm{diag}\left\{\frac{\kappa_i G_i(s)}{1 + \kappa_i G_i(s) a_{ii} \mathrm{e}^{-\tau_{ii}s}},\ i \in \overline{1,n}\right\} \\ &\triangleq \mathrm{diag}\{H_i(s),\ i \in \overline{1,n}\} \end{aligned}$$

Then, the open-loop transfer function of the system can be written as $\mathrm{diag}\{H_i(s)\}A(s)$. By Theorem 2.20, the closed-loop system has no open RHP poles if and only if the eigenloci

$$\lambda_m(\mathrm{diag}\{H_i(\mathrm{j}\omega), i \in \overline{1,n}\}A(\mathrm{j}\omega))$$

do not enclose the point $(-1, \mathrm{j}0)$ for $m = 1, \cdots, n$ as ω varies from 0 to infinity. By Lemma 2.30, the spectrum of $\mathrm{diag}\{H_i(\mathrm{j}\omega), i \in \overline{1,n}\}A(\mathrm{j}\omega)$ is contained in the union of the following discs:

$$\begin{aligned} D_i &= \left\{z \in C : |z| \leq \sum_{i=1, j\neq i}^{n} |H_i(\mathrm{j}\omega) a_{ij} e^{-\mathrm{j}\omega\tau_{ij}}|\right\} \\ &= \left\{z \in C : |z| \leq |H_i(\mathrm{j}\omega)| \sum_{i=1, j\neq i}^{n} |a_{ij}|\right\}. \end{aligned}$$

Obviously, the eigenloci $\lambda_m(\mathrm{diag}\{H_i(\mathrm{j}\omega), i \in \overline{1,n}\}A(\mathrm{j}\omega))$ do not enclose $(-1, \mathrm{j}0)$ if

$$|H_i(\mathrm{j}\omega)| \sum_{j=1, j\neq i}^{n} |a_{ij}| < 1, \quad \forall\omega \in [0, \infty), \quad \forall i \in \overline{1,n}$$

which is equivalent to (2.72). Therefore, the closed-loop system does indeed have no open RHP poles. If there is no cancelation of $\mathrm{j}\omega$-axis poles between the open-loop system and the closed-loop system, by Theorem 2.20 again, the closed-loop system is stable. □

Theorem 2.36 gives a sufficient stability condition which is independent of communication delays τ_{ij}, $i \neq j$. But it still depends on the self-delays τ_{ii}. However, if we replace $\mathrm{e}^{-\mathrm{j}\omega\tau_{ii}}$ by a unit disc $\mathcal{D} = \{d = \mathrm{e}^{\mathrm{j}\theta} : \theta \in [0, 2\pi)\}$, then a sufficient condition for (2.72) is given by the following corollary.

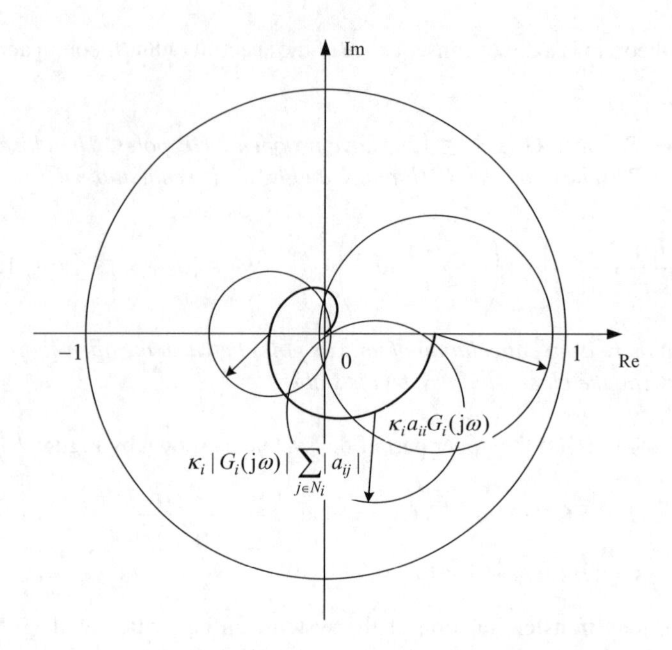

Figure 2.17 Geometrical illustration of Corollary 2.37.

Corollary 2.37 *Suppose $G_i(s)$, $i \in \overline{1,n}$, have no open RHP poles. The closed-loop system shown by Figure 2.16 has no open RHP poles if*

$$|\kappa_i a_{ii} G_i(\mathrm{j}\omega) - d| > \kappa_i |G_i(\mathrm{j}\omega)| \sum_{j \in N_i} |a_{ij}|, \ \forall d \in \mathcal{D}, \forall \omega \in [0, +\infty), \ \forall i \in \overline{1,n}. \tag{2.73}$$

Furthermore, if there is no cancelation of jω*-axis poles between the open-loop system and the closed-loop system, the closed-loop system is stable.*

This is a more conservative condition but it is independent of self-delays. The geometric meaning of condition (2.73) can be illustrated as follows: all the circle centered at the Nyquist plot of $\kappa_i a_{ii} G_i(\mathrm{j}\omega)$ with radius $\kappa_i |G_i(\mathrm{j}\omega)| \sum_{j \in N_i} |a_{ij}|$ is inside but does not touch the unit disc (see Figure 2.17).

Figure 2.17 clearly shows that the condition (2.73) can be satisfied only if $G_i(s) \in \mathbb{H}_\infty$. In the other hand, if $G_i(s) \in \mathbb{H}_\infty$, it can be shown that there always exist local proportional controls stabilizing the system.

Corollary 2.38 *Suppose $G_i(s) \in \mathbb{H}_\infty$, $i \in \overline{1,n}$. Then, the system shown by Figure 2.16 is stable if control gains κ_i, $i \in \overline{1,n}$, are sufficiently small.*

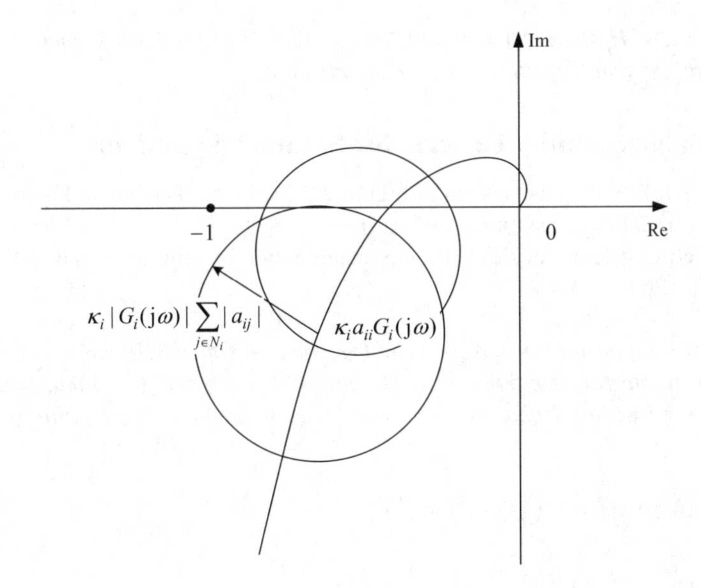

Figure 2.18 Geometrical illustration of Theorem 2.36 for the case $\tau_{ii} = 0$.

Proof. Since $G_i(s)$ has no jω-axis poles, we know that $|G_i(\mathrm{j}\omega)|$ is bounded for all $\omega \in \mathbb{R}$. Therefore,

$$\lim_{\kappa_i \to 0} \left| \frac{\kappa_i G_i(\mathrm{j}\omega)}{\mathrm{e}^{\mathrm{j}\omega\tau_{ii}} + \kappa_i a_{ii} G_i(\mathrm{j}\omega)} \right| = 0.$$

This implies that (2.72) is always satisfied if κ_i, $i \in \overline{1, n}$, are sufficiently small. □

Remark. If all self-delays are zero, i.e. $\tau_{ii} = 0$, the condition (2.72) reduces to

$$\left| \frac{\kappa_i G_i(\mathrm{j}\omega)}{1 + \kappa_i a_{ii} G_i(\mathrm{j}\omega)} \right| < \left(\sum_{j \in N_i} |a_{ij}| \right)^{-1}, \quad \forall \omega \in [0, +\infty).$$

This condition does not require $G_i(s) \in \mathbb{H}_\infty$ (see Figure 2.18 for illustration).

For the case when the transfer function of the open-loop system is stable, by applying Theorem 2.24 and the argument similar to the proof of Theorem 2.36, we can get the following scalable criterion for the test of the steady semi-stability and the marginal stability.

Theorem 2.39 *Suppose $G_i(s) \in \mathbb{H}_\infty$, $i \in \overline{1, n}$. The closed-loop system shown by Figure 2.16 is steady semi-stable if*

$$\left| \frac{\kappa_i G_i(\mathrm{j}\omega)}{\mathrm{e}^{\mathrm{j}\omega\tau_{ii}} + \kappa_i a_{ii} G_i(\mathrm{j}\omega)} \right| < \left(\sum_{j \in N_i} |a_{ij}| \right)^{-1}, \quad \forall \omega \in (0, +\infty), \quad \forall i \in \overline{1, n}. \tag{2.74}$$

Moreover, if $\rho(\text{diag}\{H_i(0)\}A(0)) = 1$*, and* $\det(I + \text{diag}\{H_i(0)\}A(0)) = 0$*, then the closed-loop system is marginally stable with* $s = 0$ *as a simple pole.*

2.5.3 Scalable Stability Criteria for Symmetric Systems

Since $\bar{A}(s) = \Lambda(s) + A(s)$, the system in Figure 2.2 can be re-sketched in Figure 2.19. Recall that the system is said to be symmetric if $\bar{A}(s) = \bar{A}^{\mathrm{T}}(-s)$.

For symmetric systems one can get some scalable stability criterion that is less conservative than Theorem 2.36.

Theorem 2.40 *Suppose* $\bar{A}(s) = \bar{A}^{\mathrm{T}}(-s)$ *and* $G_i(s)$ *has no open RHP poles. Let* γ_i *be the gain margin of the stable transfer function* $G_i(s)$*, and* $\hat{G}_i(s) = \gamma_i^{-1}G_i(s)$*. Then, the closed-loop system shown by Figure 2.19 has no open RHP poles if the following conditions are satisfied for all* $\omega \geq 0$*:*

(1) $(-1, \mathrm{j}0) \notin \mathrm{Co}\left(0 \cup \{\hat{G}_i(\mathrm{j}\omega),\ i \in \overline{1,n}\}\right)$,

(2) $\kappa_i \left(|a_{ii}| + \sum\limits_{j \in N_i} |a_{ij}| \right) < \gamma_i,\ i \in \overline{1,n}.$

Furthermore, if there is no cancelation of $\mathrm{j}\omega$*-axis poles between the open-loop system and the closed-loop system, the closed-loop system is stable.*

Proof. We just prove the semi-stability part. The stability part follows directly from Theorem 2.20.

By Theorem 2.20, the closed-loop system has no RHP poles if and only if the eigenloci

$$\lambda_m(\text{diag}\{\kappa_i G_i(\mathrm{j}\omega), i \in \overline{1,n}\}\bar{A}(\mathrm{j}\omega))$$

do not enclose the point $(-1, \mathrm{j}0)$ for $m = 1, \cdots, n$ as ω varies from 0 to infinity. Note that

$$\begin{aligned}\lambda_m(\text{diag}\{\kappa_i G_i(\mathrm{j}\omega)\}\bar{A}(\mathrm{j}\omega)) &= \lambda_m(\text{diag}\{\kappa_i \gamma_i^{-1}\hat{G}_i(\mathrm{j}\omega)\}\bar{A}(\mathrm{j}\omega)) \\ &= \lambda_m(\text{diag}\{\bar{G}_i(\mathrm{j}\omega)\}\text{diag}\{\kappa_i\gamma_i^{-1}\}\bar{A}(\mathrm{j}\omega)).\end{aligned}$$

And moreover, $\bar{A}(\mathrm{j}\omega) = \bar{A}^*(\mathrm{j}\omega)$ because $\bar{A}(s) = \bar{A}^{\mathrm{T}}(-s)$. Therefore, by Lemma 2.34, the closed-loop system has no open RHP poles if

$$(-1, \mathrm{j}0) \notin \rho(\text{diag}\{\kappa_i\gamma_i^{-1}\}\bar{A}(\mathrm{j}\omega)\mathrm{Co}\left(0 \cup \{\hat{G}_i(\mathrm{j}\omega)\}\right),\ \forall\omega \geq 0.$$

The above condition is satisfied if

$$(-1, \mathrm{j}0) \notin \mathrm{Co}\left(0 \cup \{\hat{G}_i(\mathrm{j}\omega),\ i \in \overline{1,n}\}\right), \tag{2.75}$$

and

$$\rho\left(\text{diag}\{\kappa_i\gamma_i^{-1},\ i \in \overline{1,n}\}\bar{A}(\mathrm{j}\omega)\right) < 1. \tag{2.76}$$

Since the spectral radius of any matrix is bounded by its maximum absolute row sum, (2.76) is satisfied if condition (2) of the theorem holds. Thus, the theorem is proved. □

If we use Corollary 2.32 instead of Corollary 2.31 of Gershgorin's disc lemma in the proof, we can get the following result for reducing the conservatism of Theorem 2.40.

Theorem 2.41 *Suppose $\bar{A}(s) = \bar{A}^{\mathrm{T}}(-s)$ and $G_i(s)$ has no open RHP poles. The closed-loop system shown by Figure 2.19 has no open RHP poles if there exist positive real numbers $h_i, i \in \overline{1,n}$ such that the following conditions are satisfied for all $\omega \geq 0$:*

(1) $(-1, \mathrm{j}0) \notin \mathrm{Co}\left(0 \cup \{\hat{G}_i(\mathrm{j}\omega),\ i \in \overline{1,n}\}\right)$,

(2) $\kappa_i \left(|a_{ii}| + \sum\limits_{j \in N_i} \frac{h_i}{h_j} |a_{ij}| \right) < \gamma_i,\ i \in \overline{1,n}.$

Furthermore, if there is no cancelation of $\mathrm{j}\omega$-axis poles between the open-loop system and the closed-loop system, the closed-loop system is stable.

If the condition

$$(-1, \mathrm{j}0) \notin \mathrm{Co}\left(0 \cup \{\hat{G}_i(\mathrm{j}\omega),\ i \in \overline{1,n}\}\right), \tag{2.77}$$

in the above theorem is satisfied, the second condition in Theorem 2.40 or Theorem 2.41 gives a local and hence scalable test for the stability. Therefore, condition (2.77) can be considered as a *scalability* condition. Note that the scalability condition (2.77) does not necessarily hold even if each Nyquist plot $\hat{G}_i(\mathrm{j}\omega)$ does not contain $(-1, \mathrm{j}0)$. This problem will be studied in the next chapter.

Exercise 2.42 *Let*

$$KG_i(s) = \frac{\mathrm{e}^{-\tau_i s}}{s}, \quad i \in \overline{1,4}$$

and

$$\bar{A}(s) = \begin{bmatrix} -6 & 1 & 3 & 2 \\ 1 & -7 & 2 & 4 \\ 3 & 2 & -6 & 1 \\ 2 & 4 & 1 & -7 \end{bmatrix}$$

in the system shown by Figure 2.19. Derive scalable semi-stability conditions by Theorem 2.36, Theorem 2.40 and Theorem 2.41, respectively. Compare the conservatism of the obtained conditions with the help of a simulation toolbox (e.g., MATLAB®).

2.5.4 Robust Stability in Deformity of Symmetry

Symmetry of is such a fragile property that may be easily destroyed by small perturbations. Many factors in practical systems, such as uncertainty of agent dynamics, diversity of communication delays or heterogeneousness of channel dynamics, may lead to deformity of the

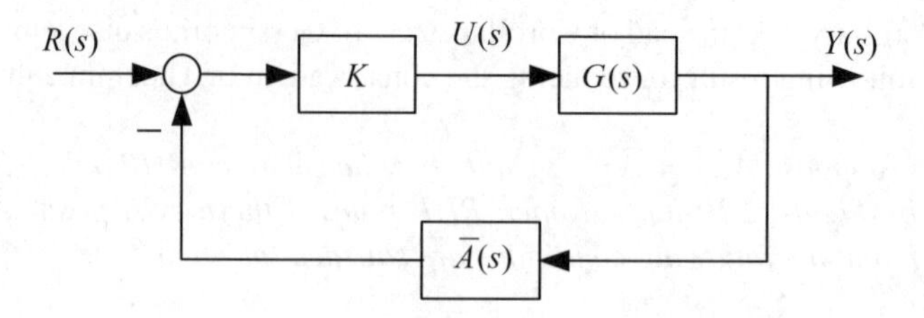

Figure 2.19 Distributed control system.

symmetry defined by Definition 2.2. Therefore, any analysis or design method based on the assumption of the system symmetry must be robust against perturbations. Fortunately, the robust control theory can help us to extend the scalable stability criteria to most cases of deformity of symmetry.

Let us consider the case when there are some uncertainties in communication channels that cause the asymmetry of the system, as shown by Figure 2.20. Here, we assume that the nominal system, i.e., the system with $\Delta_A(s) = 0$, is symmetric. Suppose that the uncertainty matrix $\Delta_A(s)$ is subject to the following structural constraint:

$$\Delta_A(s) = E_A \Delta(s) F_A, \tag{2.78}$$

where $E_A \in \mathbb{R}^{n\times r}$ and $F_A \in \mathbb{R}^{r\times n}$ are some constant matrices that describe the structure of the uncertainty, $\Delta(s)$ satisfies the following condition:

$$\Delta(s) \in \mathbb{H}_\infty^{r\times r}, \quad \text{and} \quad \bar{\sigma}(\Delta(\mathrm{j}\omega)) \leq \gamma_1(\omega), \quad \forall \omega \in \mathbb{R}, \tag{2.79}$$

where $\gamma_1(\omega) \in \mathbb{R}$ is a known function of ω. Then, by loop transformation we can equivalently transform the system into an interconnection form shown by Figure 2.21 with

$$M(s) = F_A(I + G(s)K(s)A(s))^{-1}G(s)K(s)E_A.$$

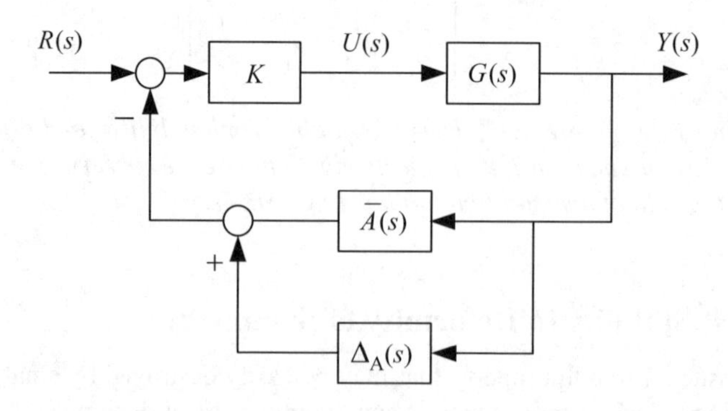

Figure 2.20 Distributed control system with uncertainty in network channels.

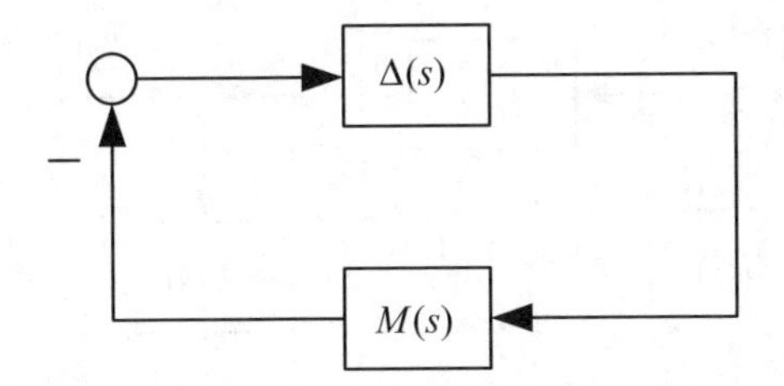

Figure 2.21 Symmetric system with asymmetric perturbations.

According to the robust control theory (see, e.g., Feng, Tian and Xin 1996, Zhou, Doyle and Glover (1996)), under the assumption that

$$F_{\mathrm{A}}(I + G(s)K(s)A(s))^{-1}G(s)K(s)E_{\mathrm{A}} \in \mathbb{H}_{\infty}^{r\times r}, \tag{2.80}$$

the asymmetric uncertain system shown by Figure 2.20 is robustly stable if and only if

$$\bar{\sigma}(F_{\mathrm{A}}(I + G(\mathrm{j}\omega)K(\mathrm{j}\omega)A(\mathrm{j}\omega))^{-1}G(\mathrm{j}\omega)K(\mathrm{j}\omega)E_{\mathrm{A}}) < (\gamma_1(\omega))^{-1}, \quad \forall\omega \in \mathbb{R}. \tag{2.81}$$

We summarize the conclusion of the above discussion in the following theorem.

Theorem 2.43 *Under the conditions (2.78), (2.79) and (2.80), the perturbed system shown by Figure 2.20 is robustly stable if and only if (2.81) holds.*

Remark. A sufficient condition for $M(s) \in \mathbb{H}_{\infty}^{r\times r}$ is that

$$(I + G(s)K(s)A(s))^{-1}G(s)K(s) \in \mathbb{H}_{\infty}^{n\times n}. \tag{2.82}$$

Note that the left-hand side of (2.82) is just the nominal symmetric system. So, its stability can be checked by Theorem 2.40 or Theorem 2.41.

We can also consider the bipartite system with an additive uncertainty $\Delta_{\mathrm{P}}(s)$ in agent plants, as shown by Figure 2.22. Note that the uncertainty $\Delta_{\mathrm{P}}(s)$ may destroy the homogeneousness assumption (2.27), and hence, destroy the symmetry of the system. Suppose that $\Delta_{\mathrm{P}}(s)$ is subject to the following structural constraint:

$$\Delta_{\mathrm{P}}(s) = E_{\mathrm{P}}\Delta(s)F_{\mathrm{P}}, \tag{2.83}$$

where $E_{\mathrm{P}} \in \mathbb{R}^{n\times 1}$ and $F_{\mathrm{P}} \in \mathbb{R}^{1\times n}$ describe the structure of the uncertainty, and $\Delta(s)$ satisfies the following condition:

$$\Delta(s) \in \mathbb{H}_{\infty}, \quad \text{and} \quad \bar{\sigma}(\Delta(\mathrm{j}\omega)) \le \gamma_2(\omega), \quad \forall\omega \in \mathbb{R}, \tag{2.84}$$

where $\gamma_2(\omega) \in \mathbb{R}$ is a known function of ω. The system can also be equivalently transformed into the interconnection form shown by Figure 2.21 with

$$M(s) = F_{\mathrm{P}}(I - \Gamma(s)A_f(s)G(s)K(s)A_b(s)P(s))^{-1}\Gamma(s)A_f(s)G(s)K(s)A_b(s)E_{\mathrm{P}}. \tag{2.85}$$

Similarly, we assume that

$$F_{\mathrm{P}}(I - \Gamma(s)A_f(s)G(s)K(s)A_b(s)P(s))^{-1}\Gamma(s)A_f(s)G(s)K(s)A_b(s)E_{\mathrm{P}} \in \mathbb{H}_{\infty}. \tag{2.86}$$

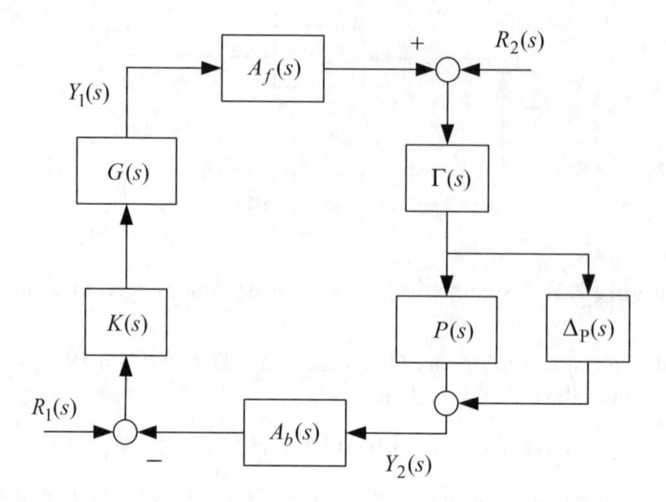

Figure 2.22 Distributed control system with uncertainty in agent plants.

Then, we have the following result.

Theorem 2.44 *Under the conditions (2.83), (2.84) and (2.86), the perturbed system shown by Figure 2.22 is robustly stable if and only if*

$$\bar{\sigma}(M(j\omega)) < (\gamma_2(\omega))^{-1}, \quad \forall \omega \in \mathbb{R}, \tag{2.87}$$

where $M(s)$ is given by (2.85).

2.6 Notes and References

Modeling of systems with forward and backward communication channels are discussed in many references on coordination control and congestion control (see, e.g., Arcak 2007; Low, Paganini and Doyle 2002; Wen and Arcak 2003). Using bipartite graphs to describe interconnection topologies of such systems and the symmetry analysis is new. The definitions of poles and zeros are given following MacFarlane and Karcanias (1976). Note that for multivariable systems there are various definitions of zeros, see MacFarlane and Karcanias (1976) and Rosenbrock (1973). For their computation, see Davison and Wang (1978) and Laub and Moore (1978). The difficulties associated with pole-zero cancelations in multivariable systems can be treated by state-space techniques using the concepts of controllability and observability. A pure algebraic approach to analyzing input-output properties including the equivalence between the systems after/before loop transformations can be found in more detail in the classic book of Desoer and Vidyasagar (1975), where general settings are used. The Nyquist stability was generalized in several ways for multivariable systems in the 1970s (Barman and Katzenelson 1974; Belletrutti and MacFarlane 1971; Callier and Desoer 1972; Davis 1972; Desoer and Vidyasagar 1975; MacFarlane and Postlethwaite 1977; Saeks 1975). Theorem 2.17 is taken from Vidyasagar (1985) while Theorem 2.19 adopts the result of Desoer and Vidyasagar (1975). Gershgorin's disc lemma has been a well-known tool for deriving scalable or decentralized stability results due to the study of frequency-domain analysis and design of multivariable systems by the so-called British school (Belletrutti and MacFarlane 1971; MacFarlane and Karcanias 1976; Rosenbrock 1970, 1974). Lemma 2.34 is a recent important contribution made by Vinnicombe (2000) to the estimation of the spectrum of complex matrices with application to the scalable stability analysis.

References

Arcak M (2007). Passivity as a design tool for group coordination. *IEEE Transactions on Automatic Control*, 52, 1380–1390.

Barman JF and Katzenelson J (1974). A generalized Nyquist-type stability criterion for multivariable feedback systems. *International Journal of Control*, 20(4), 593–622.

Belletrutti JJ and MacFarlane AGJ (1971). Characteristic loci techniques in multivariable control systems. *Proc. Inst. Elec. Eng.*, 118, 1291–1297.

Callier FM and Desoer CA (1972). A graphical test for checking the stability of a linear time-invariant feedback systems. *IEEE Transactions on Automatic Control*, 17, 773–780.

Chen CT (1984). *Linear System Theory and Design*. CBS College Publishing, New York.

Davis JH (1972). Encirclement conditions for stability and instability of feedback control systems with delays. *International Journal of Control*, 15, 793–799.

Davison EJ (1983). Some properties of minimum phase systems and "square-down" systems. *IEEE Transactions on Automatic Control*, 28, 221–222.

Davison EJ and Wang SH (1978). An algorithm for the calculation of transmission zeros of (C,A,B,D) using high-gain output feedback. *IEEE Transactions on Automatic Control*, 23, 738–741.

Desoer CA and Vidyasagar M (1975). *Feedback Systems: Input-Output Properties*, Academic Press, New York.

Desoer CA and Wang Y-T (1980). On the generalized Nyquist Stability Criterion. *IEEE Transactions on Automatic Control*, 25(2), 187–196.

Feng C-B, Tian Y-P and Xin X (1996). *Robust Control System Design* (in Chinese). Southeast University Press, Nanjing.

Gu K, Kharitonov VL and Chen J (2003). *Stability of Time-delayed Systems*. Birkhäuser, Boston.

Horn RA and Johnson CR (1985). *Matrix analysis*, 1st edn. Cambridge University Press, New York.

Horn RA and Johnson CR (1991). *Topics in Matrix analysis*, 1st edn. Cambridge University Press, New York.

Laub AJ and Moore BC (1978). Calculation of transmission zeros using QZ techniques. *Automatica*, 14, 557–566.

Lestas I and Vinnicombe G (2005). Scalable robustness for consensus protocols with heterogeneous dynamics. *Proceedings of IFAC world Congress*, Prague.

Low SH, Paganini F and Doyle JC (2002). Internet congestion control. *IEEE control systems Magazine*, 22, 28–43.

MacFarlane AGJ and Karcanias N (1976). Poles and zeros of linear multivariable systems: a survey of algebraic, geometric and complex variable theory. *International Journal of Control*, Vol.24, 33–74.

MacFarlane AGJ and Postlethwaite I (1977). Generalized Nyquist stability criterion and multivariable root loci. *International Journal of Control*, 25(1), 81–127.

Nise NS (2000). *Control Systems Engineering*. 3rd ed., John Wiley & Sons, New York.

Rosenbrock HH (1970). *State-space and Multivariable Theory*. Nelson, London.

Rosenbrock HH (1973). The zeros of a system. *International Journal of Control*, 18, 297–299.

Rosenbrock HH (1974). *Computer-Aided Control System Design*. Academic Press, New York.

Saeks R (1975). On the encirclement condition and its generalization. *IEEE Transactions on Circuits and Systems*, 22, 780–785.

Skogestad S and Postlethwaite I (1996). *Multivariable Feedback Control: Analysis And Design*. John Wiley & Sons.

Vidyasagar M (1985). *Control System Synthesis: A Factorization Approach*. MIT Press.

Vinnicombe G (2000). On the stability of end-to-end congestion control for the Internet. *Technical report CUED/F-INFENG/TR.*, No.398, 2000.

Wen JT and Arcak M (2003). A unifying passivity framework for network flow control. *Proceedings of IEEE Infocom*, San Francisco, California.

Zhou K, Doyle JC and Glover K (1996). *Robust and Optimal Control*. Prentice-Hall, Upper Saddle River, NJ.

3

Scalability in the Frequency Domain

Tao generates one, one generates two, two generates three, three generates all things.
—Lao Dan (580–500 BC), Tao Te Ching

This chapter studies the problem of the scalability of the stability results for distributed control systems. The problem is closely related to some differential geometric properties of frequency response plots of local dynamics of agents (nodes) in networks, such as clockwise property, modulus monotonicity, slope monotonicity, phase velocity, critical point of clockwise property, etc. The chapter starts from the clockwise property which plays a key role in scalability analysis. Then, detailed geometric analysis of scalability is conducted for first-order and second-order time-delayed systems which are often encountered in coordinated control systems and end-to-end congestion control systems. Finally, based on the notion of convex directions in the space of stable quasi-polynomials, a frequency sweeping method of scalability test is provided for high-order time-delayed systems.

3.1 How the Scalability Condition is Related with Frequency Responses

Denote by γ the gain margin the transfer function

$$G(s) = W(s)\mathrm{e}^{-Ts},$$

where $W(s)$ is a rational function of s, and $T > 0$ is the delay constant. Actually, γ is defined by

$$\gamma = 1/|W(\mathrm{j}\omega_c)|, \tag{3.1}$$

where $\omega_c > 0$ is the *minimal crossing frequency* that satisfies the following equation

$$\mathrm{Im}[W(\mathrm{j}\omega)\mathrm{e}^{-\mathrm{j}\omega T}] = 0. \tag{3.2}$$

Frequency-Domain Analysis and Design of Distributed Control Systems, First Edition. Yu-Ping Tian.

Note that the Nyquist plot of $\gamma W(\mathrm{j}\omega)\mathrm{e}^{-\mathrm{j}\omega T}$ crosses the real axis at $(-1, \mathrm{j}0)$ when $\omega = \omega_c$. We will call $\gamma W(s)\mathrm{e}^{-Ts}$ *the normalized transfer function* of $G(s)$ and denote it as $\hat{G}(s)$.

Given n transfer functions $G_i(s) = W_i(s)\mathrm{e}^{-T_i s}$, $i \in \overline{1, n}$, denote by $\hat{G}_i(s)$ the normalized function of $G_i(s)$. Consider a symmetric multi-agent system with $G_i(s)$ as the transfer function of the ith agent. We have shown in the last chapter that the scalability condition for the system is

$$(-1, \mathrm{j}0) \notin \kappa \mathrm{Co}\left(0 \cup \{\hat{G}_i(\mathrm{j}\omega),\ i \in \overline{1, n}\}\right),\ \forall \omega \in (-\infty, \infty), \forall \kappa \in [0, 1). \tag{3.3}$$

This condition requires us to check if the point $(-1, \mathrm{j}0)$ is contained in the convex hull of $\kappa \hat{G}_i(\mathrm{j}\omega)$ and zero.

The frequency response of the z-transfer function of a discrete-time system can also be denoted as $\hat{G}(\mathrm{j}\omega)$, where the frequency variable ω takes values from the interval $[-\pi, \pi]$. To include the analysis of discrete-time systems in a unified framework, we denote the frequency interval as $[\omega_a, \omega_b]$, and rewrite the scalability condition as

$$(-1, \mathrm{j}0) \notin \kappa \mathrm{Co}\left(0 \cup \{\hat{G}_i(\mathrm{j}\omega),\ i \in \overline{1, n}\}\right),\ \forall \omega \in [\omega_a, \omega_b], \forall \kappa \in [0, 1). \tag{3.4}$$

The following lemma shows that it suffices to check the scalability condition for the boundary of the convex hull involved.

Lemma 3.1 *Given normalized transfer functions $\hat{G}_i(s)$, $i \in \overline{1, n}$, assume that*

$$(-1, \mathrm{j}0) \notin \kappa \mathrm{Co}(0 \cup \{\hat{G}_i(\mathrm{j}\omega_0), i \in \overline{1, n}\}),\ \exists \omega_0 \in [\omega_a, \omega_b],\ \forall \kappa \in [0, 1), \tag{3.5}$$

and

$$(-1, \mathrm{j}0) \notin \kappa \mathrm{Co}(0, \hat{G}_i(\mathrm{j}\omega)),\ \forall \omega \in [\omega_a, \omega_b],\ \forall \kappa \in [0, 1),\ \forall i \in \overline{1, n}. \tag{3.6}$$

Then, the scalability condition (3.4) holds if and only if

$$(-1, \mathrm{j}0) \notin \kappa \mathrm{Co}(\hat{G}_i(\mathrm{j}\omega), \hat{G}_k(\mathrm{j}\omega)), \forall \omega \in [\omega_a, \omega_b], \forall \kappa \in [0, 1) \tag{3.7}$$

for all $i, k \in \overline{1, n}$, $i \neq k$.

Proof. Note that (3.5) indicates that the condition (3.4) holds for $\omega = \omega_0$. By continuity of the set $\kappa \mathrm{Co}(0 \cup \{G^i(\omega), i \in \overline{1, n}\})$ on ω, the point $(-1, \mathrm{j}0)$ enters $\kappa \mathrm{Co}(0 \cup \{G^i(\omega), i \in \overline{1, n}\})$ if and only if the boundary, $\kappa \mathrm{Co}(G_{i_1}(\omega), G_{i_2}(\omega))$, or $\kappa \mathrm{Co}(0, G^i(\omega))$, passes through $(-1, \mathrm{j}0)$. (3.6) has excluded the possibility that $\kappa \mathrm{Co}(0, G^i(\omega))$ passes through $(-1, \mathrm{j}0)$. Therefore, (3.4) holds if and only if (3.7) holds. □

Remark. In most cases condition (3.5) can be easily verified. For continuous-time systems, if $W_i(s)$, $i \in \overline{1, n}$, are proper, then $\hat{G}_i(\mathrm{j}\infty) = 0$. Therefore, $\kappa \mathrm{Co}(0 \cup \{G^i(\mathrm{j}\infty), i \in \overline{1, n}\}) = 0$ and hence (3.5) holds. For discrete-time systems, to verify (3.5) we usually choose $\omega_0 = \pi$ because $\hat{G}_i(\mathrm{j}\omega)$'s are real numbers. Condition (3.6) is ensured by the stability of the individual closed-loop system of each agent. When conditions (3.5) and (3.6) are satisfied, the scalability condition holds if and only if $(-1, \mathrm{j}0)$ is not contained by the Nyquist plot of the convex combination of each pair of agents.

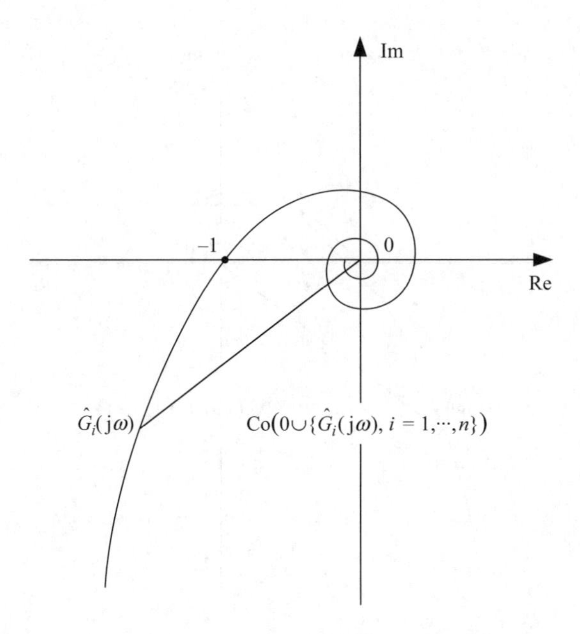

Figure 3.1 Homogeneous agents with identical delay.

Now, let us consider the following different cases.

Case 1: Homogeneous agents
In this case, the transfer functions of all the agents are assumed to be the same both in the rational part and in the delay unit, i.e.,

$$\hat{G}_i(s) = \hat{G}_0(s)\mathrm{e}^{-T_0 s}, \ \forall i \in \overline{1, n}.$$

Obviously, for such a special case we have

$$\mathrm{Co}\left(0 \cup \{\hat{G}_i(\mathrm{j}\omega),\ i \in \overline{1, n}\}\right) = \mathrm{Co}(0, \hat{G}_0(\mathrm{j}\omega)\mathrm{e}^{-\mathrm{j}\omega T_0}),$$

which is a line segment between the origin and any point at $\hat{G}_i(\mathrm{j}\omega)$ for a given frequency ω (see Figure 3.1). Obviously, in this case, *the condition (3.4) holds if and only if the Nyquist plot of* $\hat{G}_i(\mathrm{j}\omega)$ *does not contain the point* $(-1, \mathrm{j}0)$.

Case 2: Heterogeneous agents with homogeneous frequency response
Let us consider the agents given by

$$\hat{G}_i(s) = \hat{G}_0(T_i s)\mathrm{e}^{-T_i s}, \ i \in \overline{1, n}.$$

If $T_i \neq T_j$, for $i \neq j, i, j \in \overline{1, n}$, the agents generally have different dynamics. However, if there exists a common transfer function $\bar{G}(s)$ such that for any given frequency w_0 the following condition holds:

$$\hat{G}_i(\mathrm{j}\omega_0) = \bar{G}(\mathrm{j}\omega_i),$$

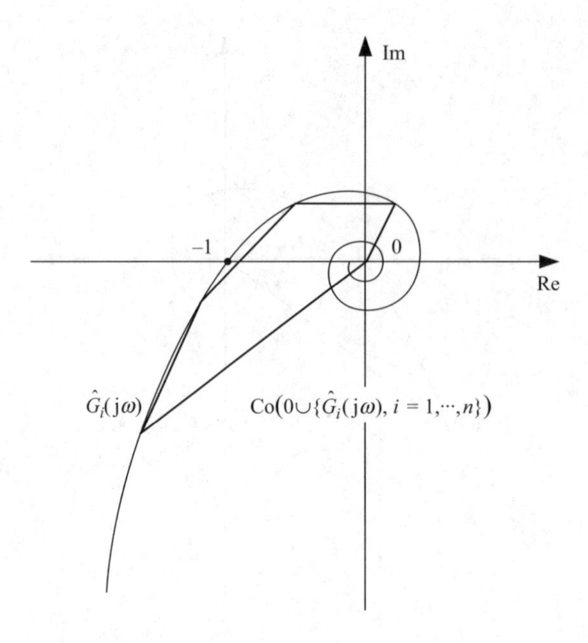

Figure 3.2 Delay-independent frequency response: convex case.

then all the agents have the same Nyquist plot which is delay-independent. For example, take $\hat{G}_i(s) = \frac{e^{-T_i s}}{T_i s}$. Then, for a given frequency ω_0, all $\hat{G}_i(j\omega_0)$, $i \in \overline{1,n}$, are distributed at different points on the same plot $\bar{G}(j\omega) = \frac{e^{-j\omega}}{j\omega}$. In this case, we are interested in the geometric property of $\bar{G}(j\omega)$ with which the following inclusion holds

$$\mathrm{Co}\left(0 \cup \{\hat{G}_i(j\omega),\ i \in \overline{1,n}\}\right) \subset \mathrm{Co}(0, \bar{G}(j\omega)). \tag{3.8}$$

Obviously, if (3.8) holds, then, as shown by Figure 3.2, $\mathrm{Co}\left(0 \cup \{\hat{G}_i(j\omega),\ i \in \overline{1,n}\}\right)$ does not contain the point $(-1, j0)$ if $\bar{G}(j\omega)$ does not contain $(-1, j0)$. Intuitively, the inclusion relationship (3.8) implies some "convexity" property of the Nyquist plot $\bar{G}(j\omega)$. Figure 3.3 gives an illustration that (3.8) may not hold when $\bar{G}(j\omega)$ does not have such convexity. In the next section we will study this property in detail.

Case 3: Heterogeneous agents with heterogeneous frequency response
In this case, the scalability problem becomes much more complicated. Since all the agents have a different frequency response, we can only expect the following inclusion

$$\mathrm{Co}\left(0 \cup \{\hat{G}_i(j\omega),\ i \in \overline{1,n}\}\right) \subset \mathrm{Co}(0, \hat{G}_i(j\omega)),\ i \in \overline{1,n}. \tag{3.9}$$

Obviously, if (3.9) holds, then $\mathrm{Co}\left(0 \cup \{\hat{G}_i(j\omega),\ i = \overline{1,n}\}\right)$ does not contain the point $(-1, j0)$ if each $\hat{G}_i(j\omega)$ does not contain $(-1, j0)$, $\forall i \in \overline{1,n}$. However, even if the convexity is satisfied by each frequency response plot, (3.9) may hold as shown by Figure 3.4(a) or may not hold as shown by Figure 3.4(b).

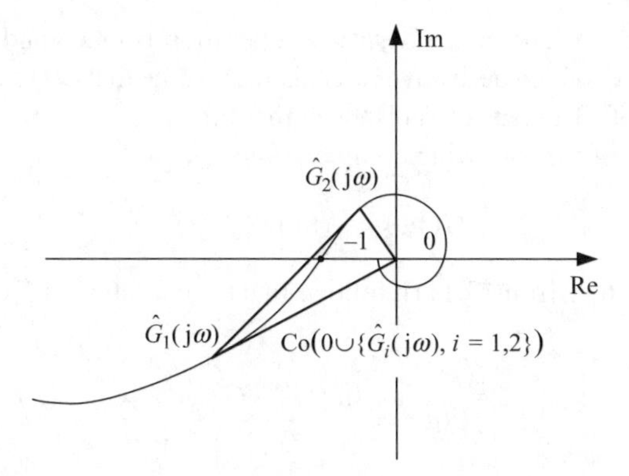

Figure 3.3 Delay-independent frequency response: non-convex case.

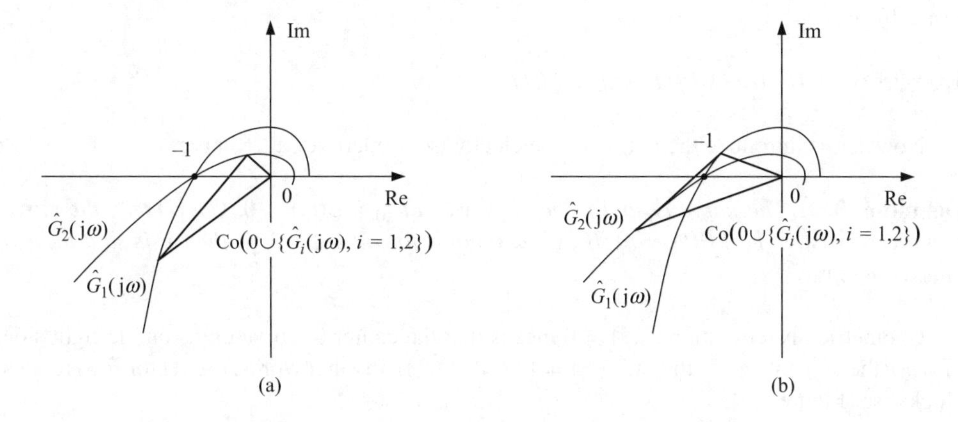

Figure 3.4 Heterogeneous frequency response.

3.2 Clockwise Property of Parameterized Curves

In this section we study the convexity of a smooth curve, which will be defined as a clockwise property. Let us first review some preliminary knowledge on the signed curvature of parameterized curves, which can be found in many textbooks of classic differential geometry such as Guggenheimer (1977).

Let $\Gamma(t)$ be a curve in $\mathbb{R}^2$ defined by two parametric equations:

$$\begin{cases} X = X(t), \\ Y = Y(t), \end{cases} \quad t \in [t_\alpha, t_\beta], \tag{3.10}$$

with $X(t), Y(t) \in C^2$, which denotes the space of functions being two-times differentiable. The curvature $\mathcal{C}(t)$ of Γ at parameter t is defined as

$$\mathcal{C}(t) := \frac{X'_t(t)Y''_{tt}(t) - X''_{tt}(t)Y'_t(t)}{(X'^2_t(t) + Y'^2_t(t))^{3/2}}, \tag{3.11}$$

where a superscript $'$ represents that a derivative operation is taken and subscripts denote the argument with respect to which derivatives are taken. Assume that $\mathcal{C}(t)$ exists for all $t \in (t_\alpha, t_\beta)$, i.e., the denominator of (3.11) never vanishes in that interval.

Let Γ be an image of the following complex function

$$G(\mathrm{j}\omega) = X(\omega) + \mathrm{j}Y(\omega) \tag{3.12}$$

in the complex plane. Then, from (3.11) it follows that the curvature of Γ can be also determined by (Gu 1994)

$$\mathcal{C}(\omega) = \frac{\mathrm{Im}(G'^{*}_{\omega} \cdot G''_{\omega\omega})}{|G'_{\omega}|^3}, \tag{3.13}$$

where G'_{ω} is the derivative of $G(\mathrm{j}\omega)$ with respect to ω, i.e., $G'_{\omega} = X'_{\omega} + \mathrm{j}Y'_{\omega}$, $G''_{\omega\omega}$ is the derivative of G'_{ω} with respect to ω, G'^{*}_{ω} and $|G'_{\omega}|$ are the conjugate and the module of G'_{ω}, respectively.

Exercise 3.2 *Derive (3.13) based on (3.11).*

Now, let us introduce the notion of the clockwise property of a C^2 curve.

Definition 3.3 *The curve is said to be clockwise at t_0 if $\mathcal{C}(t_0) < 0$. Conversely, the curve is anticlockwise at t_0 if $\mathcal{C}(t_0) > 0$. If $\mathcal{C}(t) < 0$ holds for all $t \in (t_\alpha, t_\beta)$ then Γ is said to be a clockwise curve.*

Geometrically, condition $\mathcal{C}(t_0) < 0$ means that the center of curvature is on the right side of $\vec{r}(t_0)$, the tangent vector to Γ at $\Gamma(t_0) = (X(t_0), Y(t_0))$, in other words, the vector $\vec{r}(t_0)$ rotates clockwise (Figure 3.5).

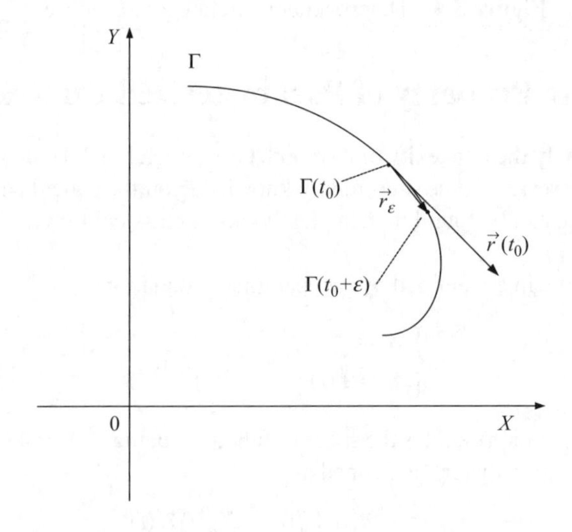

Figure 3.5 Clockwise curve.

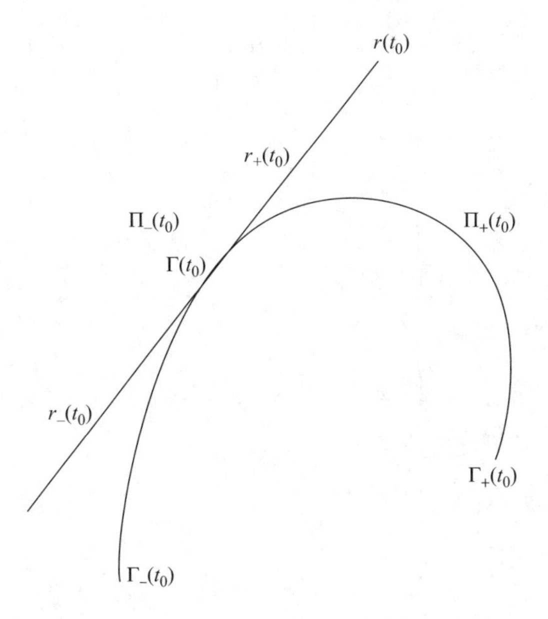

Figure 3.6 Partition of a clockwise curve.

For any $\varepsilon > 0$, denote by $\vec{r}_\varepsilon$ the vector from the point $\Gamma(t_0)$ to the point $\Gamma(t_0 + \varepsilon)$ (Figure 3.5). For convenience of statement we define the positive direction of the tangent line at $\Gamma(t_0)$ as follows. The tangent vector $\vec{r}(t_0)$ is said to point to the positive direction if the angle between $\vec{r}(t_0)$ and $\vec{r}_\varepsilon$ goes to zero when $\varepsilon \to 0$. Otherwise, $\vec{r}(t_0)$ is said to point to the negative direction if the angle between $\vec{r}(t_0)$ and $\vec{r}_\varepsilon$ goes to π when $\varepsilon \to 0$.

Then, we can split the tangent line $r(t_0)$ at $\Gamma(t_0)$ as follows:

$$r(t_0) = r_-(t_0) \cup \Gamma(t_0) \cup r_+(t_0)$$

where $r_+(t_0)$ ($r_-(t_0)$) denotes the half of the tangent line $r(t_0)$ starting at $\Gamma(t_0)$ (not including $\Gamma(t_0)$) in the positive (negative) direction (Figure 3.6).

Accordingly, for any given $t_0 \in (t_\alpha, t_\beta)$, we can also split the curve Γ as follows (Figure 3.6):

$$\Gamma = \Gamma_-(t_0) \cup \Gamma(t_0) \cup \Gamma_+(t_0)$$

where $\Gamma_+(t_0)$ ($\Gamma_-(t_0)$) denotes the part of the curve parameterized by $t \in (t_0, t_\beta]$ ($t \in [t_\alpha, t_0)$). Finally, we split the plane $\mathbb{R}^2$ as follows

$$\mathbb{R}^2 = \Pi_-(t_0) \cup r(t_0) \cup \Pi_+(t_0)$$

where $\Pi_+(t_0)$ ($\Pi_-(t_0)$) is the open-half plane containing (not containing) the center of curvature of Γ at t_0, assuming that $\mathcal{C}(t_0) \neq 0$.

The following lemma states how a clockwise curve enters and leaves the half plane $\Pi_+(t_0)$.

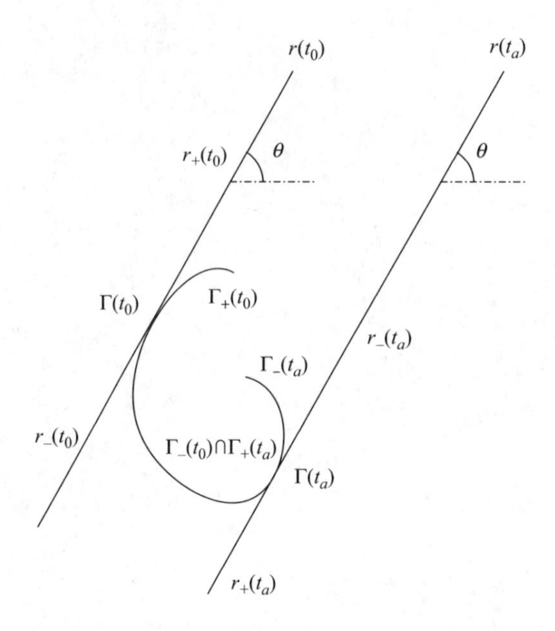

Figure 3.7 Illustration of the proof of Lemma 3.4.

Lemma 3.4 *Suppose that the curve Γ is clockwise and does not admit self-intersections. Then, the following statements are true:*

(1) If there exists a $t_\star < t_0$ such that $\Gamma(t) \in \Pi_+(t_0)$, $\forall t \in (t_\star, t_0)$ and $\Gamma(t_\star) \in r(t_0)$, i.e., Γ enters $\Pi_+(t_0)$ at $t_\star$, then $\Gamma(t_\star) \in r_+(t_0)$.

(2) If there exists a $t^\star > t_0$ such that $\Gamma(t) \in \Pi_+(t_0)$, $\forall t \in (t_0, t^\star)$ and $\Gamma(t^\star) \in r(t_0)$, i.e., Γ leaves $\Pi_+(t_0)$ at $t^\star$, then $\Gamma(t^\star) \in r_-(t_0)$.

Proof. We only give the proof of statement (1). The proof of statement (2) is similar.

Assume that the phase of the tangent vector of Γ at t_0 is $\Psi(t_0) = \theta$, and assume by contradiction that $\Gamma(t_\star) \in r_-(t_0)$. Since Γ is clockwise, the phase, $\Psi(t)$, of $\vec{r}(t)$ is strictly decreasing in t. Therefore, $\Psi(t) > \theta$, $\forall t < t_0$. So, we have

$$(2k_\star - 1)\pi + \theta < \Psi(t_\star) < 2k_\star\pi + \theta \tag{3.14}$$

for some positive integer $k_\star$. Hence, there exists a $t_a \in (t_\star, t_0)$ such that $\Psi(t_a) = \pi + \theta$, i.e., the tangent line $r(t_a)$ is parallel to $r(t_0)$ (see Figure 3.7). $\Gamma_-(t_a)$ can intersect $r_-(t_0)$ without crossing $r_+(t_0)$ and $\Gamma_-(t_0) \cap \Gamma_+(t_a)$ only if it crosses $r_-(t_a)$ for some $t_b < t_a$ (before crossing $r_+(t_a)$). By repeating the argument, it can be shown that $\Psi(t_\star) > k\pi + \theta$ for any finite positive k which can be arbitrarily large. This contradicts (3.14) and hence, $\Gamma(t_\star) \in r_+(t_0)$. □

By Lemma 3.4, a clockwise curve having no self-intersections lies on one side of its tangent line if it neither crosses the half tangent line $r_+(t_0)$ for $t < t_0$ nor crosses the half tangent line

$r_{-}(t_0)$ for $t > t_0$. For a curve lying on one side of its tangent line, the following lemma states that the convex hull of the curve together with the origin lies on one side of the tangent line if and only if the origin and the curve lie on the same side of the tangent line. This lemma addresses the concern of the second case given in the last section.

Lemma 3.5 *Given $t_0 \in (t_\alpha, t_\beta)$ and $t_i \in [t_\alpha, t_\beta], i \in \overline{1, n}$, suppose*

$$\Gamma(t) \in \Pi_{+}(t_0) \cup \Gamma(t_0), \ \ \forall t \in [t_\alpha, t_\beta]. \tag{3.15}$$

Then,

$$\kappa \mathrm{Co}(0, \{\Gamma(t_i), i \in \overline{1, n}\}) \in \Pi_{+}(t_0) \tag{3.16}$$

holds for all $\kappa \in [0, 1)$ if and only if $0 \in \Pi_{+}(t_0)$.

Proof. Necessity is obvious. We just show sufficiency. Note that $\kappa \mathrm{Co}(0, \{\Gamma(t_i), i \in \overline{1, n}\}) = \mathrm{Co}(0, \{\kappa\Gamma(t_i), i \in \overline{1, n}\})$. Since $0 \in \Pi_{+}(t_0)$ and $\Gamma(t_i) \in \Pi_{+}(t_0) \cup \Gamma(t_0)$, we have $\kappa\Gamma(t_i) \in \Pi_{+}(t_0)$ for all $\kappa \in [0, 1)$. Now, all $\kappa\Gamma(t_i)$s, $i \in \overline{1, n}$ together with the origin are inside $\Pi_{+}(t_0)$, by the separation principle of the convex analysis,

$$\kappa \mathrm{Co}(0, \{\Gamma(t_i), i \in \overline{1, n}\}) = \mathrm{Co}(0, \{\kappa\Gamma(t_i), i \in \overline{1, n}\}) \in \Pi_{+}(t_0)$$

for all $\kappa \in [0, 1)$. □

For the convex combination of two curves we have the following lemma.

Lemma 3.6 *Consider two curves $\Gamma_1(t)$ and $\Gamma_2(t)$ with $t \in [t_\alpha, t_\beta]$. Suppose $\Gamma_1(t)$ and $\Gamma_2(t)$ intersect with each other at the point $(-1, 0)$ with parameters t_c^1 and t_c^2, respectively, i.e.,*

$$\Gamma_1(t_c^1) = \Gamma_2(t_c^2) = (-1, 0). \tag{3.17}$$

Then,

$$\kappa \mathrm{Co}(\Gamma_1(t), \Gamma_2(t)) \subset \Pi_{+}^1(t_c^1) \cup \Pi_{+}^2(t_c^2), \ \ \forall \kappa \in [0, 1) \tag{3.18}$$

holds for all $t \in [t_\alpha, t_\beta]$ if

$$\Gamma_1(t), \Gamma_2(t) \in \Pi_{+}(t_c^1) \cup \Gamma_1(t_c^1),$$

or

$$\Gamma_1(t), \Gamma_2(t) \in \Pi_{+}(t_c^2) \cup \Gamma_2(t_c^2)$$

for all $t \in [t_\alpha, t_\beta]$.

Proof. For any fixed $t \in [t_\alpha, t_\beta]$, $\Gamma_1(t)$ and $\Gamma_2(t)$ simultaneously lie on the right side of the tangent line $r^1(t_c^1)$ or on the right side of $r^2(t_c^2)$. Therefore, for any $t \in [t_\alpha, t_\beta]$, we have

$$\kappa \mathrm{Co}(\Gamma_1(t), \Gamma_2(t)) \subset \Pi_{+}^1(t_c^1)$$

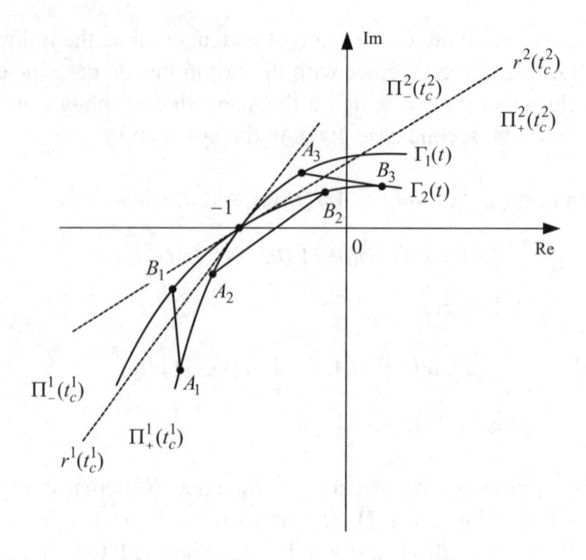

Figure 3.8 Illustration of Lemma 3.6.

or

$$\kappa \text{Co}\{\Gamma_1(t), \Gamma_2(t)\} \subset \Pi_+^2(t_c^2) \cup \Gamma_2(t_c^2)$$

for $0 \leq \kappa < 1$, which implies (3.18). □

Lemma 3.6 is illustrated by Figure 3.8, where segments $A_i B_i$, $i = 1, 2, 3$, represent possible cases of $\text{Co}(\Gamma_1(t), \Gamma_2(t))$ for different values of parameter t.

3.3 Scalability of First-Order Systems

In this section we study some global differential geometric properties of the Nyquist plot of the transfer function of the first-order time-delayed system with a single (open-loop) integrator. Many dynamic nodes in distributed control systems, such as the primal or dual algorithm of the congestion control of communication networks, ideal mobile agents (i.e., mobile agents without inertia), can be modeled by this kind of transfer function.

3.3.1 Continuous-Time System

Consider the transfer function

$$G(s) = k\frac{\mathrm{e}^{-sT}}{s}, \tag{3.19}$$

where $T > 0$ is the delay constant, and $k > 0$ is the gain. The frequency response of the system is

$$G(\mathrm{j}\omega) = \frac{k\mathrm{e}^{-\mathrm{j}T\omega}}{\mathrm{j}\omega} \tag{3.20}$$

$$= -\frac{k\sin(T\omega)}{\omega} - \mathrm{j}\frac{k\cos(T\omega)}{\omega} \tag{3.21}$$

$$\triangleq X(\omega) + \mathrm{j}Y(\omega). \tag{3.22}$$

Proposition 3.7 *The Nyquist plot $G(\mathrm{j}\omega)$ of system (3.19) is clockwise in the parameter interval* $(0, \infty)$.

Proof. From (3.20) it is easy to get

$$G_\omega'^{*} = -k\frac{\mathrm{e}^{\mathrm{j}T\omega}}{\omega^2}(-T\omega - \mathrm{j}),$$

$$G_{\omega\omega}'' = -k\frac{\mathrm{e}^{-\mathrm{j}T\omega}}{\omega^3}(2T\omega + \mathrm{j}(T^2\omega^2 - 2)).$$

It follows that

$$G_\omega'^{*}G_{\omega\omega}'' = -\frac{T^2\omega^2 - 2}{\omega^5} - \mathrm{j}\frac{T^3}{\omega^2}.$$

Therefore

$$\mathrm{Im}(G_\omega'^{*}G_{\omega\omega}) = -\frac{T^3}{\omega^2} < 0, \ \forall\omega \in (0, \infty).$$

By Definition 3.3 this implies that $G(\mathrm{j}\omega)$ is clockwise for all $\omega \in (0, \infty)$. □

Proposition 3.8 *The Nyquist plot $G(\mathrm{j}\omega)$ of system (3.19) has no self-intersection in the parameter interval* $(0, \infty)$.

Proof. The proposition is obvious from the fact that the modulus of the frequency response (see (3.20)) is strictly decreasing with respect to ω in the interval $(0, \infty)$. □

Let ω_c be the minimal crossing frequency of $G(\mathrm{j}\omega)$, i.e., the Nyquist plot of $G(\mathrm{j}\omega)$ crosses the real axis for the first time at $\omega = \omega_c$ as ω varies from 0 to ∞. It is easy to get

$$\omega_c = \frac{\pi}{2T} \tag{3.23}$$

for system (3.19). According to (3.21), straightforward calculation shows that the slope of $r(\omega_c)$, the tangent line to $G(\mathrm{j}\omega)$ at $G(\mathrm{j}\omega_c)$, is

$$k_c = \frac{Y_\omega'(\omega_c)}{X_\omega'(\omega_c)} = \frac{\pi}{2}, \tag{3.24}$$

which is independent of the delay constant T. Therefore, the angle Ψ_c between the line $r_+(\omega_c)$ and the real axis of the complex plane satisfies

$$0 < \Psi_c = \arctan\left(\frac{\pi}{2}\right) < \frac{\pi}{2}. \tag{3.25}$$

Lemma 3.9 *For all $\omega \in [0, \infty)$, the Nyquist plot of system (3.19) lies on the right side of the tangent line to $G(\mathrm{j}\omega)$ at $G(\mathrm{j}\omega_c)$, denoted by $r(\omega_c)$, i.e., $G(\mathrm{j}\omega) \in G(\mathrm{j}\omega_c) \cup \Pi_+(\omega_c)$, $\forall \omega \in [0, \infty)$.*

Proof. Firstly, we prove the lemma for all $\omega \in (0, \infty)$. By Proposition 3.7 and Proposition 3.8, the Nyquist plot of system (3.19) is clockwise and has no self-intersection in the parameter interval $(0, \infty)$. Therefore, by Lemma 3.4, the only way that the curve $G_+(\mathrm{j}\omega_c)$ can leave the half plane $\Pi_+(\omega_c)$ is by crossing the line $r_-(\omega_c)$ and the only way that $G_-(\omega_c)$ can enter the half plane $\Pi_+(\omega_c)$ is by crossing the line $r_+(\omega_c)$. Inequality (3.25) implies that the modulus of any point on $r_-(\omega_c)$ is greater than $|G(\mathrm{j}\omega_c)|$. However, the modulus of the frequency response, $\frac{k}{\omega}$, is strictly decreasing in the interval $(0, \infty)$ and hence, $|G(\mathrm{j}\omega)| < |G(\mathrm{j}\omega_c)|$, $\forall \omega \in (\omega_c, \infty)$. Therefore, it is impossible that the curve $\Gamma_+(\omega_c)$ crosses the line $r_-(\omega_c)$ for all $\omega \in (\omega_c, \infty)$. On the other hand, from (3.20) we know that for all $\omega \in (0, \omega_c)$ the phase of $G(\mathrm{j}\omega)$ is in the interval $(-\pi, -\frac{\pi}{2})$, i.e., the curve $G_-(\mathrm{j}\omega_c)$ lies inside the third quadrant of the complex plane and hence cannot cross the line $r_+(\omega_c)$ which is above the real axis of the complex plane. Hence, we have proved $G(\mathrm{j}\omega) \in G(\mathrm{j}\omega_c) \cup \Pi_+(\omega_c)$, for all $\omega \in (0, \infty)$.

When $\omega \to 0$, we can write $G(\mathrm{j}\omega)$ as

$$G(\mathrm{j}\omega) = k_1 \frac{\mathrm{e}^{-\mathrm{j}T\cdot 0}}{\varepsilon \mathrm{e}^{\mathrm{j}\theta}} = \frac{k_1}{\varepsilon \mathrm{e}^{\mathrm{j}\theta}}, \tag{3.26}$$

where ε is an infinitely small positive number and $\theta \in [0, \pi/2]$. When θ varies from 0 to $\pi/2$, $G(\mathrm{j}\omega)$ draws a clockwise infinitely large quarter-circle with phase angle in $[-\pi/2, 0]$. So it is also in $\Pi_+(\omega_c)$. □

Now we are ready to prove the following theorem.

Theorem 3.10 *For any given natural number $n \geq 2$, let*

$$G_i(\mathrm{j}\omega) = \gamma_i \frac{\mathrm{e}^{-\mathrm{j}T_i\omega}}{\mathrm{j}\omega}, \quad i \in \overline{1, n}, \tag{3.27}$$

where

$$\gamma_i = \frac{\pi}{2T_i}, \tag{3.28}$$

and $T_i \in \mathbb{R}^+$, $i \in \overline{1, n}$, are divers nonnegative delay constants. Then $\kappa\mathrm{Co}(0 \cup \{G_i(\mathrm{j}\omega), i \in \overline{1, n}\})$ does not contain the point $(-1, \mathrm{j}0)$ for all $\kappa \in [0, 1)$ and all $\omega \in (-\infty, \infty)$.

Proof. By the symmetry property of the frequency response we need only to prove the theorem for all $\omega \in [0, \infty)$.

Let

$$x_i = T_i\omega. \tag{3.29}$$

Obviously, (3.29) forms an onto map from $\omega \in [0, \infty)$ to $x_i \in [0, \infty)$. We notice that under the given gain condition (3.28), $G_i(\mathrm{j}\omega) = \frac{\pi \mathrm{e}^{-\mathrm{j}x_i}}{\mathrm{j}x_i}$, $x_i \in [0, \infty)$, $i \in \overline{1, n}$. Therefore, all of the Nyquist plots of $G_i(\mathrm{j}\omega)$, $i \in \overline{1, n}$, share a common curve, i.e., $G(\mathrm{j}x) = \frac{\pi \mathrm{e}^{-\mathrm{j}x}}{\mathrm{j}x}$, $x \in [0, \infty)$, on the complex plane. And the curve crosses the real axis for the first time at the point $(-1, \mathrm{j}0)$. By Lemma 3.9 we have $G(\mathrm{j}x) \in (-1, \mathrm{j}0) \cup \Pi_+(-1, \mathrm{j}0)$, $\forall x \in [0, \infty)$. For any given $\omega \in [0, \infty)$ we have $x_i = T_i\omega \in [0, \infty)$. So it follows that

$$\kappa \mathrm{Co}(0 \cup \{G_i(\mathrm{j}\omega), i \in \overline{1, n}\}) = \kappa \mathrm{Co}(0 \cup \{G(\mathrm{j}x_i), i \in \overline{1, n}\}). \tag{3.30}$$

By Lemma 3.5 we know that

$$\kappa \mathrm{Co}(0, \{G(\mathrm{j}x_i), i \in \overline{1, n}\}) \subset \Pi_+(-1, \mathrm{j}0), \tag{3.31}$$

holds for all $\kappa \in [0, 1)$. Therefore $\kappa \mathrm{Co}(0 \cup \{G_i(\mathrm{j}\omega), i \in \overline{1, n}\})$ does not contain the point $(-1, \mathrm{j}0)$ for all $\kappa \in [0, 1)$ and all $\omega \in [0, \infty)$ because $(-1, \mathrm{j}0) \notin \Pi_+(-1, \mathrm{j}0)$. □

3.3.2 Discrete-Time System

For a discrete-time system the frequency response is the value of its z-transfer function at the unit circle, namely, $G(\mathrm{e}^{\mathrm{j}\omega})$, $\omega \in (-\pi, \pi]$. For simplicity of notation we will write $G(\mathrm{e}^{\mathrm{j}\omega})$ as $G(\omega)$ for discrete-time systems. Let us study the first-order time-delayed system

$$G(z) = k\frac{z^{-D}}{z - 1}, \tag{3.32}$$

where D is a positive integer representing the delay, and $k > 0$ is the gain. This is a discrete-time analogue of system (3.19). Along the modified unit circle the frequency response of system (3.32) is

$$G(\omega) = k\frac{\mathrm{e}^{-\mathrm{j}D\omega}}{\mathrm{e}^{\mathrm{j}\omega} - 1} \tag{3.33}$$

$$= k\frac{\mathrm{e}^{-\mathrm{j}(\frac{\pi}{2} + \frac{2D+1}{2}\omega)}}{2\sin(\frac{\omega}{2})} \tag{3.34}$$

$$= -k\frac{\sin(\frac{2D+1}{2}\omega)}{2\sin(\frac{\omega}{2})} - \mathrm{j}k\frac{\cos(\frac{2D+1}{2}\omega)}{2\sin(\frac{\omega}{2})} \tag{3.35}$$

$$\triangleq X(\omega) + \mathrm{j}Y(\omega). \tag{3.36}$$

Since $G(z)$ has a pole $z = 1$ on the unit circle, we should modify the unit circle by adding a small half-circle around the pole $z = 1$ (see Figure 2.13). In this way the pole can be considered as inside the unit circle. The Nyquist plot of $G(\omega)$ will go to infinity when $\omega \to 0$ and draw a half-circle with an infinite radius clockwise.

Proposition 3.11 *The Nyquist plot of system $G(\omega)$ (3.32) is clockwise in the parameter interval* $[0, \pi]$.

Proof. Without loss of generality we suppose that $k = 1$ in (3.32). To prove the proposition we need to consider only the sign of the numerator of the curvature. By (3.13) we have

$$\mathcal{C}(\omega) = \frac{\mathrm{Im}(G'^{*}_{\omega} \cdot G''_{\omega\omega})}{|G'_{\omega}|^3}.$$

Straightforward calculating yields

$$G'_{\omega} = \frac{\mathrm{e}^{-\mathrm{j}\left(\frac{\pi}{2}+\frac{2D+1}{2}\omega\right)}}{\left(2\sin\left(\frac{\omega}{2}\right)\right)^2}\left(-\cos\left(\frac{\omega}{2}\right) - \mathrm{j}(2D+1)\sin\left(\frac{\omega}{2}\right)\right),$$

$$G'^{*}_{\omega} = \frac{\mathrm{e}^{\mathrm{j}\left(\frac{\pi}{2}+\frac{2D+1}{2}\omega\right)}}{\left(2\sin\left(\frac{\omega}{2}\right)\right)^2}\left(-\cos\left(\frac{\omega}{2}\right) + \mathrm{j}(2D+1)\sin\left(\frac{\omega}{2}\right)\right),$$

$$G''_{\omega\omega} = \frac{\mathrm{e}^{-\mathrm{j}\left(\frac{\pi}{2}+\frac{2D+1}{2}\omega\right)}}{\left(2\sin\left(\frac{\omega}{2}\right)\right)^3}\left(-4D(D+1)\sin^2\left(\frac{\omega}{2}\right)\right.$$
$$\left. +2\cos^2\left(\frac{\omega}{2}\right) + \mathrm{j}2(2D+1)\sin\left(\frac{\omega}{2}\right)\cos\left(\frac{\omega}{2}\right)\right).$$

Therefore, we have

$$\mathrm{Im}(G'^{*}_{\omega} \cdot G''_{\omega\omega}) = -\frac{D(D+1)(2D+1)}{2\left(2\sin\left(\frac{\omega}{2}\right)\right)^2},$$

which implies

$$\mathcal{C}(\omega) < 0, \ \forall \omega \in (0, 2\pi).$$

So, the clockwise property of the Nyquist plot of (3.32) is proved. □

Proposition 3.12 *The Nyquist plot $G(\omega)$ of system (3.32) has no self-intersection in the parameter interval* $(0, \pi]$.

Proof. The proposition is obvious from the fact that the modulus of the frequency response (see (3.34)), $k|\left(2\sin\left(\frac{\omega}{2}\right)\right)^{-1}|$, is strictly decreasing in the interval $(0, \pi]$. □

Let ω_c be the frequency at which the Nyquist plot $\Gamma(\omega)$ intersects the real axis for the first time as ω varies from 0 to π. It is easy to get

$$\omega_c = \frac{\pi}{2D+1} \tag{3.37}$$

for system (3.32). From (3.35), straightforward calculation shows that the slope of $r(\omega_c)$, the tangent line to $G(\omega)$ at $G(\omega_c)$, is

$$k_c = \frac{Y'_{\omega}(\omega_c)}{X'_{\omega}(\omega_c)} = (2D+1)\tan\left(\frac{\pi}{2(2D+1)}\right). \tag{3.38}$$

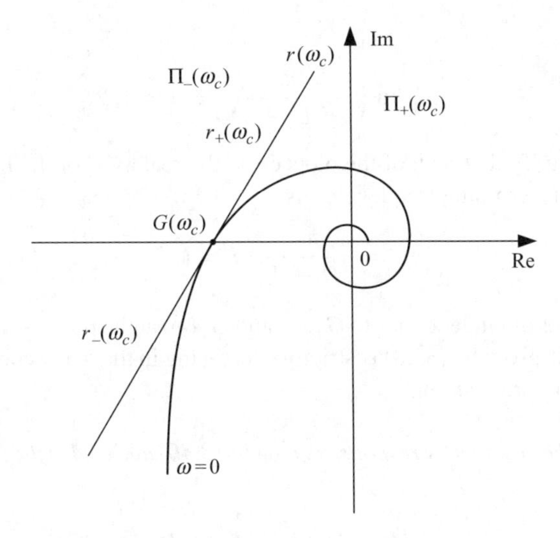

Figure 3.9 Proposition 3.13.

It is always the case that $k_c > 0$ as D is a delay constant which is a nonnegative integer for discrete-time systems. Therefore, the angle Ψ_c between the line $r_+(\omega_c)$ and the real axis of the complex plane satisfies

$$0 < \Psi_c < \frac{\pi}{2}. \tag{3.39}$$

Proposition 3.13 *For all $\omega \in (0, \pi]$, the Nyquist plot $G(\omega)$ of system (3.19) lies on the right side of $r(\omega_c)$, i.e., $G(\omega) \in G(\omega_c) \cup \Pi_+(\omega_c)$, $\forall \omega \in (0, \pi]$.*

Proof. By Proposition 3.11 and Proposition 3.12, the Nyquist plot $G(\omega)$ of system (3.32) is clockwise and has no self-intersection in the parameter interval $[0, \pi]$. Therefore, by Lemma 3.4, the only way that the curve $G_+(\omega_c)$ can leave the half plane $\Pi_+(\omega_c)$ is by crossing the line $r_-(\omega_c)$ and the only way that $G_-(\omega_c)$ can enter the half plane $\Pi_+(\omega_c)$ is by crossing the line $r_+(\omega_c)$ (see Figure 3.9). Equation (3.34) shows that the modulus of the frequency response, $k|(2\sin(\frac{\omega}{2}))^{-1}|$, is strictly decreasing in the interval $(0, \pi]$ and hence, $|G(\omega)| < |G(\omega_c)|$, $\forall \omega \in (\omega_c, \pi]$. But inequality (3.39) implies that the modulus of any point on $r_-(\omega_c)$ is greater than $|G(\omega_c)|$. Therefore, it is impossible that the curve $G_+(\omega_c)$ crosses the line $r_-(\omega_c)$ for all $\omega \in (\omega_c, \pi]$. On the other hand, from (3.34) we know that for all $\omega \in (0, \omega_c)$ the phase of $G(\omega)$ is in the interval $(-\pi, -\frac{\pi}{2})$, i.e., the curve $G_-(\omega_c)$ lies inside the third quadrant of the complex plane and hence cannot cross the line $r_+(\omega_c)$ which is above the real axis of the complex plane. □

Now, let us consider the Nyquist plots of two systems of the same form with different delay constants

$$G_i(\omega) = \gamma_i \frac{e^{-jD_i\omega}}{e^{j\omega} - 1}, \quad i = 1, 2, \tag{3.40}$$

where

$$\gamma_i = 2\sin\left(\frac{\pi}{2(2D_i+1)}\right), \quad i = 1, 2. \tag{3.41}$$

It is not difficult to verify that both of the plots cross the real axis for the first time at the same point $(-1, j0)$ with the crossing frequencies as

$$\omega_c^i = \frac{\pi}{2D_i+1}, \quad i = 1, 2. \tag{3.42}$$

Let k_c^1 and k_c^2 be slopes of tangent lines to $G_1(\omega)$ and $G_2(\omega)$ at the point $(-1, j0)$ respectively. It is easy to show that k_c given by (3.38) is strictly decreasing in the delay constant D. Therefore, we have the following proposition.

Proposition 3.14 *For frequency responses given by (3.40) and (3.41) the following inequality holds:*

$$(D_1 - D_2)(k_c^1 - k_c^2) < 0. \tag{3.43}$$

Lemma 3.15 *Let $G_i(\omega)$, $i = 1, 2$ be given by (3.40) and (3.41). Then,*

$$\kappa\mathrm{Co}(G_1(\omega), G_2(\omega)) \subset \Pi_+^1(\omega_c^1) \cup \Pi_+^2(\omega_c^2) \tag{3.44}$$

holds for all $\omega \in [-\pi, \pi]$ and $0 \le \kappa < 1$.

Proof. First of all, we note that both $G_1(\omega)$ and $G_2(\omega)$ enjoy the symmetry property with respect to the real axis of the complex plane, i.e.,

$$\overline{G_i(\omega)} = G_i(-\omega), \quad i = 1, 2, \quad \forall\omega \in [0, \pi].$$

Thus, we need only to prove (3.44) for all $\omega \in [0, \pi]$. Without loss of generality, we assume that $D_2 < D_1$. Then we have $\omega_c^1 < \omega_c^2$ by (3.42). Now, let us split the frequency interval $[0, \pi]$ into four parts, namely 0, $(0, \omega_c^1]$, (ω_c^1, ω_c^2) and $[\omega_c^2, \pi]$, and consider each case below.

Case 1: $0 < \omega \le \omega_c^1$
In this case both $G_1(\omega)$ and $G_2(\omega)$ are in the third quadrant of the complex plane because their phases are between $-\pi$ and $-\pi/2$. In this quadrant, tangent line $r^2(\omega_c^2)$ is on the right side of tangent line $r^1(\omega_c^1)$ because $k_c^1 < k_c^2$ by Proposition 3.14. And the tangent lines meet at the point $(-1, j0)$ (see Figure 3.10). So, by Proposition 3.13, both curves $G_1(\omega)$ and $G_2(\omega)$ are on the right side of $r^1(\omega_c^1)$ in this case. Moreover, only $G_1(\omega)$ intersects $r^1(\omega_c^1)$ at the point $(-1, j0)$ when $\omega = \omega_c^1$. Therefore, we have $G_1(\omega), G_2(\omega) \in \Pi_+^1(\omega_c^1) \cup (-1, j0)$, $\forall\omega \in (0, \omega_c^1]$.

Case 2: $\omega_c^2 \le \omega \le \pi$
In this case we claim that both $G_1(\omega)$ and $G_2(\omega)$ are on the right side of $r^2(\omega_c^2)$. Noticing Proposition 3.13 and Proposition 3.14, to prove this claim it suffices to show that $G_1(\omega)$ never intersects $r_-^2(\omega_c^2)$ in the third quadrant. This is indeed the case because $|G_1(\omega)| < 1$, $\forall\omega \in (\omega_c^1, \pi]$ but any point on $r_-^2(\omega_c^2)$ has modulus greater than unity (since $k_c^2 > 0$). Therefore, we have $G_1(\omega), G_2(\omega) \in \Pi_+^2(\omega_c^2) \cup (-1, j0)$, $\forall\omega \in [\omega_c^2, \pi]$.

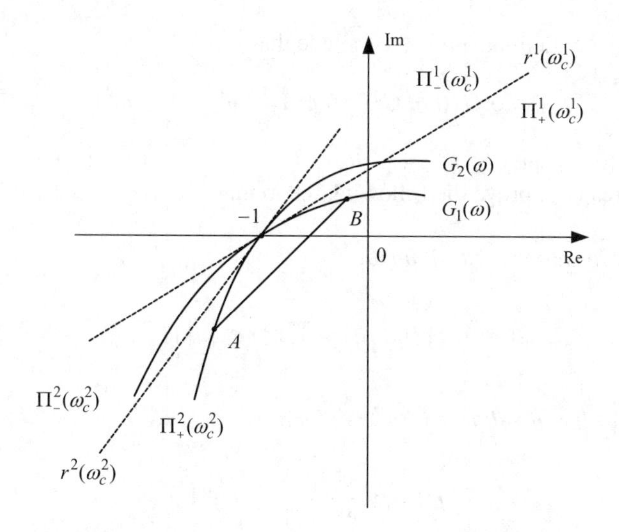

Figure 3.10 Illustration of proof of Lemma 3.15.

Case 3: $\omega_c^1 < \omega < \omega_c^2$
In this case $G_2(\omega)$ on the right side of $r^2(\omega_c^2)$ (by Proposition 3.13) and under the real axis of the complex plane (by the same argument given in Case 1), and $G_1(\omega)$ is on the right side of $r^1(\omega_c^1)$ (by Proposition 3.13) and never intersects $r_-^2(\omega_c^2)$ in the third quadrant (by the same argument given in Case 2). Therefore, both $G_1(\omega)$ and $(G_2(\omega)$ are inside $\Pi_+^1(\omega_c^1) \cap \Pi_+^2(\omega_c^2)$.

Case 4: $\omega \to 0$
In this case we can write $G_1(\omega)$ and $G_2(\omega)$ as

$$G_1(\omega) = k_1 \frac{\mathrm{e}^{-\mathrm{j}D_1 \cdot 0}}{\varepsilon \mathrm{e}^{\mathrm{j}\theta}} = \frac{k_1}{\varepsilon \mathrm{e}^{\mathrm{j}\theta}},$$
$$G_2(\omega) = k_1 \frac{\mathrm{e}^{-\mathrm{j}D_2 \cdot 0}}{\varepsilon \mathrm{e}^{\mathrm{j}\theta}} = \frac{k_2}{\varepsilon \mathrm{e}^{\mathrm{j}\theta}}$$

where ε is an infinitely small positive number and $\theta \in [0, \pi/2]$. So delays play no role when $\omega \to 0$. When θ varies from 0 to $\pi/2$ Both $G_1(\omega)$ and $G_2(\omega)$ synchronously draw a clockwise infinitely large quarter-circle with phase angle in $[-\pi/2, 0]$. Hence, $G_1(\omega), G_2(\omega) \in \Pi_+^2(\omega_c^2)$ in this case.

Summarizing all the four cases we know

$$G_1(\omega), G_2(\omega) \in \Pi_+^1(\omega_c^1) \cup (-1, \mathrm{j}0),$$

or

$$G_1(\omega), G_2(\omega) \in \Pi_+^2(\omega_c^2)$$

for any $\omega \in [0, \pi]$. By Lemma 3.6, we conclude that

$$\kappa\text{Co}(G_1(\omega), G_2(\omega)) \subset \Pi_+^1(\omega_c^1) \cup \Pi_+^2(\omega_c^2)$$

holds for all $\omega \in [0, \pi]$ and $0 \leq \kappa < 1$. □

Now, we are ready to prove the following theorem.

Theorem 3.16 *Given any natural number $n \geq 2$,*

$$\kappa\text{Co}(0 \cup \{G_i(\omega), i \in \overline{1, n}\}) \subset \bigcup_{i=1}^{n} \Pi_+^i(\omega_c^i) \tag{3.45}$$

holds for all $\kappa \in [0, 1)$ and all $\omega \in (-\pi, \pi)$, where

$$G_i(\omega) = \gamma_i \frac{e^{-jD_i\omega}}{e^{j\omega} - 1}, \quad i \in \overline{1, n},$$

$$\gamma_i = 2\sin\left(\frac{\pi}{2(2D_i + 1)}\right),$$

where D_i, $i \in \overline{1, n}$, are divers nonnegative integers. Consequently, $\kappa\text{Co}(0 \cup \{G_i(\omega), i \in \overline{1, n}\})$ does not contain the point $(-1, \text{j}0)$ for all $\omega \in [-\pi, \pi]$ and all $\kappa \in [0, 1)$.

Proof. By the symmetry property of the frequency response we need only to prove the lemma for all $\omega \in [0, \pi]$.

We first note that

$$\kappa\text{Co}(0, G_i(\omega)) \subset \Pi_+^i(\omega_c^i) \tag{3.46}$$

holds for all $\omega \in [0, \pi]$ since both $G_i(\omega)$ and the origin of the complex plane lie in $\Pi_+^i(\omega_c^i) \cup G_i(\omega_c^i)$. We also note that

$$\kappa\text{Co}(0 \cup \{G_i(\omega), i \in \overline{1, n}\}) \subset \bigcup_{i=1}^{n} \Pi_+^i(\omega_c^i)$$

holds for $\omega = \pi$ since $G_i(\pi) = -\cos(D_i\pi)/2 > -1$. Then, by continuity of the set $\kappa\text{Co}(0 \cup \{G_i(\omega), i \in \overline{1, n}\})$ on ω, $\kappa\text{Co}(0 \cup \{G_i(\omega), i \in \overline{1, n}\})$ goes out of the region $\bigcup_{i=1}^{n} \Pi_+^i(\omega_c^i)$ implies its boundary, $\kappa\text{Co}(G_{i_1}(\omega), G_{i_2}(\omega))$, or $\kappa\text{Co}\{0, G_i(\omega)\}$, goes out of the same region. But Lemma 3.15 and (3.46) have excluded this possibility. Therefore, we conclude that

$$\kappa\text{Co}(0 \cup \{G_i(\omega), i \in \overline{1, n}\}) \subset \bigcup_{i=1}^{n} \Pi_+^i(\omega_c^i)$$

for all $\omega \in [-\pi, \pi]$. Therefore, $\kappa\text{Co}(0 \cup \{G_i(\omega), i \in \overline{1, n}\})$ does not contain the point $(-1, \text{j}0)$ for all $\omega \in [-\pi, \pi]$ because $(-1, \text{j}0) \notin \bigcup_{i=1}^{n} \Pi_+^i(\omega_c^i)$. □

3.4 Scalability of Second-Order Systems

In this section we study the second-order time-delayed systems of two types. In accordance with textbooks on classic control theory, it will be called *a system of type I* if its (open-loop) transfer function contains a single integrator, or *a system of type II* if its transfer function contains double integrators.

3.4.1 System of Type I

Many dynamic nodes of distributed control systems, such as the primal-dual algorithm of the congestion control of communication networks, and position feedback control of a mobile agent with inertia, can be modeled as a second-order system with a single integrator. The time-delayed form of such a system is given by

$$G(s) = k\frac{\mathrm{e}^{-Ts}}{s(s+\alpha)}, \tag{3.47}$$

where $k > 0$ is the gain, $T > 0$ is the delay constant, and $-\alpha < 0$ is the LHP pole of the system.

The frequency response of system (3.47) is given by

$$G(\mathrm{j}\omega) = k\frac{\mathrm{e}^{-\mathrm{j}T\omega}}{(\mathrm{j}\omega)(\mathrm{j}\omega+\alpha)}. \tag{3.48}$$

Denote by $\omega_c > 0$ *the minimal crossing frequency* of $G(\mathrm{j}\omega)$, i.e., $G(\mathrm{j}\omega)$ crosses the real axis of the complex plane for the first time when $\omega = \omega_c$. Actually, $\omega_c > 0$ is the minimal frequency satisfying the following equation:

$$T\omega_c = \frac{\pi}{2} - \arctan(\omega_c/\alpha). \tag{3.49}$$

Denote by γ the gain margin of $G(s)$, which is defined by

$$\gamma = 1/|G(\mathrm{j}\omega_c)|. \tag{3.50}$$

From (3.49) we have

$$\frac{\alpha}{\omega_c} = \tan(T\omega_c).$$

Using this equality, we get

$$\begin{aligned}
\gamma^{-1} &= \frac{1}{\omega_c\sqrt{\omega_c^2 + \omega_c^2\tan^2(T\omega_c)}} \\
&= \frac{T\sin(T\omega_c)}{\alpha T\omega_c} \\
&< \frac{T}{\alpha}.
\end{aligned} \tag{3.51}$$

Now, let

$$x = T\omega, \tag{3.52}$$

$$A = T\alpha. \tag{3.53}$$

Then, $G(\mathrm{j}\omega)$ can be rewritten as

$$G(\mathrm{j}\omega) = kT^2 \frac{\mathrm{e}^{-\mathrm{j}x}}{(\mathrm{j}x)(\mathrm{j}x + A)} \triangleq kT^2 G(\mathrm{j}x), \tag{3.54}$$

by abusing the notation $G(\cdot)$ just for convenience. Obviously, for a given system $G(s)$, the curve

$$G(\mathrm{j}x) = \frac{\mathrm{e}^{-\mathrm{j}x}}{(\mathrm{j}x)(\mathrm{j}x + A)} \tag{3.55}$$

is just a linear zoom of the Nyquist plot of $G(\mathrm{j}\omega)$ and hence preserves almost all of the geometric properties of $G(\mathrm{j}\omega)$. For example, the Nyquist plot of $G(\mathrm{j}\omega)$ is clockwise in the frequency interval $[\omega_1, \omega_2]$ if and only if the curve of $G(\mathrm{j}x)$ is clockwise in the parameter interval $[T\omega_1, T\omega_2]$. Similarly, $G(\mathrm{j}\omega)$ does not admit self-intersection in the interval $[\omega_1, \omega_2]$ if and only if $G(\mathrm{j}x)$ does not admit self-intersections in the interval $[T\omega_1, T\omega_2]$.

Proposition 3.17 *The curve $G(\mathrm{j}x)$ does not admit self-intersections for all $x \geq 0$.*

Proof. The assertion is obvious from the fact that the modulus of $G(\mathrm{j}x)$, i.e.,

$$|G(\mathrm{j}x)| = \frac{1}{x\sqrt{A^2 + x^2}},$$

is strictly decreasing in x in the interval $[0, \infty)$. □

Proposition 3.18 *The curve $G(\mathrm{j}x)$ is clockwise in the parameter interval* $[0, \infty)$.

Proof. To prove the proposition, we only need to check the sign of the numerator of the curvature. By (3.13) we know that the curvature of $G(\mathrm{j}x)$ is given by

$$\mathcal{C}(x) = \frac{\mathrm{Im}(G_x'^* \cdot G_{xx}'')}{|G_x'|^3}.$$

Denote

$$G(\mathrm{j}x) = \frac{g_1(x)}{g_2(x)},$$

where

$$g_1(x) = \mathrm{e}^{-\mathrm{j}x}, \tag{3.56}$$

$$g_2(x) = \mathrm{j}x(\mathrm{j}x + A). \tag{3.57}$$

Then, we have

$$G'_x = \frac{g'_{1x}g_2 - g_1 g'_{2x}}{g^2} \triangleq \frac{G_1(x)}{G_2(x)}. \tag{3.58}$$

Therefore,

$$G'^*_x = \frac{(G_1 G_2^*)^*}{|G_2|^2} = \frac{G_2 G_1^*}{|G_2|^2},$$
$$G''_{xx} = \frac{G'_{1x}G_2 - G_1 G'_{2x}}{G_2^2}.$$

Straightforward calculation yields

$$G'^*_x G''_{xx} = \frac{|G_1|^2}{|G_2|^2}\left(\frac{G'_{1x}}{G_1} - \frac{G'_{2x}}{G_2}\right). \tag{3.59}$$

Taking (3.58) into account, we get

$$G'^*_x G''_{xx} = \frac{|G_1|^2}{|G_2|^2}\left(\frac{g''_{1xx}g_2 - g_1 g''_{2xx}}{g'_{1x}g_2 - g_1 g'_{2x}} - \frac{2g'_{2x}}{g_2}\right). \tag{3.60}$$

From (3.56) and (3.57), we get

$$g'_{1x} = -\mathrm{j}e^{-\mathrm{j}x},$$
$$g''_{1xx} = -e^{-\mathrm{j}x},$$
$$g'_{2x} = -2x + \mathrm{j}A,$$
$$g''_{2xx} = -2.$$

So, we have

$$\frac{g''_{1xx}g_2 - g_1 g''_{2xx}}{g'_{1x}g_2 - g_1 g'_{2x}} - \frac{2g'_{2x}}{g_2} \triangleq \frac{Q}{P},$$

where

$$Q = -x^4 + 6x^2 + x^2A^2 + 6x^2A - 2A^2 + \mathrm{j}(2x^3A + 4x^3 - 6xA - 2xA^2), \tag{3.61}$$
$$P = -2x^3A - 2x^3 + xA^2 + \mathrm{j}(x^2A^2 + 3x^2A - x^4). \tag{3.62}$$

Therefore,

$$\mathrm{Im}(G'^*_x \cdot G''_{xx}) = \frac{|G_1|^2}{|G_2|^2|P|^2}\mathrm{Im}(QP^*), \tag{3.63}$$

where

$$\mathrm{Im}(QP^*) = -x^8 - 2x^6A^2 - 3x^6A - 2x^6 - 6x^4A^2 - 6x^4A - x^4A^4 - 3x^4A^3. \tag{3.64}$$

Obviously, $\mathrm{Im}(QP^*) < 0$, for all $x > 0$ and all $A \geq 0$. This implies

$$\mathcal{C}(x) < 0, \quad \forall x \in (0, \infty).$$

Thus, the clockwise property of the curve of $G(\mathrm{j}x)$ in the interval $[0, \infty)$ is proved. □

Given ω_c the minimal crossing frequency of $G(\mathrm{j}\omega)$, it is obvious that $G(\mathrm{j}x)$ crosses the real axis of the complex plane for the first time when

$$x_c = T\omega_c. \tag{3.65}$$

According to (3.49), the minimal crossing parameter x_c satisfies

$$\arctan(x_c/A) = \frac{\pi}{2} - x_c. \tag{3.66}$$

Differentiating both sides of the above equation with respect to A, we get

$$\frac{1}{1 + (x_c/A)^2} \cdot \frac{x_c' A - x_c}{A^2} = -x_c',$$

which gives

$$x_c' = \frac{x_c}{x_c^2 + A^2 + A} > 0, \tag{3.67}$$

where $x_c' = \frac{\mathrm{d}x_c}{\mathrm{d}A}$. So, we have the following proposition.

Proposition 3.19 *The minimal crossing point x_c is a strictly monotonically increasing function of parameter A.*

Denote by $r(x_c)$ the tangent line of $G(\mathrm{j}x)$ at the minimal crossing point x_c. Then, we can prove the following result.

Proposition 3.20 *The slope of $r(x_c)$, denoted by T_c, is a strictly monotonically increasing function of parameter A. Moreover, $T_c > 0$ for all $A > 0$.*

Proof. The slope of the tangent line $r(x_c)$ is given by

$$T_c = \left.\frac{\mathrm{Im}(G_x')}{\mathrm{Re}(G_x')}\right|_{x=x_c}. \tag{3.68}$$

Note that

$$\frac{\partial T_c}{\partial x_c} = \frac{\mathrm{Im}(G_x'^* G_{xx}'')}{(\mathrm{Re}(G_x'))^2},$$

$$\frac{\partial T_c}{\partial A} = \frac{\mathrm{Im}(G_x'^* G_{xA}'')}{(\mathrm{Re}(G_x'))^2}.$$

So, we get

$$\begin{aligned}\frac{\mathrm{d}T_c}{\mathrm{d}A} &= \frac{\partial T_c}{\partial x_c}\frac{\partial x_c}{\partial A} + \frac{\partial T_c}{\partial A} \\ &= \frac{\mathrm{Im}(G_x'^{*}G_{xx}'')}{(\mathrm{Re}(G_x'))^2}\frac{\mathrm{d}x_c}{\mathrm{d}A} + \frac{\mathrm{Im}(G_x'^{*}G_{xA}'')}{(\mathrm{Re}(G_x'))^2},\end{aligned} \tag{3.69}$$

where $\mathrm{Im}(G_x'^{*}G_{xx}'')$ has been obtained in the proof of Proposition 3.18 (see (3.63) and (3.64)), and $\frac{\mathrm{d}x_c}{\mathrm{d}A}$ is given by (3.67). Now, we calculate $\mathrm{Im}(G_x'^{*}G_{xA}'')$. Denote

$$G(\mathrm{j}x) = \frac{g_1(x)}{g_2(x)},$$

where $g_1(x)$ and $g_2(x)$ are given by (3.56) and (3.57), respectively. Then, we have

$$G_x' = \frac{g_{1x}'g_2 - g_1g_{2x}'}{g^2} \triangleq \frac{G_1(x)}{G_2(x)}. \tag{3.70}$$

Therefore,

$$\begin{aligned}G_x'^{*} &= \frac{(G_1G_2^*)^*}{|G_2|^2} = \frac{G_2G_1^*}{|G_2|^2}, \\ G_{xA}'' &= \frac{G_{1A}'G_2 - G_1G_{2A}'}{G_2^2}.\end{aligned}$$

Straightforward calculation yields

$$G_x'^{*}G_{xA}'' = \frac{|G_1|^2}{|G_2|^2}\left(\frac{G_{1A}'}{G_1} - \frac{G_{2A}'}{G_2}\right). \tag{3.71}$$

Taking (3.70) into account, we get

$$G_x'^{*}G_{xA}'' = \frac{|G_1|^2}{|G_2|^2}\left(\frac{g_{1xA}''g_2 + g_{1x}'g_{2A}' - g_{1A}'g_{2x}' - g_1g_{2xA}''}{g_{1x}'g_2 - g_1g_{2x}'} - \frac{2g_{2A}'}{g_2}\right).$$

From (3.56) and (3.57), we get

$$\begin{aligned}g_{1x}' &= -\mathrm{j}e^{-\mathrm{j}x}, \quad & g_{1A}' &= 0, \quad & g_{1xA}'' &= 0, \\ g_{2x}' &= -2x + \mathrm{j}A, \quad & g_{2A}' &= \mathrm{j}x, \quad & g_{2xA}'' &= \mathrm{j}.\end{aligned}$$

So, we have

$$\frac{g_{1xA}''g_2 + g_{1x}'g_{2A}' - g_{1A}'g_{2x}' - g_1g_{2xA}''}{g_{1x}'g_2 - g_1g_{2x}'} - \frac{2g_{2A}'}{g_2} \triangleq \frac{N}{P},$$

where

$$N = x^3 - xA + \mathrm{j}(-x^3A - 3x^2),$$

and P is given by (3.62). Therefore,

$$\mathrm{Im}(G_x'^* \cdot G_{xA}'') = \frac{|G_1|^2}{|G_2|^2|P|^2}\mathrm{Im}(NP^*), \tag{3.72}$$

where

$$\mathrm{Im}(NP^*) = x^7 + 4x^5A + x^5A^2 + 6x^5. \tag{3.73}$$

Now, substituting (3.67), (3.63), (3.64), (3.72) and (3.73) into (3.69) yields

$$\frac{\mathrm{d}T_c}{\mathrm{d}A} = \frac{|G_1|^2}{|G_2|^2|P|^2(\mathrm{Re}(G_x'))^2}W,$$

where

$$W = 4x^7 + 2x^7A + 4x^5A^2 + 2x^5A^3 > 0.$$

Therefore, $T_c(A)$ is a strictly monotonically increasing function of the parameter A. Finally, we note that $T_c = 0$ when $A = 0$. This implies that $T_c > 0$ for all $A > 0$. The proposition is thus proved. □

Now, using Lemma 3.4, we can prove the following result.

Proposition 3.21 *For all $x \in [0, \infty)$, the curve $G(\mathrm{j}x)$ lies on the right side of $r(x_c)$, i.e., $G(\mathrm{j}x) \in G(\mathrm{j}x_c) \cup \Pi_+(x_c)$, $\forall x \in [0, \infty)$.*

Proof. For simplicity of statement, we denote the curve of $G(\mathrm{j}x)$ in the parameter interval $[0, \infty)$ as $G[0, \infty)$. Also, we denote the curve $G(\mathrm{j}x)$ in the parameter interval (x_c, ∞) as $G_+(x_c)$ and the curve $G(\mathrm{j}x)$ in the parameter interval $[0, x_c)$ as $G_-(x_c)$. By Proposition 3.17 and Proposition 3.18, $G[0, \infty)$ is clockwise and has no self-intersections. Therefore, by Lemma 3.4, the only way that the curve $G_+(x_c)$ can leave the half plane $\Pi_+(x_c)$ is by crossing the line $r_-(x_c)$ and the only way that $G_-(x_c)$ can enter the half plane $\Pi_+(\omega_c)$ is by crossing the line $r_+(\omega_c)$. In the proof of Proposition 3.17, we have mentioned that the modulus $|G(\mathrm{j}x)|$ is strictly decreasing in x, and hence, $|G(\mathrm{j}x)| < |G(\mathrm{j}x_c)|$, $\forall x \in (x_c, \infty)$. But, from Proposition 3.20, we know that the slope of the tangent line at $x = x_c$ is greater than zero. This implies that the modulus of any point on $r_-(x_c)$ is greater than $|G(\mathrm{j}x_c)|$. Therefore, it is impossible that the curve $G_+(x_c)$ crosses the line $r_-(x_c)$ for all $x \in (x_c, \infty)$. On the other hand, for any $x \in [0, x_c)$, the curve $G_-(x_c)$ lies inside the third quadrant of the complex plane and hence cannot cross the line $r_+(x_c)$, which is located above the real axis of the complex plane. The proposition is thus proved. □

Next, let us consider the Nyquist plots of two systems of the same form with different delay constants:

$$G_i(\mathrm{j}\omega) = \gamma_i \frac{\mathrm{e}^{-\mathrm{j}T_i\omega}}{(\mathrm{j}\omega)(\mathrm{j}\omega + \alpha_i)}, \quad i = 1, 2, \tag{3.74}$$

where γ_i is the gain margin of the transfer function

$$W_i(s) = \frac{e^{-T_i s}}{s(s + \alpha_i)}.$$

Lemma 3.22 *If the following condition*

$$(T_1\alpha_1 - T_2\alpha_2)(\omega_c^1 - \omega_c^2) \geq 0 \tag{3.75}$$

holds for the frequency responses of systems given by (3.74), then

$$\kappa \mathrm{Co}(G_1(j\omega), G_2(j\omega)) \subset \Pi_+^1(\omega_c^1) \cup \Pi_+^2(\omega_c^2) \tag{3.76}$$

holds for all real numbers $\kappa \in [0, 1)$ *and all* $\omega \in [0, \infty)$.

Proof. First of all, we note that both of the two plots of the functions in the form (3.74) cross the real axis for the first time at the same point $(-1, j0)$, with minimal crossing frequencies

$$\omega_c^i = x_c^i / T_i, \quad i = 1, 2, \tag{3.77}$$

where x_c^i is given by (3.66). Without loss of generality, assume that $A_1 < A_2$, which by (3.75) also implies that

$$\omega_c^1 \leq \omega_c^2.$$

Since when $\omega < \omega_c^1$ both $G_1(j\omega)$ and $G_2(j\omega)$ are in the third quadrant, the conclusion of the lemma obviously holds in the frequency interval $[0, \omega_c^1)$. We need to prove the proposition for the frequency interval $[\omega_c^1, \infty)$. To do so, let us split this frequency interval into two parts, namely $[\omega_c^1, \omega_c^2)$ and $[\omega_c^2, \infty)$. We consider each case below.

Case 1: $\omega_c^1 \leq \omega < \omega_c^2$
In this case, $G_2(j\omega)$ is on the right side of $r^2(\omega_c^2)$ (by Proposition 3.21) and under the real axis of the complex plane (since $\omega_c^2 > \omega_c^1$), and $G_1(j\omega)$ is on the right side of $r^1(\omega_c^1)$ (by Proposition 3.21). Moreover, $G_1(j\omega)$ never intersects $r_-^2(\omega_c^2)$ in the third quadrant because $|G_1(j\omega)| < 1, \forall \omega \in (\omega_c^1, \infty)$, but any point on $r_-^2(\omega_c^2)$ has modulus greater than unity (since $T_c^2 > 0$ by Proposition 3.20). Therefore, both $G_1(j\omega)$ and $G_2(j\omega)$ are inside $\Pi_+^1(\omega_c^1) \cap \Pi_+^2(\omega_c^2)$, which is a convex set (since $T_c^1 < T_c^2$ by Proposition 3.20). Thus, for any given $\kappa \in [0, 1)$, we have $\kappa \mathrm{Co}(G_1(j\omega), G_2(j\omega)) \subset \Pi_+^1(\omega_c^1) \cap \Pi_+^2(\omega_c^2)$, $\forall \omega \in [\omega_c^1, \omega_c^2)$.

Case 2: $\omega_c^2 \leq \omega < \infty$
In this case, both $G_1(j\omega)$ and $G_2(j\omega)$ have already crossed the real axis for the first time. We claim that both $G_1(j\omega)$ and $G_2(j\omega)$ are on the right side of $r^2(\omega_c^2)$. Noticing Proposition 3.21, to prove this claim it suffices to show that $G_1(j\omega)$ never intersects $r_-^2(\omega_c^2)$ in the third quadrant. This is indeed the case as we have shown in *Case 1*. Therefore, both $G_1(j\omega)$ and $G_2(j\omega)$ are inside $\Pi_+^2(\omega_c^2) \cup G_2(\omega_c^2)$, which is obviously a convex set. Note that only $G_2(j\omega)$ intersects $r^2(\omega_c^2)$ at the point $(-1, j0)$ when $\omega = \omega_c^2$. Thus, we have $\kappa \mathrm{Co}(G_1(j\omega), G_2(j\omega)) \subset \Pi_+^2(\omega_c^2)$, $\forall \omega \in [\omega_c^2, \infty)$, for any given $0 \leq \kappa < 1$.

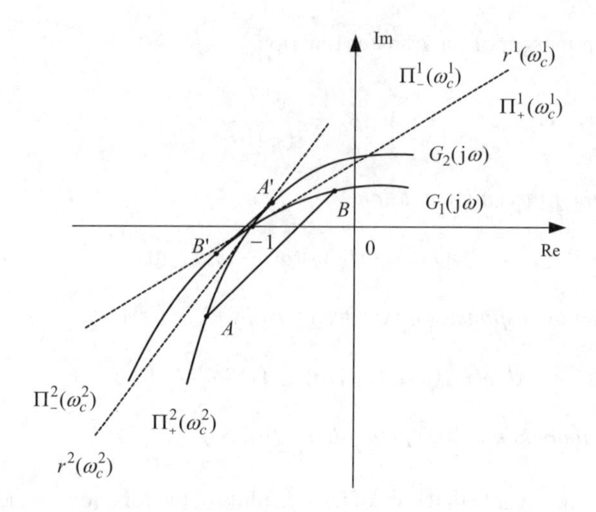

Figure 3.11 Illustration of Lemma 3.22.

Summarizing the conclusion obtained in the above two cases, we complete the proof of the lemma. □

A graphical illustration of Lemma 3.22 is given by Figure 3.11. Note that condition (3.75) is necessary for the conclusion of Lemma 3.22. Otherwise, the less curved $G_2(j\omega)$ may cross the real axis and enter the second quadrant before the more curved $G_1(j\omega)$ does. In this case, $\mathrm{Co}(G_1(j\omega), G_2(j\omega))$ will be out of the region $\Pi_+^1(\omega_c^1) \cup \Pi_+^2(\omega_c^2)$, as illustrated as $A'B'$ in Figure 3.11.

Now, we are ready to prove the following theorem.

Theorem 3.23 *Suppose that the following condition*

$$(T_i\alpha_i - T_j\alpha_j)(\omega_c^i - \omega_c^j) \geq 0, \ \forall i, j \in \overline{i, n}, \ i \neq j. \tag{3.78}$$

holds for the frequency response of a family of systems described by

$$G_i(j\omega) = \gamma_i \frac{e^{-jT_i\omega}}{(j\omega)(j\omega + \alpha_i)}, \ i \in \overline{1, n}. \tag{3.79}$$

where γ_i is the gain margin of the transfer function

$$W_i(s) = \frac{e^{-T_i s}}{s(s + \alpha_i)}.$$

Then, $\kappa\mathrm{Co}(0 \cup \{G_i(j\omega), i \in \overline{1, n}\})$ does not contain the point $(-1, j0)$ for all real numbers $\kappa \in [0, 1)$ and all $\omega \in [0, \infty)$.

Proof. By Lemma 3.22, we know that

$$\kappa\mathrm{Co}(G_i(j\omega), G_k(j\omega)) \subset \Pi_+^i(\omega_c^i) \cup \Pi_+^k(\omega_c^k) \tag{3.80}$$

holds for all $i, k \in \overline{1, n}$. Note that

$$\kappa \text{Co}(0, G_i(j\omega)) \subset \Pi^i_+(\omega^i_c) \tag{3.81}$$

holds for all $\omega \in [0, \infty)$ since both $G_i(j\omega)$ and the origin of the complex plane lie in $\Pi^i_+(\omega^i_c) \cup G_i(\omega^i_c)$. Also, note that

$$\kappa \text{Co}(0 \cup \{G_i(j\omega), i \in \overline{1, n}\}) \subset \bigcup_i^n \Pi^i_+(\omega^i_c)$$

holds when $\omega \to \infty$ since $G_i(j\infty)$ goes to the origin of the complex plane for all $i \in \overline{1, n}$. Therefore, by continuity of the set $\kappa \text{Co}(0 \cup \{G_i(j\omega), i \in \overline{1, n}\})$ on ω, if $\kappa \text{Co}(0 \cup \{G_i(j\omega), i \in \overline{1, n}\})$ goes out of the region $\bigcup_i^n \Pi^i_+(\omega^i_c)$, then its boundary,

$$\kappa \text{Co}(G_i(j\omega), G_k(j\omega)),$$

or

$$\kappa \text{Co}(0, G_i(j\omega)),$$

will go out of the same region at some frequency. But (3.80) and (3.81) have excluded this possibility. Therefore, we conclude that

$$\kappa \text{Co}(0 \cup \{G_i(j\omega), i \in \overline{1, n}\}) \subset \bigcup_{i=1}^n \Pi^i_+(\omega^i_c).$$

Since $(-1, j0) \notin \bigcup_{i=1}^n \Pi^i_+(\omega^i_c)$, $\kappa \text{Co}(0 \cup \{G_i(j\omega), i \in \overline{1, n}\})$ does not contain the point $(-1, j0)$ for all $\omega \in [0, \infty)$. The theorem is thus proved. □

For the family of systems given by (3.79), denote by $\check{i}$ the index of the system with minimal value of parameter A_i, i.e.,

$$\check{i} = \arg \min_{i \in \overline{1,n}} A_i. \tag{3.82}$$

Let $T^{\check{i}}_c$ be the slope of the tangent line to the Nyquist plot of $W_{\check{i}}(j\omega)$ at the intersection point C, as illustrated by Figure 3.12. It is easy to see that $T^{\check{i}}_c = |OE|/|EC|$. An analytic formula of $T^{\check{i}}_c$ is given by

$$T^{\check{i}}_c = \left. \frac{\text{Im} G'_{\check{i}}(\omega)}{\text{Re} G'_{\check{i}}(\omega)} \right|_{\omega = \omega^{\check{i}}_c}. \tag{3.83}$$

Define

$$\mu_i = \begin{cases} 1, & \text{if } i = \check{i}; \\ \sqrt{\frac{(T^{\check{i}}_c)^2}{1 + (T^{\check{i}}_c)^2}}, & \text{otherwise.} \end{cases} \tag{3.84}$$

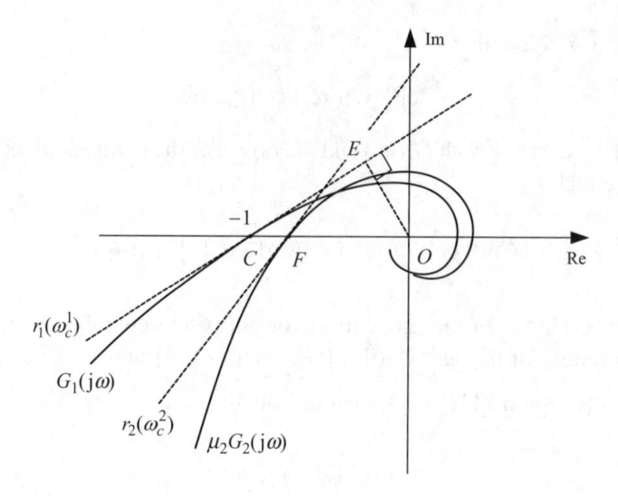

Figure 3.12 Definition of μ_i and illustration of the proof of Theorem 3.24.

Now, we are ready to give an alternative theorem for checking if $(-1, j0)$ is contained by $\kappa\text{Co}\{0 \cup \{G_i(j\omega), i \in \overline{1, n}\}\}$.

Theorem 3.24 *Let μ_i be defined by (3.84) and $G_i(j\omega)$ be defined by (3.79). Then, $\kappa\text{Co}(0 \cup \{\mu_i G_i(j\omega), i \in \overline{1, n}\})$ does not contain the point $(-1, j0)$ for all real numbers $\kappa \in [0, 1)$ and all $\omega \in [0, \infty)$.*

Proof. For simplicity, we only show the validity of the theorem for the case of two systems. But the proof can be directly extended to the general case.

Without loss of generality, assume $A_1 < A_2$, i.e., $\check{i} = 1$. Then, we have $\mu_1 = 1$ and

$$\mu_2 = \sqrt{\frac{(T_c^1)^2}{(T_c^1)^2 + 1}} < 1,$$

where T_c^1 is the slope of the tangent line $r_1(\omega_c^1)$ (see Figure 3.12). To prove the theorem, it suffices to show that $\mu_2 G_2(j\omega) \subset \Pi_+^1(\omega_c^1)$.

Note that $G_1(j\omega)$ intersects the real axis of the complex plane at $(-1, j0)$ while $\mu_2 G_2(j\omega)$ intersects the real axis at point F with coordinate $(\mu_2, j0)$. It is easy to see that $|OF| = |OE| = \mu_2$, where $|OE|$ is the shortest distance to the origin from the tangent line $r_1(\omega_c^1)$. Since the modulus of $\mu_2 G_2(j\omega)$ is strictly monotonically decreasing in ω, we have

$$|\mu_2 G_2(j\omega)| < |OE| = |OF|, \ \forall \omega > \omega_c^2,$$

which implies that $\mu_2 G_2(j\omega)$ never touches $r_1(\omega_c^1)$ for $\omega > \omega_c^2$. To show that $\mu_2 G_2(j\omega)$ never touches $r_1(\omega_c^1)$ for $\omega < \omega_c^2$, we only need to notice the fact that $\mu_2 < 1$ and $T_c^2 > T_c^1$ (by Proposition 3.20), where T_c^1 and T_c^2 are the slopes of the tangent lines $r_1(\omega_c^1)$ and $r_1(\omega_c^2)$, respectively (see Figure 3.12). □

3.4.2 System of Type II

Many dynamic nodes in distributed control systems, such as the second-order active queue management (AQM) algorithm, force-controlled ideal mobile agents with both position and velocity feedback, can be modeled as a second-order system with double integrators, which is referred to as the second-order system of type II in the literature of classic feedback control. In this subsection, we give detailed analysis of the geometric properties of the frequency response of the time-delayed system

$$G(s) = k\frac{(s+\alpha)\mathrm{e}^{-Ts}}{s^2}, \tag{3.85}$$

where $k > 0$ is the gain, $T > 0$ is the delay constant, and $-\alpha < 0$ is the zero of the system. The frequency response of system (3.85) is

$$G(\mathrm{j}\omega) = k\frac{(\mathrm{j}\omega+\alpha)\mathrm{e}^{-\mathrm{j}T\omega}}{(\mathrm{j}\omega)^2}. \tag{3.86}$$

Let $x = T\omega$. Then, $G(\mathrm{j}\omega)$ can be rewritten as

$$G(\mathrm{j}\omega) = kT\frac{(\mathrm{j}x+A)\mathrm{e}^{-\mathrm{j}x}}{(\mathrm{j}x)^2} \triangleq kTG(\mathrm{j}x), \tag{3.87}$$

where

$$A = T\alpha \tag{3.88}$$

is a key parameter in the geometric analysis for this system. (3.87) shows that for a given system $G(s)$, the curve

$$G(\mathrm{j}x) = \frac{(\mathrm{j}x+A)\mathrm{e}^{-\mathrm{j}x}}{(\mathrm{j}x)^2} \tag{3.89}$$

is just a linear zoom of the Nyquist plot of $G(\mathrm{j}\omega)$. Hence, $G(\mathrm{j}x)$ preserves all of the geometric properties of $G(\mathrm{j}\omega)$. For example, the Nyquist plot of $G(\mathrm{j}\omega)$ is clockwise in the frequency interval $[\omega_1, \omega_2]$ if and only if the curve of $G(\mathrm{j}x)$ is clockwise in the parameter interval $[T\omega_1, T\omega_2]$. Similarly, $G(\mathrm{j}\omega)$ does not admit self-intersection in the interval $[\omega_1, \omega_2]$ if and only if $G(\mathrm{j}x)$ does not admit self-intersections in the interval $[T\omega_1, T\omega_2]$. Therefore, next we will mainly study the geometric property of $G(\mathrm{j}x)$ instead of $G(\mathrm{j}\omega)$.

Proposition 3.25 *The curve $G(\mathrm{j}x)$ does not admit self-intersections for all $x \geq 0$.*

Proof. The proposition is obvious from the fact that the modulus of $G(\mathrm{j}x)$, i.e.,

$$|G(\mathrm{j}x)| = \frac{\sqrt{A^2+x^2}}{x^2}$$

is strictly decreasing in x in the interval $[0, \infty)$. □

Proposition 3.26 *If $A < 1$, then the curve $G(\mathrm{j}x)$ is clockwise in the parameter interval $[x_0, \infty)$, where*

$$x_0 = \sqrt{\left(3A - A^2 + \sqrt{(3A - A^2)^2 + 8A(1 - A)}\right)/2}. \tag{3.90}$$

Proof. To prove the proposition we only need to check the sign of the numerator of the curvature

$$\mathcal{C}(x) = \frac{\mathrm{Im}(G_x'^* \cdot G_{xx}'')}{|G_x'|^3}.$$

Straightforward calculating yields

$$G_x' = -\frac{\mathrm{e}^{-\mathrm{j}x}}{x^3}(x(x - \mathrm{j}) + A(-2 - \mathrm{j}x)), \tag{3.91}$$

$$G_x'^* = -\frac{\mathrm{e}^{\mathrm{j}x}}{x^3}(x(x + \mathrm{j}) + A(-2 + \mathrm{j}x)), \tag{3.92}$$

$$G_{xx}'' = -\frac{\mathrm{e}^{-\mathrm{j}x}}{x^4}\left(x(-2x + \mathrm{j}(2 - x^2)) + A(6 - x^2 + 4x\mathrm{j})\right). \tag{3.93}$$

Therefore,

$$\mathrm{Im}(G_x'^* \cdot G_{xx}'') = -(x^4 + (A^2 - 3A)x^2 + 2A^2 - 2A)/x^6. \tag{3.94}$$

Denote

$$p(x) = x^4 + (A^2 - 3A)x^2 + 2A^2 - 2A. \tag{3.95}$$

When $0 \le A \le 1$, it is easy to get the largest root of $p(x)$ as

$$x_0 = \sqrt{\left(3A - A^2 + \sqrt{(3A - A^2)^2 + 8A(1 - A)}\right)/2}. \tag{3.96}$$

When $x > x_0$ we have $p(x) > 0$. This implies

$$\mathcal{C}(x) < 0, \ \forall x \in (x_0, \infty).$$

So, the clockwise property of the curve of $G(\mathrm{j}x)$ in the interval $[x_0, \infty)$ is proved. □

We call x_0 *the maximal critical parameter* of $G(\mathrm{j}x)$, and correspondingly, $\omega_0 = x_0/T$ *the maximal critical frequency* of $G(\mathrm{j}\omega)$.

Let ω_c be the minimal crossing frequency of $G(\mathrm{j}\omega)$, i.e, the Nyquist plot of $G(\mathrm{j}\omega)$ crosses the real axis for the first time at $\omega = \omega_c$ as ω varies from 0 to ∞. It is easy to prove that ω_c is the minimal frequency satisfying

$$\omega_c = \alpha \tan(\omega_c T). \tag{3.97}$$

Similarly, we have the minimal crossing parameter of $G(\mathrm{j}x)$ at

$$x_c = T\omega_c.$$

According to (3.97), the minimal crossing point x_c satisfies

$$x_c = A\tan(x_c). \tag{3.98}$$

Both x_0 and x_c can be regarded as functions of A, an adjustable parameter. The following proposition gives some important properties of $x_0(A)$ and $x_c(A)$ of $G(jx)$.

Proposition 3.27 *If $A \in [0, 1)$, then the following claims are true:*

(1) The maximal critical parameter $x_0(A)$ is a strictly monotonically increasing function of the parameter A.

(2) The minimal crossing point $x_c(A)$ is a strictly monotonically decreasing function of the parameter A.

(3) There exists $\bar{A} \in (0, 1)$ such that $x_0(A) \le x_c(A)$ for all $A \in (0, \bar{A}]$, and the equality holds if and only if $A = \bar{A}$.

Proof. Denote $y = x_0^2$. Then from (3.90) we have

$$y = \frac{1}{2}\left(3A - A^2 + \sqrt{(3A - A^2)^2 + 8A(1 - A)}\right).$$

Thus,

$$\frac{\mathrm{d}y}{\mathrm{d}A} = \frac{F(A)}{\sqrt{(3A - A^2)^2 + 8A(1 - A)}} > 0 \tag{3.99}$$

as $A \in [0, 1)$, where

$$F(A) = (3 - 2A)\sqrt{(3A - A^2)^2 + 8A(1 - A)} + (3A - A^2)(3 - 2A) + 4 - 8A > 0.$$

Therefore, y is a strictly monotonically increasing function of A. Hence, so is $x_0(A)$. Claim (1) of the proposition is proved.

According to the definition of x_c we have

$$x_c = A\tan(x_c).$$

Obviously, the equation has a solution $x_c \in [0, \pi/2)$ if $A \in (0, 1]$. $x_c = 0$ if and only if $A = 1$. By the formula for derivatives of implicit functions, it is easy to get

$$\frac{\mathrm{d}x_c}{\mathrm{d}A} = \frac{\sin(x_c)}{\sin(x_c)\cos(x_c) - x_c} < 0$$

for $x_c \in (0, \pi/2)$. So $x_c(A)$ is a strictly monotonically decreasing function of A. Claim (2) of the proposition is proved.

Notice that $x_c = \pi/2$, $x_0 = 0$ when $A = 0$ and $x_c = 0$, $x_0 = \sqrt{2}$ when $A = 1$. According to Claim (1) and Claim (2), this fact directly results in Claim (3) of the proposition. □

Letting $x_c = x_0$, we can get the value of $\bar{A}$. When $A = \bar{A}$, the maximal critical point meets the minimal crossing point. Substituting (3.98) into (3.90), we obtain an equation with respect to the parameter A, the solution of which is

$$\bar{A} = 0.4495. \tag{3.100}$$

So, the following proposition is straightforward from the foregoing discussion.

Proposition 3.28 *If $A \in [0, \bar{A})$, then the curve $G(\mathrm{j}x)$ is inside the third quadrant of the complex plane for the parameter interval $[0, x_0]$, and is clockwise for the parameter interval $[x_0, \infty)$, where x_0 is the maximal critical parameter of $G(\mathrm{j}x)$.*

Denote the tangent line of $G(\mathrm{j}x)$ at the minimal crossing point x_c as $r(x_c)$. Then we can prove the following proposition.

Proposition 3.29 *The slope, denoted by T_c, of $r(x_c)$ is a strictly monotonically decreasing function of the parameter A, and moreover, $T_c > 0$ for all $A \in (0, \bar{A})$.*

Proof. The slope of the tangent line $r(x_c)$ is given by

$$T_c = \left.\frac{\mathrm{Im}(G'_x)}{\mathrm{Re}(G'_x)}\right|_{x=x_c} = \frac{x_c^3 - Ax_c + A^2 x_c}{x_c^2 + 2A^2}. \tag{3.101}$$

To show $T_c > 0$ for all $A \in (0, \bar{A})$ it suffices to show

$$x_c^2 > A - A^2, \ \forall A \in (0, \bar{A}). \tag{3.102}$$

By Proposition 3.27, we know that for all $A \in (0, \bar{A})$

$$\begin{aligned} x_c &> x_0 \\ &= \sqrt{\left(3A - A^2 + \sqrt{(3A - A^2)^2 + 8A(1 - A)}\right)/2} \\ &> \sqrt{3A - A^2}. \end{aligned}$$

So $x_c^2 > 3A - A^2 > A - A^2$, and hence, $T_c > 0$ for all $A \in (0, \bar{A})$.

Now we show that T_c is a strictly monotonically decreasing function of A. Note that

$$\frac{\partial T_c}{\partial x_c} = \frac{\mathrm{Im}(G'^*_x G''_{xx})}{(\mathrm{Re}(G'_x))^2}, \quad \frac{\partial T_c}{\partial A} = \frac{\mathrm{Im}(G'^*_x G''_{xA})}{(\mathrm{Re}(G'_x))^2}$$

where G''_{xA} represents $\frac{\partial G'_x}{\partial A}$ and all the remaining notations are the same as given in the *Proof of Proposition 3.26*. So from (3.101) it follows that

$$\begin{aligned} \frac{\mathrm{d}T_c}{\mathrm{d}A} &= \frac{\partial T_c}{\partial x_c}\frac{\partial x_c}{\partial A} + \frac{\partial T_c}{\partial A} \\ &= \frac{\mathrm{Im}(G'^*_x G''_{xx})}{(\mathrm{Re}(G'_x))^2}\frac{\partial x_c}{\partial A} + \frac{\mathrm{Im}(G'^*_x G''_{xA})}{(\mathrm{Re}(G'_x))^2}. \end{aligned}$$

Straightforward computing gives

$$G''_{xA} = -\frac{e^{-jx}}{x^3}(-jx - 2). \tag{3.103}$$

In *Proof of Proposition 3.27* we have shown

$$\frac{\partial x_c}{\partial A} = \frac{\sin x_c}{\sin x_c \cos x_c - x_c} = \frac{x_c}{A - (A^2 + x_c^2)} < 0. \tag{3.104}$$

So, using (3.91)–(3.94) and (3.103), we get

$$\frac{dT_c}{dA} = -\frac{1}{x_c^6(\mathrm{Re}(G'_x))^2}\left(\Pi(x_c)\frac{\partial x_c}{\partial A} + (x_c^3 + 2x_c)\right)$$

where the polynomial $\Pi(x_c)$ is defined by (3.95). Substituting (3.95) and (3.104) into the above equation yields

$$\frac{dT_c}{dA} = \frac{2x_c^2(A+1)}{x_c^6(\mathrm{Re}(G'_x))^2} \cdot \frac{x}{A - (A^2 + x_c^2)} < 0.$$

Therefore, $T_c(A)$ is a strictly monotonically decreasing function of A. The proposition is thus proved. □

Now, using Lemma 3.4, we can prove the following proposition.

Proposition 3.30 *For all $x \in [x_0, \infty)$, the curve $G(jx)$ lies on the right side of $r(x_c)$, i.e., $G(jx) \in G(jx_c) \cup \Pi_+(x_c)$, $\forall x \in [x_0, \infty)$.*

Proof. For simplicity of statement we denote the curve of $G(jx)$ in the parameter interval $[x_0, \infty)$ as $G[x_0, \infty)$. Also, we denote the curve $G(jx)$ in the parameter interval (x_c, ∞) as $G_+(x_c)$ and the curve $G(jx)$ in the parameter interval $[x_0, x_c)$ as $G_-(x_c)$. By Proposition 3.25 and Proposition 3.26, $G[x_0, \infty)$ is clockwise and has no self-intersections. Therefore, by Lemma 3.4, the only way that the curve $G_+(x_c)$ can leave the half plane $\Pi_+(x_c)$ is by crossing the line $r_-(x_c)$ and the only way that $G_-(x_c)$ can enter the half plane $\Pi_+(\omega_c)$ is by crossing the line $r_+(\omega_c)$ (see Figure 3.13). In the proof of Proposition 3.25 we have mentioned that the modulus $|G(jx)|$ is strictly decreasing in x, and hence, $|G(jx)| < |G(jx_c)|$, $\forall x \in (x_c, \infty)$. But, from Proposition 3.29, we know the slope of the tangent line at $x = x_c$ is greater than zero. This implies that the modulus of any point on $r_-(x_c)$ is greater than $|G(jx_c)|$. Therefore, it is impossible that the curve $G_+(x_c)$ crosses the line $r_-(x_c)$ for all $x \in (x_c, \infty)$. On the other hand, from Proposition 3.28 we know that for all $x \in [x_0, x_c)$ the curve $G_-(x_c)$ lies inside the third quadrant of the complex plane and hence cannot cross the line $r_+(x_c)$ which is above the real axis of the complex plane. The proposition is thus proved. □

Now, let us consider the Nyquist plots of two systems of the same form with different delay constants:

$$G_i(j\omega) = \gamma_i \frac{(j\omega + \alpha)e^{-jT_i\omega}}{(j\omega)^2}, \quad i = 1, 2, \tag{3.105}$$

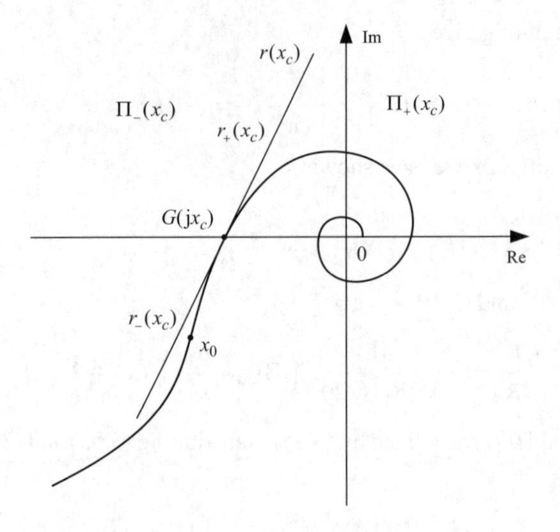

Figure 3.13 Graphic illustration of Proposition 3.30.

where γ_i is the gain margin of the transfer function

$$W_i(s) = \frac{(s+\alpha)\mathrm{e}^{-T_i s}}{s^2}. \tag{3.106}$$

Lemma 3.31 *If the frequency responses of systems (3.105) satisfy the following conditions:*

(1) $A_i < 0.4495, \forall i = 1, 2,$

(2) $\omega_0(i) \le T_{\hat{i}}^{-1} \arctan\left(\frac{\omega_0(i)}{\alpha}\right), \forall i = 1, 2,$

where

$$\hat{i} = \arg\max_{i=1,2} T_i, \tag{3.107}$$

then

$$\kappa \mathrm{Co}(G_1(\mathrm{j}\omega), G_2(\mathrm{j}\omega)) \subset \Pi_+^1(\omega_c^1) \cup \Pi_+^2(\omega_c^2) \tag{3.108}$$

holds for any given real number $\kappa \in [0, 1)$ *and any* $\omega \in [\min\{\omega_c^1, \omega_c^2\}, \infty)$.

Proof. First of all, we note that both of the two plots of the functions in the form (3.105) cross the real axis for the first time at the same point $(-1, \mathrm{j}0)$ with the minimal crossing frequencies as

$$\omega_c^i = x_c^i / T_i, \quad i = 1, 2 \tag{3.109}$$

where x_c^i is given by (3.98). Without loss of generality, we assume that $A_2 < A_1$. This also implies $T_2 < T_1$. Then, by Proposition 3.27, we have

$$x_c^1 < x_c^2$$

which also implies that

$$\omega_c^1 < \omega_c^2$$

since $T_2 < T_1$.

Now, we consider the lemma in the following two cases separately.

Case 1: $\omega_c^1 \leq \omega < \omega_c^2$

In this case $G_1(j\omega)$ has crossed the real axis while $G_2(j\omega)$ is still under the real axis of the complex plane. We note that condition (2) of the lemma implies the maximal critical frequencies of both $G_1(j\omega)$ and $G_2(j\omega)$ are less than $\min\{\omega_c^1, \omega_c^2\} = \omega_c^1$. Therefore, by Proposition 3.28, the Nyquist plots of $G_1(j\omega)$ and $G_2(j\omega)$ are both clockwise for the frequency interval $[\omega_c^1, \infty)$ under condition (1) of the lemma. Hence, by Proposition 3.30, $G_1(j\omega)$ is on the right side of $r^1(\omega_c^1)$ and $G_2(j\omega)$ is on the right side of $r^2(\omega_c^2)$. Moreover, $G_1(j\omega)$ never intersects $r_-^2(\omega_c^2)$ in the third quadrant because $|G_1(j\omega)| < 1, \forall\omega \in (\omega_c^1, \infty)$ but any point on $r_-^2(\omega_c^2)$ has modulus greater than unity (since the slope of the tangent line $r^2(\omega_c^2)$ is greater than zero by Proposition 3.29). Now, from Proposition 3.29 we know that under the assumption $A_2 < A_1$, the slope of the tangent line $r^1(\omega_c^1)$ is less than the slope of the tangent line $r^2(\omega_c^2)$. This implies that both $G_1(j\omega)$ and $G_2(j\omega)$ lie in the intersection of Π_+^1 and Π_+^2 for each $\omega \in [\omega_c^1, \omega_c^2]$. Hence, by Lemma 3.6, for any given $0 \leq \kappa < 1$ and for each $\omega \in [\omega_c^1, \omega_c^2)$, the convex combination $\kappa\text{Co}(G_1(j\omega), G_2(j\omega))$ lies in $\Pi_+^1(\omega_c^1) \cap \Pi_+^2(\omega_c^2)$.

Case 2: $\omega_c^2 \leq \omega < \infty$

In this case both $G_1(j\omega)$ and $G_2(j\omega)$ have crossed the real axis for the first time. We claim that both $G_1(j\omega)$ and $G_2(j\omega)$ are on the right side of $r^2(\omega_c^2)$. Noticing Proposition 3.28 and Proposition 3.29, to prove this claim it suffices to show that $G_1(j\omega)$ never intersects $r_-^2(\omega_c^2)$ in the third quadrant. This is indeed the case as we have shown in *Case 1*. Therefore, both $G_1(j\omega)$ and $G_2(j\omega)$ are inside $\Pi_+^2(\omega_c^2) \cup G_2(\omega_c^2)$ which is obviously a convex set. Note that only $G_2(j\omega)$ intersects $r^2(\omega_c^2)$ at the point $(-1, j0)$ when $\omega = \omega_c^2$. Thus, by Lemma 3.6, we have $\kappa\text{Co}(G_(j\omega), G_2(j\omega)) \subset \Pi_+^2(\omega_c^2), \forall\omega \in [\omega_c^2, \infty)$, for any given $0 \leq \kappa < 1$.

So, we conclude that the lemma is true. □

Now, we are ready to prove the following theorem.

Theorem 3.32 *Suppose that the frequency responses of a family of systems are described by*

$$G_i(j\omega) = \gamma_i \frac{(j\omega + \alpha)e^{-jT_i\omega}}{(j\omega)^2}, \quad i \in \overline{1, n}, \tag{3.110}$$

where γ_i is the gain margin of the transfer function

$$W_i(s) = \frac{(s+\alpha)\mathrm{e}^{-T_i s}}{s^2}.$$

If the following conditions hold:

(1) $A_i < 0.4495,\ \forall i \in \overline{1,n},$

(2) $\omega_0(i) \leq T_{\hat{i}}^{-1} \arctan\left(\frac{\omega_0(i)}{\alpha}\right),\ \forall i \in \overline{1,n},$

where

$$\hat{i} = \arg\max_{i\in\overline{1,n}} T_i, \tag{3.111}$$

then, $\kappa\mathrm{Co}(0 \cup \{G_i(\mathrm{j}\omega), i \in \overline{1,n}\})$ *does not contain the point* $(-1, \mathrm{j}0)$ *for any given real number* $\kappa \in [0, 1)$ *and any* $\omega \in [0, \infty)$.

Proof. Let ω_c^i be the minimal crossing frequency of $G_i(\mathrm{j}\omega)$. By Proposition 3.27, $x_c^i = T_i\omega_c^i$ is strictly decreasing in A_i, and hence in T_i. Therefore, we have

$$\min_{i\in\overline{1,n}} x_c(i) = x_c(\hat{i}),$$

where $\hat{i}$ is given by (3.111). Since

$$T_{\hat{i}} = \max_{i\in\overline{1,n}} T_i,$$

we have

$$\omega_c(\hat{i}) = \min_{i\in\overline{1,n}} \omega_c^i.$$

Now, we split the frequency interval $[0, \infty)$ into two parts: $[0, \omega_c(\hat{i}))$ and $[\omega_c(\hat{i}), \infty)$, and prove the theorem for the following two cases separately.

Case 1: $0 \leq \omega < \omega_c(\hat{i}))$
In this case, all the plots $G_i(\mathrm{j}\omega), i \in \overline{1,n}$ are in the third quadrant under condition (1) of the theorem. So $\kappa\mathrm{Co}(0 \cup \{G_i(\mathrm{j}\omega), i \in \overline{1,n}\})$ does not contain the point $(-1, \mathrm{j}0)$.

Case 2: $\omega_c(\hat{i}) \leq \omega < \infty$
Note that in this case all the plots $G_i(\mathrm{j}\omega), i \in \overline{1,n}$ have passed their maximal critical frequencies and are thus clockwise because condition (2) of the theorem guarantees that

$$\omega_0(i) \leq \omega_c(\hat{i}),\ \forall i \in \overline{1,n}.$$

By Lemma 3.31 we know that

$$\kappa\mathrm{Co}(G_i(\mathrm{j}\omega), G_k(\mathrm{j}\omega)) \subset \Pi_+^i(\omega_c^i) \cup \Pi_+^k(\omega_c^k) \tag{3.112}$$

holds for any $i, k \in \overline{1, n}$. Now, we note that

$$\kappa \mathrm{Co}(0, G_i(\mathrm{j}\omega)) \subset \Pi_+^i(\omega_c^i) \tag{3.113}$$

holds for all $\omega \in [\omega_c(\hat{i}), \infty)$ since both $G_i(\mathrm{j}\omega)$ and the origin of the complex plane lie in $\Pi_+^i(\omega_c^i) \cup G_i(\omega_c^i)$. We also note that $\kappa \mathrm{Co}(0 \cup \{G_i(\mathrm{j}\omega), i \in \overline{1, n}\}) \subset \bigcup_{i=1}^{n} \Pi_+^i(\omega_c^i)$ holds when $\omega \to \infty$ since $G_i(\mathrm{j}\infty)$ goes to the origin of the complex plane for all $i \in \overline{1, n}$. Therefore, by continuity of the set $\kappa \mathrm{Co}(0 \cup \{G_i(\mathrm{j}\omega), i \in \overline{1, n}\})$ on ω, $\kappa \mathrm{Co}(0 \cup \{G_i(\mathrm{j}\omega), i \in \overline{1, n}\})$ goes out of the region $\bigcup_{i=1}^{n} \Pi_+^i(\omega_c^i)$ implies that its boundary, i.e., $\kappa \mathrm{Co}(G_i(\mathrm{j}\omega), G_k(\mathrm{j}\omega))$ or $\kappa \mathrm{Co}(0, G_i(\mathrm{j}\omega))$, will go out of the same region. But (3.112) and (3.113) have excluded this possibility. Therefore, we conclude that

$$\kappa \mathrm{Co}(0 \cup \{G_i(\mathrm{j}\omega), i \in \overline{1, n}\}) \subset \bigcup_{i=1}^{n} \Pi_+^i(\omega_c^i).$$

Since $(-1, \mathrm{j}0) \notin \bigcup_{i=1}^{n} \Pi_+^i(\omega_c^i$, $\kappa \mathrm{Co}(0 \cup \{G_i(\mathrm{j}\omega), i \in \overline{1, n}\})$ does not contain the point $(-1, \mathrm{j}0)$ for $\omega \in [\omega_c(\hat{i}), \infty)$.

Therefore, summarizing the above two cases, the theorem is proved. □

3.5 Frequency-Sweeping Condition

Lemma 3.1 shows that the scalability condition for symmetric multi-agent system can be verified by checking if the point $(-1, \mathrm{j}0)$ is touched by the convex combination of Nyquist plots of each pair of agents. In this section we will further convert the latter into the problem of the stability test for a convex combination of two stable quasi-polynomials.

3.5.1 Stable Quasi-Polynomials

First let us recall the concept of stability of a quasi-polynomial.

A quasi-polynomial is an entire function of the form

$$\begin{aligned} f(s) &= p_0(s)\mathrm{e}^{\tau_0 s} + p_1(s)\mathrm{e}^{\tau_1 s} + \cdots + p_m(s)\mathrm{e}^{\tau_m s} \\ &= \sum_{i=0}^{m}\sum_{k=0}^{n} a_{ik} s^k \mathrm{e}^{\tau_i s}, \end{aligned} \tag{3.114}$$

where $p_i(s), i = 0, 1, \cdots, m$, are polynomials with coefficients $a_{ik} \in \mathbb{R}$ and $\tau_0 < \tau_1 < \cdots < \tau_m$ are real numbers representing delays.

For $n \geq 0, m > 0$ and any real vector $\tau = [\tau_0, \tau_1, \cdots, \tau_m]^{\mathrm{T}}$ with ordered components $\tau_0 < \tau_1 < \cdots < \tau_m$, let $\mathcal{Q}_n^{m,\tau}(R)$ denote the set of all quasi-polynomials defined by (3.114) with coefficients in $\mathbb{R}$.

A non-constant quasi-polynomial $f(s) \in \mathcal{Q}_n^{m,\tau}(R)$ is called *Hurwitz stable* or simply *stable* if all its roots belong to the open left half plane. The set of all Hurwitz stable quasi-polynomials in $\mathcal{Q}_n^{m,\tau}(R)$ is denoted by $\mathcal{H}_n^{m,\tau}(R)$.

Definition 3.33 *A quasi-polynomial $g(s) \in \mathcal{Q}_{n-1}^{m,\tau}(R)$ is called a global convex direction for the set $\mathcal{H}_n^{m,\tau}(R)$ if, for all stable quasi-polynomials $f(s) \in \mathcal{H}_n^{m,\tau}(R)$, the stability of $f(s) + g(s)$ implies the stability of the whole segment of quasi-polynomials $[f, f+g] = \{f(s) + \mu g(s); \mu \in [0,1]\}$, i.e., if $g(s)$ satisfies, for all $f(s) \in \mathcal{H}_n^{m,\tau}(R)$,*

$$f(s),\ f(s) + g(s) \in \mathcal{H}_n^{m,\tau}(R) \Rightarrow [f, f+g] \in \mathcal{H}_n^{m,\tau}(R).$$

The following lemma due to Kharitonov and Zhabko (1994) shows that the global convex direction can be verified through a frequency-sweeping test of the phase velocity.

Lemma 3.34 *A quasi-polynomial $g(s) \in \mathcal{Q}_{n-1}^{m,\tau}(R)$ is a global convex direction for the set $\mathcal{H}_n^{m,\tau}(R)$ if and only if for all $\omega \in \{\omega > 0 \mid g(\mathrm{j}\omega) \neq 0\}$ the following condition is satisfied:*

$$\frac{\partial \arg(g(\mathrm{j}\omega))}{\partial \omega} \leq \frac{\tau_0 + \tau_m}{2} + \left| \frac{\sin(2\arg(g(\mathrm{j}\omega)) - (\tau_0 + \tau_m)\omega)}{2\omega} \right|. \tag{3.115}$$

Note that the global convex direction is defined for the set of all Hurwitz stable quasi-polynomials. To check the stability of a convex combination of two specific quasi-polynomials, we need the following definition of local convex direction.

Definition 3.35 *A quasi-polynomial $g(s) \in \mathcal{Q}_{n-1}^{m,\tau}(R)$ is called a local convex direction for a Hurwitz quasi-polynomial $f(s)$ if the stability of $f(s) + g(s)$ implies the stability of the whole segment of quasi-polynomials $[f, f+g] = \{f(s) + \mu g(s); \mu \in [0,1]\}$, i.e., if $g(s)$ satisfies*

$$f(s),\ f(s) + g(s) \in \mathcal{H}_n^{m,\tau}(R) \Rightarrow [f, f+g] \in \mathcal{H}_n^{m,\tau}(R).$$

By Definition 3.35, the following proposition is true.

Proposition 3.36 *Given $f_0(s), f_1(s) \in \mathcal{Q}_n^{m,\tau}(R)$. Then, the following two statements are equivalent:*

(1) $\mathrm{Co}(f_0(s), f_1(s)) \in \mathcal{H}_n^{m,\tau}(R)$*;*

(2) $f_0(s), f_1(s) \in \mathcal{H}_n^{m,\tau}(R)$, and $f_1(s) - f_0(s)$ is a local convex direction for $f_0(s)$.

The following lemma gives a sufficient condition for the local convex direction.

Lemma 3.37 *A quasi-polynomial $g(s) \in \mathcal{Q}_{n-1}^{m,\tau}(R)$ is a local convex direction for the quasi-polynomial $f(s)$ if for all $\omega \in \{\omega > 0 :\ g(\mathrm{j}\omega) \neq 0\}$ the following condition is satisfied:*

$$\frac{\partial \arg(g(\mathrm{j}\omega))}{\partial \omega} \leq \min\left\{ \frac{\partial \arg(f(\mathrm{j}\omega))}{\partial \omega}, \frac{\partial \arg((f+g)(\mathrm{j}\omega))}{\partial \omega} \right\}. \tag{3.116}$$

Proof. The lemma can be proved based on the fact that the phase of any Hurwitz quasi-polynomial is monotonically increasing when s goes along the imaginary axis from $-j\infty$ to $j\infty$. We leave the completion of the proof to the reader as an exercise. □

The following lemma gives a necessary and sufficient condition of the local convex direction. It is directly from the fact that the instability of one of the polynomials $f_{\mu_1}(s)$, $\mu_1 \in [0, 1]$, implies that there is a quasi-polynomial $f_{\mu^\star}(s)$, $\mu^\star \in [0, 1]$, with at least one zero on the imaginary axis (Gu, Kharitonov and Chen 2003).

Lemma 3.38 *A quasi-polynomial $g(s) \in \mathcal{Q}_{n-1}^{m,\tau}(R)$ is a local convex direction for the quasi-polynomial $f(s)$ if and only if the complex curve*

$$r(j\omega) = \frac{f(j\omega)}{f(j\omega) + g(j\omega)}, \ \omega \in (-\infty, +\infty),$$

does not touch the negative real semi-axis of the complex plane.

3.5.2 Frequency-Sweeping Test

Given the normalized transfer function of time-delayed system $\hat{G}_i(s) = \gamma_i W_i(s)$ as

$$\hat{G}_i(s) = \frac{N_i(s)e^{-T_i s}}{s^\eta D_i(s)}, i = 1, 2 \tag{3.117}$$

where η, l, n are non-negative integers satisfying $l \leq n + \eta$; $N_i(s)$ and $D_i(s)$ are l-order and n-order real-coefficient polynomials of s, respectively; $D_1(s)$ and $D_2(s)$ are coprime. Note that $D_i(s)$ can be a constant which can be considered as a zero-order polynomial. Without loss of generality, it is assumed that $0 \leq T_1 \leq T_2$. Denote

$$f_0(s) = s^\eta D_1(s)D_2(s) + N_1(s)D_2(s)e^{-T_1 s}, \tag{3.118}$$

$$f_1(s) = s^\eta D_1(s)D_2(s) + N_2(s)D_1(s)e^{-T_2 s}, \tag{3.119}$$

$$g(s) = N_2(s)D_1(s)e^{-T_2 s} - N_1(s)D_2(s)e^{-T_1 s}, \tag{3.120}$$

$$\tau = [-T_2, -T_1, 0]^T. \tag{3.121}$$

The following lemma relates the stability of a convex hull of two time-delayed systems to the robust stability of a convex combination of two Hurwitz polynomials.

Theorem 3.39 *Suppose normalized transfer functions are given by (3.117), and $D_i(s)$, $i = 1, 2$, are coprime Hurwitz stable polynomials or positive constants. Then, the following conditions are equivalent:*

(1) $(-1, j0) \notin \mathrm{Co}\left(\hat{G}_1(j\omega), \hat{G}_2(j\omega)\right), \forall\omega \in [0, \infty)$;

(2) $\mathrm{Co}\left(f_0(s), f_1(s)\right) \in \mathcal{H}_{2n+\eta}^{3,\tau}(R)$;

(3) $f_0(s)$ and $f_1(s)$ are stable, $g(s)$ is a local convex direction for $f_0(s)$.

Proof. "(1) $\Leftrightarrow$ (2)". Write $\text{Co}\left(\hat{G}_1(\text{j}\omega), \hat{G}_2(\text{j}\omega)\right)$ as

$$\begin{aligned}
&\text{Co}\left(\hat{G}_1(\text{j}\omega), \hat{G}_2(\text{j}\omega)\right) \\
&= \frac{\lambda N_1(\text{j}\omega)D_2(\text{j}\omega)\text{e}^{-\text{j}T_1\omega} + (1-\lambda)N_2(\text{j}\omega)D_1(\text{j}\omega)\text{e}^{-\text{j}T_2\omega}}{(\text{j}\omega)^\eta D_1(\text{j}\omega)D_2(\text{j}\omega)} \\
&\triangleq G(\text{j}\omega, \lambda),
\end{aligned} \tag{3.122}$$

where $\lambda \in [0, 1]$. Since $D_1(s)$, $D_2(s)$ are Hurwitz stable polynomials or positive constants, by the Nyquist criterion of stability we know that the system with characteristic equation as

$$\det(I + G(s, \lambda)) = 0 \tag{3.123}$$

is robustly stable for all $\lambda \in [0, 1]$ if and only if

$$(-1, \text{j}0) \notin \text{Co}\left(\hat{G}_1(\text{j}\omega), \hat{G}_2(\text{j}\omega)\right), \forall \omega \in [0, \infty).$$

From (3.122) it is easy to see that the characteristic quasi-polynomial associated with (3.123) is given by

$$f_\lambda(s) = s^\eta D_1(s)D_2(s) + \lambda N_1(s)D_2(s)\text{e}^{-T_1 s} + (1-\lambda)N_2(s)D_1(s)\text{e}^{-T_2 s}. \tag{3.124}$$

Rewrite $f_\lambda(s)$ as

$$\begin{aligned}
f_\lambda(s) = &\lambda(s^\eta D_1(s)D_2(s) + N_1(s)D_2(s)\text{e}^{-T_1 s}) \\
&+(1-\lambda)(s^\eta D_1(s)D_2(s) + N_2(s)D_1(s)\text{e}^{-T_2 s}).
\end{aligned}$$

Then, the equivalence between (1) and (2) is clear.

"(2) $\Leftrightarrow$ (3)". It follows directly from Proposition 3.36. □

Note that if we replace statement (1) in Theorem 3.39 by

$$(-1, \text{j}0) \notin \kappa\text{Co}\left(\hat{G}_1(\text{j}\omega), \hat{G}_2(\text{j}\omega)\right), \forall \omega \in [0, \infty), \forall \kappa \in [0, 1), \tag{3.125}$$

then, it is not equivalent to statement (2) or (3) in Theorem 3.39. Actually, straightforward calculation will show that (3.125) is equivalent to saying

$$f_{\lambda,\kappa}(s) = s^\eta D_1(s)D_2(s) + \lambda\kappa N_1(s)D_2(s)\text{e}^{-T_1 s} + (1-\lambda)\kappa N_2(s)D_1(s)\text{e}^{-T_2 s} \tag{3.126}$$

is Hurwitz stable. By the Nyquist criterion of stability it is equivalent to say that the system with characteristic equation as

$$\det(I + \kappa G(s, \lambda)) = 0 \tag{3.127}$$

is robustly stable for all $\lambda \in [0, 1]$. Rewrite $f_{\lambda,\kappa}(s)$ as

$$\begin{aligned}
f_{\lambda,\kappa}(s) = &\lambda(s^\eta D_1(s)D_2(s) + \kappa N_1(s)D_2(s)\text{e}^{-T_1 s}) \\
&+(1-\lambda)(s^\eta D_1(s)D_2(s) + \kappa N_2(s)D_1(s)\text{e}^{-T_2 s}).
\end{aligned}$$

Note that the Nyquist plots of normalized transfer-functions $\hat{G}_i(s)$, $i = 1, 2$, cross the real axis at the point $(-1, \text{j}0)$. Therefore, $\kappa\hat{G}_i(\text{j}\omega)$, $i = 1, 2$, do not contain the point $(-1, \text{j}0)$ for any $\omega \in [0, \infty)$ and any $\kappa \in [0, 1)$. Since $D_i(s)$, $i = 1, 2$, are Hurwitz stable polynomials or positive constants, by the Nyquist criterion of stability again, we know the systems with the following characteristic equations

$$1 + \kappa \frac{N_i(s)}{s^\eta D_i(s)} \mathrm{e}^{-T_i s} = 0, \ i = 1, 2,$$

are stable. This implies that quasi-polynomials

$$f_{0,\kappa}(s) = s^\eta D_1(s) D_2(s) + \kappa N_1(s) D_2(s) \mathrm{e}^{-T_1 s},$$
$$f_{1,\kappa}(s) = s^\eta D_1(s) D_2(s) + \kappa N_2(s) D_1(s) \mathrm{e}^{-T_2 s}$$

are Hurwitz stable. Therefore, by Definition 3.33 we know that $f_{\lambda,\kappa}(s)$ is Hurwitz stable if and only if the quasi-polynomial

$$g(s) = N_2(s) D_1(s) \mathrm{e}^{-T_2 s} - N_1(s) D_2(s) \mathrm{e}^{-T_1 s}$$

is a local convex direction for $f_{0,\kappa}(s)$, or simply if it is a convex direction. The following lemma summarizes the above discussion.

Theorem 3.40 *Suppose normalized transfer functions are given by (3.117), and $D_i(s)$, $i = 1, 2$, are coprime Hurwitz stable polynomials or positive constants. Then,*

$$(-1, \mathrm{j}0) \notin \kappa \mathrm{Co}\left(\hat{G}_1(\mathrm{j}\omega), \hat{G}_2(\mathrm{j}\omega)\right), \forall \omega \in [0, \infty), \forall \kappa \in [0, 1)$$

if and only if $g(s) = N_2(s) D_1(s) \mathrm{e}^{-T_2 s} - N_1(s) D_2(s) \mathrm{e}^{-T_1 s}$ is a local convex direction for $f_{0,\kappa}(s)$, or simply if $g(s)$ is convex direction.

Now, we can generalize the result of Theorem 3.40 to a family of n normalized systems.

Theorem 3.41 *Consider a family of normalized systems described by*

$$\hat{G}_i(s) = \frac{N_i(s)}{s^\eta D_i(s)} \mathrm{e}^{-T_i s}, \ i \in \overline{1, n}. \tag{3.128}$$

Suppose that

$$\lim_{\omega \to \infty} |\hat{G}_i(\mathrm{j}\omega)| = 0, \ \forall i \in \overline{1, n}. \tag{3.129}$$

Then,

$$(-1, \mathrm{j}0) \notin \kappa \mathrm{Co}\left(0 \cup \{\hat{G}_i(\mathrm{j}\omega), i \in \overline{1, n}\}\right), \ \forall \omega \in [0, \infty), \ \forall \kappa \in [0, 1)$$

if for any $i, k \in \overline{1, n}$, the quasi-polynomial

$$N_k(s) D_i(s) \mathrm{e}^{-T_k s} - N_i(s) D_k(s) \mathrm{e}^{-T_i s}$$

is a convex direction.

Proof. Suppose $N_k(s)D_i(s)\mathrm{e}^{-T_k s} - N_i(s)D_k(s)\mathrm{e}^{-T_i s}$ is a convex direction. Then, by Theorem 3.40, we know that

$$(-1, \mathrm{j}0) \notin \kappa\mathrm{Co}\left(\hat{G}_i(\mathrm{j}\omega), \hat{G}_k(\mathrm{j}\omega)\right), \ \forall\omega \in [0, \infty), \ \forall\kappa \in [0, 1) \tag{3.130}$$

holds for any $i, k \in \overline{1, n}$. Note that

$$(-1, \mathrm{j}0) \notin \kappa\mathrm{Co}\left(0, \hat{G}_i(\mathrm{j}\omega)\right), \ \forall\omega \in [0, \infty), \ \forall\kappa \in [0, 1) \tag{3.131}$$

also holds for any $i \in \overline{1, n}$ since $\kappa\hat{G}_i(\mathrm{j}\omega)$ does not contain $(-1, \mathrm{j}0)$ and the origin is inside the region enclosed by $\kappa\hat{G}_i(\mathrm{j}\omega)$. Also, note that

$$(-1, \mathrm{j}0) \notin \kappa\mathrm{Co}\left(0 \cup \{\hat{G}_i(\mathrm{j}\omega), i \in \overline{1, n}\}\right)$$

holds when $\omega \to \infty$ since $\lim_{\omega\to\infty} |\hat{G}_i(\mathrm{j}\omega)| = 0, \forall i \in \overline{1, n}$. Therefore, by continuity of the set $\kappa\mathrm{Co}\left(0 \cup \{\hat{G}_i(\mathrm{j}\omega), i \in \overline{1, n}\}\right)$ on ω, if $\kappa\mathrm{Co}\left(0 \cup \{\hat{G}_i(\mathrm{j}\omega), i \in \overline{1, n}\}\right)$ intersects the point $(-1, \mathrm{j}0)$ then its boundary, $\kappa\mathrm{Co}\left(\hat{G}_i(\mathrm{j}\omega), \hat{G}_k(\mathrm{j}\omega)\right)$ or $\kappa\mathrm{Co}\left(0, \hat{G}_i(\mathrm{j}\omega)\right)$ will definitely intersects the point $(-1, \mathrm{j}0)$ for some $\omega^* \in [0, \infty)$. But (3.130) and (3.131) have excluded the possibility. Therefore, we conclude that

$$(-1, \mathrm{j}0) \notin \kappa\mathrm{Co}\left(0 \cup \{\hat{G}_i(\mathrm{j}\omega), i \in \overline{1, n}\}\right), \ \forall\omega \in [0, \infty), \ \forall\kappa \in [0, 1).$$

Theorem is thus proved. □

3.6 Notes and References

The concept and expression of the signed curvature for parametric equation a curve can be found in many textbooks on classical differential geometry such as Guggenheimer (1977).

The clockwise property of the Nyquist plot of transfer functions attracted attention of control researchers as early as the 1980s (Horowitz and Ben-Adam 1989). It was found useful in the study of absolute stability (Tesi *et al.* 1992). Lemma 3.5 was initially given in Tesi *et al.* (1992) and extended to the current version in Tian and Yang (2004a).

In 1990s, convexity, a notion very closely related to the clockwise property, was proved for the frequency response arc of stable polynomials as a by-product of the study of robust stability against parametric uncertainties (Gu 1994; Hamann and Barmish 1993). Rantzer (1992) studied the relationship between the phase velocity of the frequency response plot of polynomials and the robust stability of a convex set of stable polynomials, and proposed the notion of convex direction in the space of stable polynomials. Kharitonov and Zhabko (1994) extended the notion of convex direction to the space of stable quasi-polynomials.

The relationship between the scalability of a distributed congestion control algorithm and differential geometric properties of the frequency response of the control system was uncovered in Tian and Yang (2004a). In this chapter it was shown that the clockwise property of the frequency response curve (Nyquist plot) of a congestion control system plays a key role in the scalability of the stability criterion for a network with diverse round-trip delays. Besides, some other differential geometric properties, such as phase monotonicity and/or modulus monotonicity, critical point of clockwise property, phase velocity, etc., are also shown to be very important in the scalability analysis for distributed congestion control systems

(Tian 2005a,b; Tian and Chen 2006). In Tian and Liu (2008) and Tian and Liu (2009) it was revealed that the scalability of consensus criteria for multi-agent systems with diverse input delays are also closely related to differential geometric properties of node dynamics of networks.

The material of Section 3.3 is taken from Tian and Yang (2004a,b). Section 3.4 is based on Tian (2005a), Tian and Chen (2006) while Section 3.5 is based on Kharitonov and Zhabko (1994) and Tian (2005b).

References

Gu K (1994). Comments on "Convexity of frequency response arcs associated with a stable polynomials". *IEEE Transactions on Robotics and Automation*, 39, 2262–2265.

Gu K, Kharitonov VL and Chen J (2003). *Stability of Time-delayed Systems*. Birkhäuser, Boston.

Guggenheimer HW (1977). *Differential Geometry*. Dover, New York.

Hamann JC and Barmish BR (1993). Convexity of frequency response arcs associated with a stable polynomial. *IEEE Transactions on Bobotics and Automation*, 38, 904–915.

Horowitz I and Ben-Adam S (1989). Clockwise nature of Nyquist locus of stable transfer functions. *International Journal Control*, 49, 1433–1436.

Kharitonov VL and Zhabko AP (1994). Robust stability of time-delayed systems. *IEEE Transactions on Automatic Control*, 39, 2388–2397.

Rantzer A (1992). Stability conditions for polytopes of polynomials. *IEEE Transactions on Automatic Control*, 37, 79–84.

Tesi A, Vicino A and Zappa G (1992). Clockwise property of the Nyquist plot with implications for absolute stability. *Automatica*, 28, 71–80.

Tian Y-P and Yang H-Y (2004a). Stability of the Internet congestion control with diverse delays. *Automatica*, 40, 1533–1541.

Tian Y-P and Yang H-Y (2004b). Stability of distributed congestion control with diverse communication delays. *Proceedings of the World Congress on Intelligent Control and Automation* 2, 1438–1442.

Tian Y-P (2005a). Stability analysis and design of the second-order congestion control for networks with heterogeneous delays. *IEEE/ACM Transactions on Networking*, 13, 1082–1093.

Tian Y-P (2005b). A general stability criterion for congestion control with diverse communication delays. *Automatica*, 41, 1255–1262.

Tian Y-P and Chen G (2006). Stability of the primal-dual algorithm for congestion control. *International Journal of Control*, 79, 662–676.

Tian Y-P and Liu C-L (2008). Consensus of multi-agent systems with diverse input and communication delays. *IEEE Transactions on Automatic Control*, 53, 2122–2128.

Tian Y-P and Liu C-L (2009). Robust consensus of multi-agent systems with diverse input delays and asymmetric interconnection perturbations. *Automatica*, 45, 1347–1353.

[illegible]

[illegible]

References

[illegible]

4

Congestion Control: Model and Algorithms

To which sources are the sun and the moon linked? In what formation are the stars deployed?

– Qu Yuan (340–278 BC*), 'Heavenly Questions'*

Congestion avoidance mechanisms serve as a keystone of a huge communication network like the Internet. This chapter introduces basic notions, algorithms and system models of end-to-end congestion control. The congestion control problem is modeled as an optimization problem of resource allocation in the network under link capacity constraints, and the congestion control algorithms are treated as distributed real-time gradient-descent algorithms of seeking for the optimal solution of the resource allocation problem.

4.1 An Introduction to Congestion Control

Consider a network containing S sources each of which is identified as an origin. Denote by $\overline{S} = \{1, \cdots, S\}$ the set of source nodes. It is supposed that each user uses a fixed route between its origin and destination. Denote by $\overline{L} = \{1, \cdots, L\}$ the set of all the links contained in the network, and by $L_i \subset \overline{L}$ the set of links used by the user of source i. Note that each link may be used by multiple sources. Denote by $S_l \subset \overline{S}$ the set of sources using link l. From the viewpoint of graph theory, the interconnection topology of such a network is a bipartite graph with vertex classes $\overline{S}$ and $\overline{L}$. L_i is the neighbor set of source node i, and S_l is the neighbor set of link node l.

Each source $i \in \overline{S}$ transmits data at rate x_i (bits/sec or bps) from the origin to its destination. Each link l in the network has a *capacity* c_l bps, which implies that the sum of the transmission rates of all the users sharing link l is less than or equal to c_l, i.e.,

$$\sum_{i \in S_l} x_i \leq c_l, \quad l \in \overline{L}. \tag{4.1}$$

Frequency-Domain Analysis and Design of Distributed Control Systems, First Edition. Yu-Ping Tian.

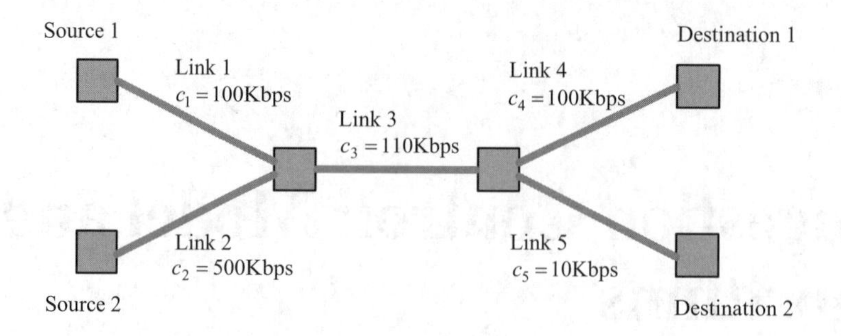

Figure 4.1 A simple network exhibiting inefficiency.

4.1.1 Congestion Collapse

Since the capacity of all the links is limited, the network may suffer *congestion collapse*. Now, we explain the reason of congestion collapse with an example. Firstly, we assume that if the offered traffic on link l exceeds the capacity c_l of the link, then all sources sharing link l see their traffic reduced in proportion to their offered traffic. This assumption is approximately true if the queueing is first-in first-out (FIFO) in the network node.

Consider a simple network with two sources as shown in Figure 4.1. There are five links labeled 1 through 5 with capacities shown on the figure. Assume that sources are limited only by their first link, without feedback from the network. Then, source 1 and source 2 may send packets at rate 100 Kbps and 500 Kbps, respectively, according to their first links. But, source 1 can actually only send at 10 Kbps because it is competing with source 2 on link 3, which sends at a higher rate on that link; however, source 2 is limited to 10 Kbps because of link 5. Therefore, the outgoing rates of source 1 and source 2 are both 10 Kbps, and the total throughput is only 20 Kbps! If source would be aware of the global situation, and if it were to cooperate, then it would send at 10 Kbps only initially, which would allow source 1 to send at 100 Kbps, without any penalty for source 2. Then, the total throughput of the network would become 110 Kbps.

The above example show some inefficiency in network communication. In complex network scenarios, such inefficiency may lead to a form of instability known as congestion collapse. Let us illustrate this by using the network shown in Figure 4.2. This network topology is a ring which is commonly used in network design. Suppose there are n nodes and links numbered $1, \cdots, n$. Source i enters node i using links i and $[i \bmod n] + 1$, and leaves at node $[(i+1) \bmod n] + 1$. Assume that source i sends as much as x_i, without feedback from the network. Denote by x'_i the achieved rate of source i through link i at node $[i \bmod n] + 1$, and by x''_i the achieved rate of source i through link $[i \bmod n] + 1$ at node $[(i+1) \bmod n] + 1$. In the rest of the example we omit "mod n" when the context is clear. Then, we have

$$\begin{cases} x'_i = \min\left(x_i, \frac{c_i}{x_i + x'_{i-1}} x_i\right), \\ x''_i = \min\left(x'_i, \frac{c_{i+1}}{x'_i + x_{i+1}} x'_i\right). \end{cases} \tag{4.2}$$

For simplicity we assume that $c_i = c$ and $x_i = x$ for $i \in \overline{1, n}$. Then, we have obviously $x'_i = x'$ and $x''_i = x''$ for some values of x' and x'' which are computed below. If $x \leq c/2$ then there

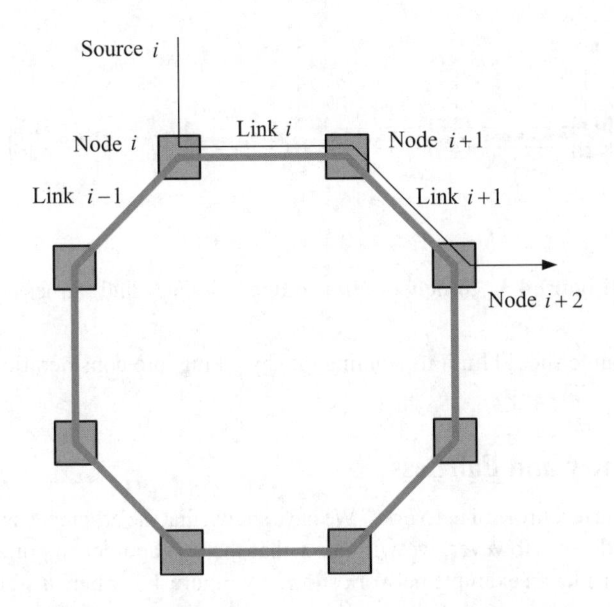

Figure 4.2 A ring network illustrating congestion collapse.

is no loss and $x' = x'' = x$ and the throughput of the network is nx. Otherwise, from (4.2) it follows that

$$x' = \frac{cx}{x + x'},$$

which yields

$$x' = \frac{x}{2}\left(-1 + \sqrt{1 + 4\frac{c}{x}}\right). \tag{4.3}$$

From (4.2) it also follows that

$$x'' = \frac{cx'}{x + x'}. \tag{4.4}$$

Combining (4.3) and (4.4) gives

$$x'' = c - \frac{x}{2}\left(\sqrt{1 + 4\frac{c}{x}} - 1\right). \tag{4.5}$$

By using the Taylor expansion, from (4.5) it is easy to get

$$x'' = \frac{c^2}{x} + o(\frac{1}{x}), \tag{4.6}$$

where $o(\frac{1}{x})$ denotes the terms with order equal to or greater than two. When the offered rate x goes to ∞, the outgoing rate approaches zero, and the throughput of the network nx'' also approaches zero. This is the so-called congestion collapse. Therefore, to avoid congestion

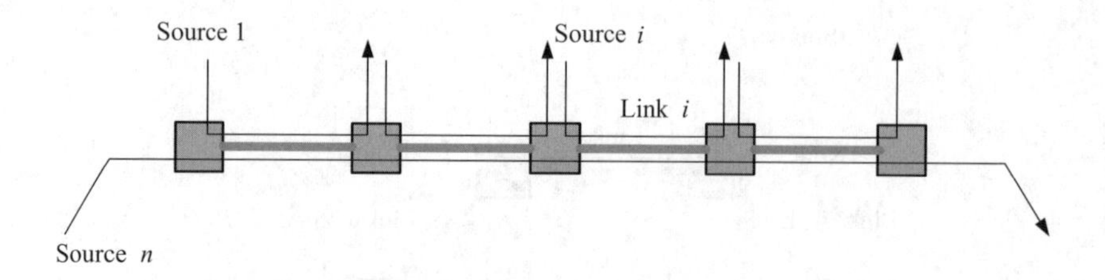

Figure 4.3 A network illustrating efficiency and fairness.

collapse, each source should limit its sending rate by taking into consideration the state of the network.

4.1.2 Efficiency and Fairness

How to allocate the resources of networks? We have shown that the efficiency is a very important issue to be considered. However, we will show that there is another important factor in this design. Let us consider an example network shown by Figure 4.3, where n sources share $n-1$ links. All links have a capacity equal to c. We assume that some congestion avoidance strategy has been implemented in the network and there is negligible packet loss. Source i sends data at rate x_i, $i \in \overline{1, n}$. To achieve the maximum efficiency, the following equality should hold

$$x_n + x_i = c, \quad i \in \overline{1, (n-1)}.$$

So, the total throughput of the network is

$$\theta = (n-1)(c - x_n) + x_n = (n-1)c - (n-2)x_n.$$

Obviously, if we want to maximize the total throughput of the network, we should set $x_n = 0$. This strategy is, of course, unfair to source n. In summary, the objective of congestion control is to provide both efficiency and some form of fairness.

4.1.3 Optimization-Based Resource Allocation

To optimize the resource allocation in the network, we suppose that each user $i \in \overline{S}$ generates a *utility* $U_i(x_i)$ when a data rate x_i is allocated to it. Then, the network resource can be allocated by solving the following optimization problem

$$\max_{x_i \geq 0} \sum_{i \in \overline{S}} U_i(x_i) \tag{4.7}$$

subject to (4.1). From optimization theory (Bertsekas 1995), the above optimization problem has a unique solution if $U_i(x_i)$, $i \in \overline{S}$, are strictly concave functions. It is also reasonable to assume that $U_i(x_i)$, $i \in \overline{S}$, are non-decreasing because more resources generate higher utility. Therefore, we make the following assumption for the utility function.

Assumption 4.1 *For each $i \in \overline{S}$, $U_i(x_i)$ is a continuously differentiable, non-decreasing, strictly concave function.*

Denote by $\{x_i^\star, i \in \overline{S}\}$ the optimal solution to the resource allocation problem (4.7). From a well-known property of convex functions it follows that

$$\sum_{i \in \overline{S}} U_i'(x_i^\star)(x_i - x_i^\star) \leq 0, \tag{4.8}$$

where $U_i'(x_i^\star)$ is the derivative of $U_i(x_i)$ at $x_i^\star$, $\{x_i, i \in \overline{S}\}$ is any other set of transmission rates satisfying the constraints (4.1). Rewrite (4.8) as

$$\sum_{i \in \overline{S}} U_i'(x_i^\star) x_i^\star \frac{x_i - x_i^\star}{x_i^\star} \leq 0. \tag{4.9}$$

As $\frac{x_i - x_i^\star}{x_i^\star}$ is the relative change of the source i's rate, $U_i'(x_i^\star)x_i^\star$ can be regarded as a fairness weight for source i. Then, (4.9) implies that the weighted sum of the changes in each user's rate is less than or equal to zero. Next, we show that different choices of the utility function lead to different kind of fairness.

Proportional fairness

Let us choose, for example,

$$U_i(x_i) = w_i \log x_i, \tag{4.10}$$

where $w_i > 0$. Then, the inequality (4.8) becomes

$$\sum_{i \in \overline{S}} w_i \frac{x_i - x_i^\star}{x_i^\star} \leq 0. \tag{4.11}$$

The above inequality implies that the weighted sum of the proportional changes in each user's rate is less than or equal to zero. Hence, the resource allocation corresponding to the choice of the utility function $U_i(x_i) = w_i \log x_i$ is said to be weighted proportionally fair. In particular, if all w_i's are equal to one, then the allocation is simply said to be proportionally fair.

Minimum potential delay fairness

Let us choose, for example,

$$U_i(x_i) = -w_i/x_i, \tag{4.12}$$

where $w_i > 0$. Then, the inequality (4.8) becomes

$$\sum_{i \in \overline{S}} \frac{w_i}{x_i^\star} \frac{x_i - x_i^\star}{x_i^\star} \leq 0. \tag{4.13}$$

Suppose that w_i represents the size of the file that source i is attempting to transmit, then w_i/x_i is the amount of time that it would take to transfer the file, and $\sum_{i \in \overline{S}} w_i/x_i^\star$ is the minimum of the sum of the amount of time. Hence, the resulting resource allocation under this choice of utility function is said to be minimum potential delay fair.

4.2 Distributed Congestion Control Algorithms

To solve the optimization problem (4.7), one has to know the utility functions and routes of all the sources in the network. In a huge network such as the Internet, this information is not available centrally. Thus, it is important to develop distributed solutions, where each source adapts its transmission rate based only on local information.

4.2.1 Penalty Function Approach and Primal Algorithm

Instead of trying to obtain the optimal solution to problem (4.7), let us consider the following problem where the constraints are added to the objective using penalty function:

$$\max_{x\geq 0} \sum_{i\in\overline{S}} U_i(x_i) - \sum_{l\in\overline{L}} \int_0^{\sum_{i\in S_l} x_i} p_l(x)\mathrm{d}x. \tag{4.14}$$

The function $p_l(x)$ denotes the penalty function corresponding to the capacity c_l of link l. It is also referred to as the *price* at link l. Note that the price at link l is a function of the aggregate arrival source rate $\sum_{i\in S_l} x_i$ at link l, and can be regarded as a measure of congestion at link l. We make the following assumption on the price function.

Assumption 4.2 *For each $l \in \overline{L}$, $p_l(\cdot)$ is a non-decreasing, continuous function such that*

$$\int_0^y p_l(x)\mathrm{d}x \to \infty \quad \text{as } y \to \infty.$$

The first-order necessary condition of the optimization problem (4.14) is given by

$$U_i'(x_i) - \sum_{l\in L_i} p_l(\sum_{r\in S_l} x_r) = 0, \quad i \in \overline{S}.$$

Thus, the optimization problem (4.14) can be solved by using the following continuous-time version of gradient-ascent algorithm:

$$\dot{x}_i(t) = k_i(x_i)\left(U_i'(x_i(t)) - \sum_{l\in L_i} p_l(\sum_{r\in S_l} x_r(t)) \right)_{x_i}^{+}, \quad i \in \overline{S}, \tag{4.15}$$

where the step size $k_i(x_)$ is any non decreasing, continuous function such that $k_i(x_) > 0$; $(z)_x^+$ is defined as

$$(z)_x^+ = \begin{cases} z, & \text{if } x > 0, \\ \max(z, 0), & \text{if } x = 0, \end{cases}$$

which ensures the non-negativeness of variable x.

Obviously, (4.15) is a distributed control algorithm as each source node uses only the price information of the links used by this source, and each link needs only the rate information of the sources sharing this link.

Since the optimization problem (4.14) has unique optimal solution under Assumption 4.1, the equilibrium point of the congestion control equation (4.15) solves (4.14). As shown by Srikant (2004), the objective in (4.14) is a Lyapunov function for the system of equations given by (4.15), and the derivative of the Lyapunov function is negatively definite if the price functions satisfy Assumption 4.2. Hence, the system given by (4.15) is globally asymptotically stable with respect to the optimal solution of (4.14).

In the literature on congestion control, the algorithm (4.15) is referred to as the *primal algorithm*. If we apply the utility function given by (4.10) for the proportional fairness and take a linear step size function, i.e.,

$$k_i(x_i) = \kappa_i x_i,$$

where $\kappa_i > 0$, then, the primal algorithm (4.15) becomes

$$\dot{x}_i(t) = \kappa_i \left(w_i - x_i(t) \sum_{l \in L_i} p_l (\sum_{r \in S_l} x_r(t)) \right)_{x_i}^{+}, \quad i \in \overline{S}. \tag{4.16}$$

Kelly, Maulloo and Tan (1998) first proposed this primal algorithm and proved the global convergence of the algorithm. So, this algorithm is also referred to as Kelly's primal algorithm.

4.2.2 Dual Approach and Dual Algorithm

Instead of solving the optimization problem (4.7) using a penalty function, one can also solve the problem exactly using Lagrange multipliers. Low and Lapsley (1999) considered the associated dual problem:

$$\min_{\lambda_l \geq 0} D(p) := \min_{\lambda_l \geq 0} \sum_{i \in \overline{S}} \max_{x_i \geq 0} \left(U_i(x_i) - x_i \sum_{l \in L_i} p_l \right) + \sum_{l \in L} p_l c_l. \tag{4.17}$$

Given the Lagrange multipliers p_l's which are also referred to as prices, the maximization over x_i can be carried out by each source based on the aggregate price of its route as follows:

$$x_i(t) = U_i'^{-1} \left(\sum_{l \in L_i} p_l(t) \right) \tag{4.18}$$

where $U_i'^{-1}$ denotes the inverse of the derivative of U_i. Note that $U_i'^{-1}$ is a decreasing function when $U_i(\cdot)$ is convex.

The continuous-time version of the gradient projection algorithm for the dual problem (4.17) is given by

$$\dot{p}_l = -\gamma_l \frac{\partial D}{\partial p_l}$$

where $\gamma_l > 0$ is a step size parameter (also referred to as a control gain) at link l. It is easy to get that

$$\frac{\partial D}{\partial p_l} = c_l - \sum_{i \in S_l} x_i.$$

Thus, one can get an algorithm to compute the price at each link as follows:

$$\dot{p}_l(t) = \gamma_l \left(\sum_{i \in S_l} x_i(t) - c_l \right)^+_{p_l}, \quad l \in \overline{L}. \tag{4.19}$$

The algorithm (4.19) together with (4.18) is called a *dual algorithm* in the literature on congestion control. The interpretation of the algorithm is very clear: if the demand $\sum_{i \in S_l} x_i$ exceeds capacity c_l, increase the price; otherwise, decrease it. The algorithm is distributive because each link uses only the total arrival rate into it to compute its price.

4.2.3 Primal-Dual Algorithm

The dual solution (or the corresponding dual algorithm) solves the resource allocation problem (4.7) exactly, whereas the penalty function approach (or the corresponding primal algorithm) solves it only approximately. From (4.19) it is clear that at the equilibrium it holds

$$\sum_{i \in S_l} x_i = c_l.$$

This implies that the dual algorithm can achieve very high utilization of link capacity. However, the dual algorithm requires the existence of the inverse of the derivative of the utility function. So, it is restricted to a specific class of utility functions, and hence, does not allow arbitrary fairness in rate allocation. To achieve both high utilization and arbitrary fairness, a so-called *primal-dual algorithm* was proposed in Alpcan and Başar (2003), Wen and Arcak (2003) and generalized in Liu *et al.* (2003). The main idea of the primal-dual algorithm is to adopt dynamical adaptations both at links and at sources, which can be described as follows:

$$\dot{x}_i(t) = k_i(x_i) \left(U_i'(x_i(t)) - \sum_{l \in L_i} p_l(t) \right)^+_{x_i(t)}, \quad i \in \overline{S}, \tag{4.20}$$

$$\dot{p}_l(t) = \gamma_l \left(\sum_{i \in S_l} x_i(t) - c_l \right)^+_{p_l}, \quad l \in \overline{L}. \tag{4.21}$$

4.2.4 REM: A Second-Order Dual Algorithm

Generally, the congestion control algorithms for Internet-like networks can be classified into two types: one is the algorithms implemented in its transmission control protocol (TCP), such as TCP Reno and TCP Vegas, which adjust transmission rates of sources based on available feedback congestion marks (Jacobson 1988). The other is the algorithms, such as DropTail and RED, used as the active queue management (AQM) at the link nodes, which decide how to drop arriving packets when the network is overloaded (Floyd and Jacobson 1993). As we have shown in this chapter, the primal algorithm has a dynamical law for adjusting source rate and a static law for generating link price, and such an algorithm properly describes currently used congestion avoidance protocols for TCP rate control at sources. Similarly, the dual algorithm has a dynamical law for adjusting link price and a static law for generating source rate, and such an algorithm can be used for the AQM mechanism at links.

When the dual algorithm is used for AQM, the price p_l, the measure of congestion at link l, is usually given some practical meaning. For example, the queue length in the buffer is used as the congestion measure in RED (Floyd and Jacobson 1993). However, when the dual algorithm (4.19) is used for adjusting the queue length, we find that at the equilibrium the queue length may be far away from the target value. To keep the queue length around a desired value, Athuraliya, Low and Yin (2001) proposed a kind of the second-order AQM control algorithm: random exponential marking (REM). The continuous-time version of the REM algorithm is given by

$$\dot{b}_l(t) = \left(\sum_{i \in S_l} x_i(t) - c_l \right)_{b_l}^{+}, \tag{4.22}$$

$$\dot{p}_l(t) = \left(\gamma_l(\alpha_l(b_l(t) - e_l) + \sum_{i \in S_l} x_i(t) - c_l) \right)_{p_l}^{+}, \tag{4.23}$$

where $\gamma_l > 0, \alpha_l > 0, l \in \overline{L}$, are the control gains. Obviously, REM reduces to the (first-order) dual algorithm if $\alpha_l = 0$.

If the queue length is taken as the congestion measure, then $b_l(t)$ can be understood as the queue length at link l and e_l is the desired value of the queue length. Under the algorithm given by (4.22) and (4.23), the equilibrium point guarantees not only the tight utilization of the link capacity, i.e., $\sum_{i \in S_l} x_i = c_l$, but also the desired value the queue length, i.e., $b_l = e_l$.

4.3 A General Model of Congestion Control Systems

4.3.1 Framework of End-to-End Congestion Control under Diverse Round-Trip Delays

Recall that the network under our consideration has S sources which share L links. We have supposed that there is only one fixed route between each user's origin and its destination, and the set of links contained in source i's route is denoted by L_i. By the sets L_i's we can define an $L \times S$ routing matrix $R = \{R_{li}\}$, where

$$R_{li} = \begin{cases} 1, & \text{if } l \in L_i \\ 0, & \text{otherwise.} \end{cases} \tag{4.24}$$

For each link l, we have the following:

- aggregate rate of all sources which use link l, denoted by y_l (bps);
- price p_l, used to indicate the link congestion.

The vectors $y, p \in \mathbb{R}^L$ are defined by the above components across the set of links. For each source i, we have the following:

- source rate x_i (bps);
- aggregate price of all links used by source i, denoted by q_i.

The vectors $x, q \in \mathbb{R}^S$ are defined by the above components across the set of sources. The following equations reflect the interconnection between links and sources:

$$y_l(t) = \sum_{i \in S_l} x_i(t), \tag{4.25}$$

$$q_i(t) = \sum_{l \in L_i} p_l(t). \tag{4.26}$$

For a given link l and each source $i \in S_l$, let $d_{li}^{\rightarrow}$ denote the forward delay of delivering packets from source i to link l, and $d_{li}^{\leftarrow}$ the backward delay of sending the feedback signal from link l to source i. Then, each route is subject to a *round-trip delay* denoted by D_i. We assume that for any $i \in \overline{S}$ the following relationship is true (Johari and Tan 2001)

$$d_{li}^{\rightarrow} + d_{li}^{\leftarrow} = D_i, \;\; \forall l \in L_i \subseteq \overline{L}. \tag{4.27}$$

The assumption (4.27) is quite reasonable for currently used TCP. It implies that the forward delay from source i to any link l in its route plus the backward delay from the link is the same for any l. An illustration of the round-trip delay is given by Figure 4.4.

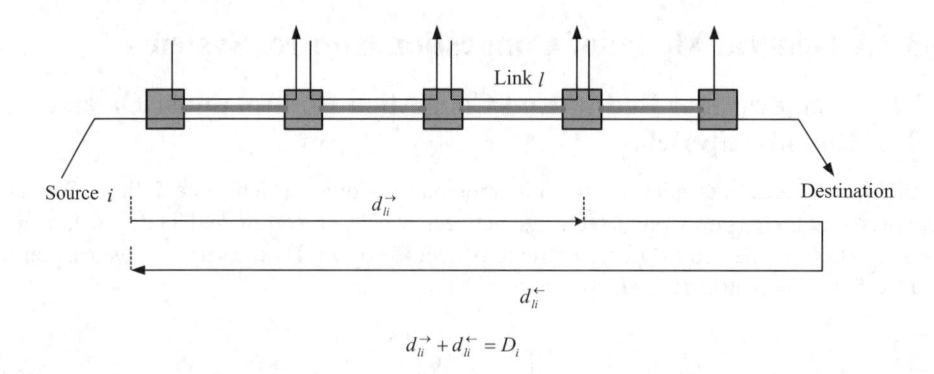

Figure 4.4 An illustration of round-trip delay.

When propagation delays are considered, the following relationships are immediate from (4.25) and (4.26):

$$y_l(t) = \sum_{i \in S_l} x_i(t - d_{li}^{\rightarrow}),$$

$$q_i(t) = \sum_{l \in L_i} p_l(t - d_{li}^{\leftarrow}).$$

Now, we can summarize the end-to-end congestion control algorithms previously introduced under diverse round-trip delays as follows.

Primal algorithm

At source $i \in \overline{S}$:

$$\dot{x}_i(t) = k_i(x_i) \left(U_i'(x_i(t)) - q_i(t) \right)_{x_i}^{+}, \tag{4.28}$$

$$q_i(t) = \sum_{l \in L_i} p_l(t - d_{li}^{\leftarrow}). \tag{4.29}$$

At link $l \in \overline{L}$:

$$p_l(t) = p_l\left(y_l(t)\right), \tag{4.30}$$

$$y_l(t) = \sum_{i \in S_l} x_i(t - d_{li}^{\rightarrow}). \tag{4.31}$$

Dual algorithm

At source i:

$$x_i(t) = U_i'^{-1}\left(q_i(t)\right), \tag{4.32}$$

$$q_i(t) = \sum_{l \in L_i} p_l(t - d_{li}^{\leftarrow}). \tag{4.33}$$

At link l:

$$\dot{p}_l(t) = \gamma_l \left(y_l(t) - c_l\right)_{p_l}^{+}, \tag{4.34}$$

$$y_l(t) = \sum_{i \in S_l} x_i(t - d_{li}^{\rightarrow}). \tag{4.35}$$

Primal dual algorithm

At source i:

$$\dot{x}_i(t) = k_i(x_i) \left(U_i'(x_i(t)) - q_i(t) \right)_{x_i}^{+}, \tag{4.36}$$

$$q_i(t) = \sum_{l \in L_i} p_l(t - d_{li}^{\leftarrow}). \tag{4.37}$$

At link l:

$$\dot{p}_l(t) = \gamma_l \left(y_l(t) - c_l\right)^+_{p_l}, \tag{4.38}$$

$$y_l(t) = \sum_{i \in S_l} x_i(t - d^{\rightarrow}_{li}). \tag{4.39}$$

Second-order dual algorithm

At source i:

$$x_i(t) = U_i'^{-1}\left(q_i(t)\right), \tag{4.40}$$

$$q_i(t) = \sum_{l \in L_i} p_l(t - d^{\leftarrow}_{li}). \tag{4.41}$$

At link l:

$$\dot{b}_l(t) = \left(y_l(t) - c_l\right)^+_{b_l}, \tag{4.42}$$

$$\dot{p}_l(t) = \left(\gamma_l(\alpha_l(b_l(t) - e_l) + y_l(t) - c_l)\right)^+_{p_l}, \tag{4.43}$$

$$y_l(t) = \sum_{i \in S_l} x_i(t - d^{\rightarrow}_{li}). \tag{4.44}$$

4.3.2 General Primal-Dual Algorithm

Let $\alpha_i(t) \in \mathbb{R}^{m_i}$ be some internal variable at source i, which is subject to a dynamical adjustment. The source rate $x_i(t)$ is adjusted according to a function F_i based on $\alpha_i(t)$ and $q_i(t)$. So the source dynamics are described as follows:

$$\dot{\alpha}_i(t) = H_i(\alpha_i(t), q_i(t)), \tag{4.45}$$

$$x_i(t) = F_i(\alpha_i(t), q_i(t)). \tag{4.46}$$

Similarly, let $\beta_l(t) \in \mathbb{R}^{m_l}$ be some internal variable at link l, which is subject to a dynamical adjustment. The link congestion measure $p_l(t)$ is adjusted according to a function E_l based on $\beta_l(t)$ and $y_l(t)$. So the link dynamics are described as follows:

$$\dot{\beta}_l(t) = V_l(\beta_l(t), y_l(t)), \tag{4.47}$$

$$p_l(t) = E_l(\beta_l(t), y_l(t)). \tag{4.48}$$

Sources and links are interconnected by the following equations:

$$y_l(t) = \sum_{i \in S_l} x_i(t - d^{\rightarrow}_{li}), \tag{4.49}$$

$$q_i(t) = \sum_{l \in L_i} p_l(t - d^{\leftarrow}_{li}). \tag{4.50}$$

Equations (4.45)–(4.50) constitute a dynamical feedback system with structure shown by Figure 4.5. For the statement convenience, we call the system "system $\sum$" in the rest of this chapter.

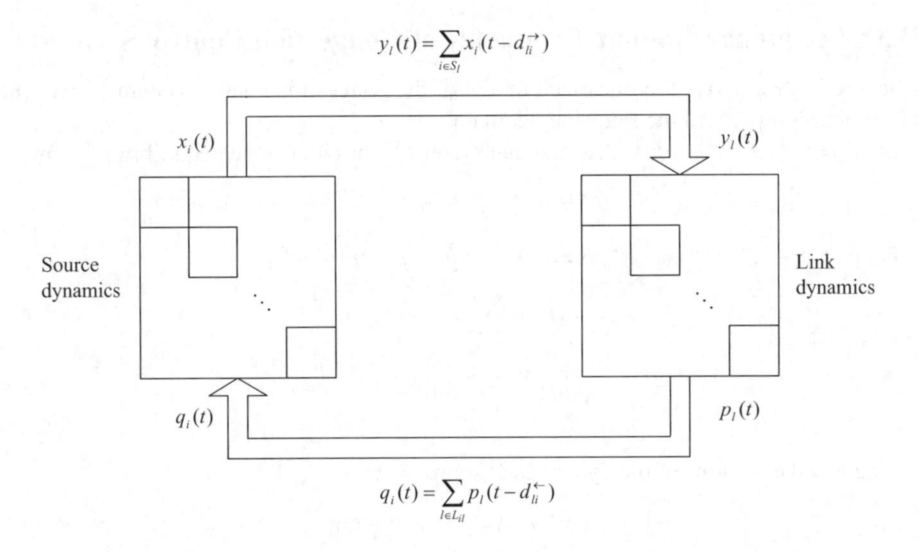

Figure 4.5 Feedback structure of distributed congestion control.

Remark 1. The model (4.45)–(4.48) unifies different congestion control algorithms. For example, one can set $x_i(t) = F_i(\alpha_i(t), q_i(t)) = \alpha_i(t)$ in the source dynamics, and $V_l(\cdot, \cdot) \equiv 0$, $\beta_l(t) \equiv 0$ in the link dynamics, then get the primal algorithm. Table 4.1 summarizes the relationship of the unified model with other algorithms.

Remark 2. For the second-order case, the only algorithm considered in this book is the REM algorithm. In the unified model, by setting $\beta_l(t) = [b_l(t), p_l(t)]^T$, where $b_l(t)$ is a scalar internal variable at link l, and $E_l(\cdot, \cdot) = [0,\ I]\beta_l(t)$, then we get the REM algorithm.

Table 4.1 Different congestion control algorithms in a unified model

	first-order	second-order
Primal algorithm	$\beta_l(t) \equiv 0$, $V_l(\cdot, \cdot) \equiv 0$, $F_i(\alpha_i(t), q_i(t)) = \alpha_i(t)$.	$\beta_l(t) \equiv 0$, $V_l(\cdot, \cdot) \equiv 0$.
Dual algorithm	$\alpha_i(t) \equiv 0$, $H_i(\cdot, \cdot) \equiv 0$, $E_l(\beta_l(t), y_l(t)) = \beta_l(t)$.	$\alpha_i(t) \equiv 0$, $H_i(\cdot, \cdot) \equiv 0$.
Primal-dual algorithm	$F_i(\alpha_i(t), q_i(t)) = \alpha_i(t)$, $E_l(\beta_l(t), y_l(t)) = \beta_l(t)$.	

4.3.3 Frequency-Domain Symmetry of Congestion Control Systems

In this section we derive the linearized model of the congestion control system $\sum$ and show its symmetry property in the frequency domain.

Let $\theta = (\alpha^{\mathrm{T}}, x^{\mathrm{T}}, \beta^{\mathrm{T}}, p^{\mathrm{T}})^{\mathrm{T}}$. Assume that system $\sum$ has an isolated equilibrium point

$$\theta^\star = (\alpha^{\star\mathrm{T}}, x^{\star\mathrm{T}}, \beta^{\star\mathrm{T}}, p^{\star\mathrm{T}})^{\mathrm{T}}.$$

Let $\hat{\alpha} = \alpha - \alpha^\star$, $\hat{x} = x - x^\star$, $\hat{\beta} = \beta - \beta^\star$, $\hat{p} = p - p^\star$, and

$$\hat{y}_l(t) = \sum_{i \in S_l} \hat{x}_i(t - d_{li}^{\rightarrow}), \tag{4.51}$$

$$\hat{q}_i(t) = \sum_{l \in L_i} \hat{p}_l(t - d_{li}^{\leftarrow}). \tag{4.52}$$

Linearized equations of the dynamics at source i are

$$\dot{\hat{\alpha}}_i(t) = \tilde{A}_i \hat{\alpha}_i(t) + \tilde{B}_i \hat{q}_i(t), \tag{4.53}$$
$$\hat{x}_i(t) = \tilde{C}_i \hat{\alpha}_i(t) + \tilde{D}_i \hat{q}_i(t), \tag{4.54}$$

where $\tilde{A}_i = \left.\frac{\partial H_i}{\partial \alpha_i(t)}\right|_{\theta=\theta^\star}$, $\tilde{B}_i = \left.\frac{\partial H_i}{\partial q_i}\right|_{\theta=\theta^\star}$, $\tilde{C}_i = \left.\frac{\partial F_i}{\partial \alpha_i}\right|_{\theta=\theta^\star}$, and $\tilde{D}_i = \left.\frac{\partial F_i}{\partial q_i}\right|_{\theta=\theta^\star}$. In the Laplace domain, the above equations can be written as

$$X_i(s) = [\tilde{C}_i(sI - \tilde{A}_i)^{-1}\tilde{B}_i + \tilde{D}_i]Q_i(s), \tag{4.55}$$

where $X_i(s)$ and $Q_i(s)$ are the Laplace transformation of $\hat{x}_i(t)$ and $\hat{q}_i(t)$, respectively. Denote

$$G_i^{(1)}(s) = \tilde{C}_i(sI - \tilde{A}_i)^{-1}\tilde{B}_i + \tilde{D}_i. \tag{4.56}$$

Let κ_i be the control gain at source i, which is defined as

$$\kappa_i = \lim_{s \to 0} s^\nu G_i^{(1)}(s), \tag{4.57}$$

where ν is the order of the pole $s = 0$ contained in $G_i^{(1)}(s)$. Define

$$W_i^{(1)}(s) = G_i^{(1)}(s)/\kappa_i. \tag{4.58}$$

Then, we have

$$G_i^{(1)}(s) = \kappa_i W_i^{(1)}(s). \tag{4.59}$$

Similarly, linearized equations of the dynamics at link l are

$$\dot{\hat{\beta}}_l(t) = \bar{A}_l \hat{\beta}_l(t) + \bar{B}_l \hat{y}_l(t), \tag{4.60}$$
$$\hat{p}_l(t) = \bar{C}_l \hat{\beta}_l(t) + \bar{D}_l \hat{y}_l(t), \tag{4.61}$$

where $\bar{A}_l = \left.\frac{\partial V_l}{\partial \beta_l}\right|_{\theta=\theta^\star}$, $\bar{B}_l = \left.\frac{\partial V_l}{\partial y_l}\right|_{\theta=\theta^\star}$, $\bar{C}_l = \left.\frac{\partial E_l}{\partial \beta_l}\right|_{\theta=\theta^\star}$, and $\bar{D}_l = \left.\frac{\partial E_l}{\partial y_l}\right|_{\theta=\theta^\star}$. The equivalent equation in the Laplace domain is

$$P_l(s) = [\bar{C}_l(sI - \bar{A}_l)^{-1}\bar{B}_l + \bar{D}_l]Y_l(s), \tag{4.62}$$

where $P_l(s)$ and $Y_l(s)$ are the Laplace transformation of $\hat{p}_l(t)$ and $\hat{y}_l(t)$, respectively. Denote

$$G_l^{(2)}(s) = \bar{C}_l(sI - \bar{A}_l)^{-1}\bar{B}_l + \bar{D}_l. \tag{4.63}$$

Let γ_l be the control gain at link l, which is defined as

$$\gamma_l = \lim_{s \to 0} s^{\mu} G_l^{(2)}(s), \tag{4.64}$$

where μ is the order of the pole $s = 0$ contained in $G_l^{(2)}(s)$. Define

$$W_l^{(2)}(s) = G_l^{(2)}(s)/\gamma_l. \tag{4.65}$$

Then we have

$$G_l^{(2)}(s) = \gamma_l W_l^{(2)}(s). \tag{4.66}$$

Assume that the system is semi-homogeneous by links, which implies that all the links used by the same source have the same dynamics with possibly different gains, i.e.,

$$W_{l_1}^{(2)}(s) = W_{l_2}^{(2)}(s) \triangleq W_i^{(2)}(s), \quad \forall l_1, l_2 \in L_i.$$

Then, we can write

$$G_l^{(2)}(s) = \gamma_l W_i^{(2)}(s), \ \ \forall l \in L_i. \tag{4.67}$$

Denote

$$W_i(s) = W_i^{(1)}(s) W_i^{(2)}(s). \tag{4.68}$$

We refer to $W_i(s)$ as the *round-trip transfer function* of the route associated with source i.

In the Laplace domain, equations (4.51) and (4.52) can be written as

$$Y_l(s) = \sum_{s \in S_l} X_i(s) \mathrm{e}^{-d_{li}^{\rightarrow} s}, \tag{4.69}$$

$$Q_i(t) = \sum_{l \in L_i} P_l(s) \mathrm{e}^{-d_{li}^{\leftarrow} s}. \tag{4.70}$$

Define the delayed forward routing matrix $R_f(s)$ and the backward routing matrix $R_b(s)$ as

$$\begin{aligned} [R_f(s)]_{li} &= \begin{cases} \mathrm{e}^{-d_{li}^{\rightarrow} s}, & \text{if } l \in L_i \\ 0, & \text{otherwise;} \end{cases} \\ [R_b(s)]_{li} &= \begin{cases} \mathrm{e}^{-d_{li}^{\leftarrow} s}, & \text{if } l \in L_i \\ 0, & \text{otherwise.} \end{cases} \end{aligned} \tag{4.71}$$

Then, from the assumption on the round-trip delay (4.27) it follows that

$$R_b(s) = R_f(-s) \mathrm{diag}\{\mathrm{e}^{-T_i s}, i \in \overline{S}\}. \tag{4.72}$$

Combining (4.55), (4.62), (4.69) and (4.70) we get

$$\begin{aligned} X_i(s) &= \kappa_i W_i^{(1)}(s) Q_i(s) \\ &= \kappa_i W_i^{(1)}(s) \sum_{l \in L_i} P_l(s) \mathrm{e}^{-s d_{li}^{\leftarrow}} \\ &= \kappa_i W_i^{(1)}(s) \sum_{l \in L_i} \gamma_l W_i^{(2)}(s) Y_l(s) \mathrm{e}^{-s d_{li}^{\leftarrow}} \\ &= \kappa_i W_i(s) \sum_{l \in L_i} \gamma_l \mathrm{e}^{-s d_{li}^{\leftarrow}} \sum_{r \in S_l} X_r(s) \mathrm{e}^{-s d_{lr}^{\rightarrow}} . \end{aligned} \tag{4.73}$$

Denote

$$X(s) = (X_1(s), \cdots, X_S(s))^T, \tag{4.74}$$

$$T(s) = \mathrm{diag}\left\{ \mathrm{e}^{-T_i s} W_i(s),\ i \in \overline{S} \right\}, \tag{4.75}$$

$$K = \mathrm{diag}\{\kappa_i, i \in \overline{S}\}, \tag{4.76}$$

$$\Gamma = \mathrm{diag}\{\gamma_l, l \in \overline{L}\}. \tag{4.77}$$

Then, by using relationship (4.72), equation (4.73) can be rewritten in the following matrix form:

$$X(s) = -KT(s) R_f^T(-s) \Gamma R_f(s) X(s). \tag{4.78}$$

Then, we get the open-loop transfer function of system $\sum$ as

$$L(s) = KT(s) R_f^T(-s) \Gamma R_f(s). \tag{4.79}$$

Now, we see that system $\sum$ is symmetric in the frequency domain by Definition 2.2.

4.4 Notes and References

In 1988, Jacobson (1988) proposed a remarkable congestion avoidance algorithm, which is now regarded as a significant technological breakthrough for the growth of the Internet from a small-scale research network to today's interconnection of tens of millions of users and links. The optimization problem (4.7) was first formulated in Kelly (1997). The dual optimization problem (4.17) was first considered in Low and Lapsley (1999). Related ideas also appeared in Yaiche, Mazumdar and Rosenberg (2000). Kar, Sarkar and Tassiulas (2001) extended these ideas to multicast networks.

The mathematical treatment of congestion control problems and algorithms in this chapter follows the classic literature on congestion control theory, see, for example, Srikant (2004), Kelly (2003), Low, Paganini and Doyle (2002). The formulation of the general model of congestion control algorithms given in Section 4.3.2 and its linearization with symmetry analysis given in Section 4.3.3 are mainly taken from Tian (2005). For the connection of the algorithms with currently used Internet protocols such as TCP Reno and TCP Vegas (Brakmo and Peterson 1995; Jacobson 1988), DropTail and RED (Floyd and Jacobson 1993), etc., the reader is referred to Low, Paganini and Doyle (2002).

References

Bertsekas D (1995). *Nonlinear Programming*. Athena Scientific, Belmont, MA.

Srikant R (2004). *The Mathematics of Internet Congestion Control*. Birkhäuser, Boston.

Alpcan T and Başar T (2003). A utility-based congestion control scheme for internet-style networks with delay. *Proceedings of IEEE Infocom*, San Francisco, California.

Athuraliya S, Low SH and Yin Q (2001). REM: Active Queue Management. *IEEE Network*, 15, 48–53.

Brakmo LS and Peterson LL (1995). TCP Vegas: end to end congestion avoidance on a global Internet. *IEEE Journal on Selected Area in Communications* 13, 1465–1480.

Floyd S and Jacobson V (1993). Random early detection gateways for congestion avoidance. *IEEE/ACM Transactions on Networking*, 1, 397–413.

Jacobson V (1988). Congestion avoidance and control. *Proceedings of ACM SIGCOMM '88*, Stanford, CA, 314–329.

Johari R and Tan D (2001). End to end congesion control for the internet: delays and stability. *IEEE/ACM Transactions on Networking*, 9, 818–832.

Kar K, Sarkar S and Tassiulas L (2001). Optimization based rate control for multirate multicast sessions. *Proceedings of IEEE Infocom*.

Kelly FP (1997). Charging and rate control for elastic traffic. *European transactions on Telecommunication*, 8, 33–37.

Kelly FP, Maulloo A and Tan D (1998). Rate control for communication networks: shadow prices proportional fairness and stability. *Journal of the Operational Research Society*, 49, 237–252.

Kelly FP (2003). Fairness and stability of end-to-end congestion control. *European Journal of Control*, 9, 149–165.

Liu S, Başar T and Srikant R (2003). Controlling the Internet: A survey and some new results. *Proceedings of IEEE Conference on Decision and Control*, Maui, Hawaii.

Low SH and Lapsley DE (1999). Optimization flow control, I: basic algorithm and convergence. *IEEE/ACM Transactions on Networking*, 7, 861–874.

Low SH, Paganini F and Doyle JC (2002). Internet congestion control. *IEEE Control Systems Magazine*, 22, 28–43.

Tian Y-P (2005). A general stability criterion for congestion control with diverse communication delays. *Automatica*, 41, 1255–1262.

Wen JT and Arcak M (2003). A unifying passivity framework for network flow control. *Proceedings of IEEE Infocom*, San Francisco, California.

Yaiche H, Mazumdar RR and Rosenberg C (2000). A game theoretical framework for bandwidth allocation and pricing in broadband networks. *IEEE/ACM Transactions on Networking*, 8, 2–14.

5

Congestion Control: Stability and Scalability

The net of Heaven has large meshes but yet it lets nothing through.
—Lao Dan (580–500 BC), Tao Te Ching

A distributed congestion control system with diverse propagation delays enjoys symmetry property in the frequency domain. This chapter applies the stability and scalability results presented in Chapter 2 and Chapter 3 to derive scalable stability criteria for the congestion control algorithms introduced in Chapter 4. The mechanism of time-delayed feedback control in the stabilization of distributed congestion control systems is also investigated.

5.1 Stability of the Primal Algorithm

5.1.1 Johari–Tan Conjecture

With the choice of utility function $U_i(x_i) = w_i \log x_i$, and the step size function $k_i(x_i) = \kappa_i x_i$, the primal algorithm (4.15) has the form (Kelly, Maulloo and Tan 1998)

$$\dot{x}_i(t) = \kappa_i \left(w_i - x_i(t) \sum_{l \in L_i} q_l(t) \right)^+_{x_i}, \quad i \in \overline{S} \tag{5.1}$$

$$q_l(t) = p_l \left(\sum_{r \in S_l} x_r(t) \right), \quad l \in \overline{L} \tag{5.2}$$

where $x_i(t) \in \mathbb{R}^+$ is the transmission rate at source i, $\kappa_i > 0$ is the control gain, w_i is some desired value of the rate of marked packets received back at source i, and the notation $(z)^+_x$ means

$$(z)^+_x = \begin{cases} z, & \text{if } x > 0, \\ \max(z, 0), & \text{if } x = 0. \end{cases}$$

Frequency-Domain Analysis and Design of Distributed Control Systems, First Edition. Yu-Ping Tian.

The first equation (5.1) describes the time evolution of the transmission rate $x_i(t)$ of source i. The second equation (5.2) describes the generation of congestion signal $q_l(t)$ at link l, by means of a congestion indication function $p_l(y)$, which is assumed to be increasing, nonnegative, and not identically zero. A candidate of $p(y)$ is $p(y) = 1 - e^y(1 - y)$, which is positive and strictly increasing for all $y > 0$ (Johari and Tan 2001).

It was shown by (Kelly, Maulloo and Tan 1998) the system of equations (5.1) and (5.2) has a unique equilibrium point $x^\star = [x_1^\star, \ldots, x_S^\star]^T$ given by

$$x_i^\star = \frac{w_i}{\sum_{l\in L_i} p_l\left(\sum_{r\in S_l} x_r^\star\right)}, \tag{5.3}$$

and this equilibrium is globally asymptotically stable. Johari and Tan (2001) studied the influence of propagation delays on the stability of congestion control algorithms by considering the following delay difference equations analogous to the equations (5.1) and (5.2):

$$x_i(t+1) = x_i(t) + \kappa_i\left(w_i - x_i(t - D_i)\sum_{l\in L_i} q_l(t - d_{li}^{\leftarrow})\right)_{x_i}^{+}, \quad i \in \overline{S} \tag{5.4}$$

$$q_l(t) = p_l\left(\sum_{r\in S_l} x_r(t - d_{lr}^{\rightarrow})\right), \quad l \in \overline{L} \tag{5.5}$$

where the forward delay $d_{li}^{\rightarrow}$ and the backward delay $d_{li}^{\leftarrow}$ are subject to (4.27). Note that the equilibrium of system (5.4)–(5.5) has the same expression as given by (5.3).

Johari and Tan (2001) proposed the following conjecture on the local stability of the equilibrium point of system (5.4)–(5.5).

Conjecture 5.1 *Let $D_i, i \in \overline{S}$, be diverse round-trip delays. System (5.4)–(5.5) is locally stable at the equilibrium point $x^\star$, if*

$$\kappa_i\left(\sum_{l\in L_i} p_l + \sum_{l\in L_i} p_l' \sum_{r\in S_l} x_r^\star\right) < 2\sin\left(\frac{\pi}{2(2D_i+1)}\right) \tag{5.6}$$

holds for all $i \in \overline{S}$, where

$$p_l = p_l\left(\sum_{i\in S_l} x_i^\star\right) \tag{5.7}$$

and p_l' is the derivative of p_l evaluated at $y = \sum_{r\in S_l} x_r^\star$, i.e.,

$$p_l' = p_l'\left(\sum_{r\in S_l} x_r^\star\right). \tag{5.8}$$

The significance of this conjecture can be explained as follows. If it is correct, then the network stability under diverse round-trip delays can be guaranteed by implementation of a simple distributed congestion control algorithm called the primal algorithm, and the stability test for such a network uses only local information of each user of the network. Hence, such a control scheme fully satisfies the scalability requirement for huge networks such as the Internet. Moreover, the conjecture looks so elegant because it gives a necessary and sufficient condition of stability for system (5.4)–(5.5) when only one source is contained in the network.

Johari and Tan (2001) proved the correctness of the conjecture for a special case when all sources share the same round-trip delay parameter, i.e., $D_i = D$ for all $i \in \overline{S}$. They also generated 10 000 "random" networks for the case of diverse round-trip delays and found no counterexamples to Conjecture 5.1. In this section we will prove that the conjecture is correct and the conservatism of its stability condition can be further reduced.

5.1.2 Scalable Stability Criterion for Discrete-Time Systems

We will show the correctness of Conjecture 5.1 by proving the following results.

Theorem 5.2 *Let $D_i, i \in \overline{S}$, be diverse delays. System (5.4)–(5.5) is locally stable at the equilibrium point $x_i^\star$, if there exists a diagonal positive real number matrix $H = \mathrm{diag}\{h_i > 0,\ i \in \overline{S}\}$ such that*

$$\kappa_i \left(\sum_{l \in L_i} p_l + \sum_{l \in L_i} p_l' \sum_{r \in S_l} \frac{h_i}{h_r} x_r^\star \right) < 2 \sin \left(\frac{\pi}{2(2D_i + 1)} \right) \tag{5.9}$$

holds for all $i \in \overline{S}$.

Obviously, Conjecture 5.1 is just a special case of Theorem 5.2 because when H is taken as the identity matrix, i.e. $H = I$, Theorem 5.2 reduces to Conjecture 5.1. This criterion of stability preserves the elegant property being locally implemented of Conjecture 5.1. Moreover, Theorem 5.2 enlarges the stability region for control gains and admissible delay constants.

Proof of Theorem 5.2. Define

$$\hat{x}_i(t) = \frac{x_i(t) - x_i^\star}{\sqrt{\kappa_i x_i^\star}}, \quad i = \overline{S}.$$

Then, linearizing system (5.4)–(5.5) about the equilibrium point $x^\star$ yields

$$\begin{aligned} \hat{x}_i(t+1) = {} & \hat{x}_i(t) - \kappa_i w_i x_i^{\star -1} \hat{x}_i(t - D_i) \\ & - \sqrt{\kappa_i x_i^\star} \sum_{l \in \overline{L}} R_{li} p_l' \sum_{r \in \overline{S}} R_{lr} \sqrt{\kappa_r x_r^\star} \hat{x}_r(t - d_{lr}^{\rightarrow} - d_{li}^{\leftarrow}). \end{aligned} \tag{5.10}$$

Assume the initial state of the linearized system is zero, $\hat{x}_i(t) = 0$ for $t \in \overline{-D_i, 0}$. Then, taking the z-transform, we obtain

$$\begin{aligned} z\hat{X}_i(z) = {} & \hat{X}_i(z) \\ & - \left(\kappa_i w_i x_i^{\star -1} \hat{X}_i(z) + \sqrt{\kappa_i x_i^\star} \sum_{l \in \overline{L}} R_{li} p_l' z^{d_{li}^{\rightarrow}} \sum_{r \in \overline{S}} R_{lr} \sqrt{\kappa_r x_r^\star} z^{-d_{lr}^{\rightarrow}} \hat{X}_r(z) \right) z^{-D_i}. \end{aligned} \tag{5.11}$$

where $\hat{X}_i(z) = \mathcal{Z}(\hat{x}_i(t))$ is the z-transform of $\hat{x}_i(t)$. Define the following matrices:

$$\kappa = \text{diag}\{\kappa_i,\ i \in \overline{S}\}, \tag{5.12}$$
$$X^\star = \text{diag}\{x_i^\star,\ i \in \overline{S}\}, \tag{5.13}$$
$$W = \text{diag}\{w_i,\ i \in \overline{S}\}, \tag{5.14}$$
$$P' = \text{diag}\{p_l',\ l \in \overline{L}\}, \tag{5.15}$$
$$R_f(z) = \{R_{li} z^{-d_{li}^{\rightarrow}},\ i \in \overline{S},\ l \in \overline{L}\}, \tag{5.16}$$
$$M(z) = \kappa^{1/2} W X^{\star-1} \kappa^{1/2} + \kappa^{1/2} X^{\star 1/2} R_f^{\mathrm{T}}(z^{-1}) P' R_f(z) X^{\star 1/2} \kappa^{1/2}. \tag{5.17}$$

Then, (5.11) can be rewritten as

$$(z-1)\hat{X}(z) = -M(s)\text{diag}\{z^{-D_i},\ i \in \overline{S}\}\hat{X}(z). \tag{5.18}$$

The characteristic equation of the above closed-loop system is

$$\det\left(I + M(z)\text{diag}\left\{\frac{z^{-D_i}}{z-1},\ i \in \overline{S}\right\}\right) = 0. \tag{5.19}$$

Now, we see that system (5.4)–(5.5) is locally stable at the equilibrium point, if all the roots of the characteristic equation have modulus less than unity. Note that

$$M(z)\text{diag}\left\{\frac{z^{-D_i}}{z-1},\ i \in \overline{S}\right\}$$

has no poles outside the unit disk in the complex plane (the pole $z = 1$ can be considered as inside the modified unit disk as shown by Figure 2.13). Therefore, by the generalized Nyquist criterion (Theorem 2.22), system (5.4)–(5.5) is locally stable, if the eigenloci of $M(z)\text{diag}\left\{\frac{z^{-D_i}}{z-1}\right\}$, i.e., $\lambda\left(M(\mathrm{e}^{j\omega})\text{diag}\left\{\frac{\mathrm{e}^{-j\omega D_i}}{\mathrm{e}^{j\omega}-1},\ i \in \overline{S}\right\}\right)$, do not enclose the point $(-1, \mathrm{j}0)$ for all $\omega \in [-\pi, \pi]$.

Now, let us rewrite $M(z)\text{diag}\left\{\frac{z^{-D_i}}{z-1},\ i \in \overline{S}\right\}$ as

$$M(z)\text{diag}\left\{\frac{z^{-D_i}}{z-1},\ i \in \overline{S}\right\} = M(z)K^{1/2}\text{diag}\left\{k_i\frac{z^{-D_i}}{z-1},\ i \in \overline{S}\right\}K^{1/2}$$

where

$$k_i = 2\sin\left(\frac{\pi}{2(2D_i+1)}\right),$$
$$K = \text{diag}\{k_i^{-1},\quad i \in \overline{S}\}.$$

Obviously,

$$\begin{aligned}
&\{0\} \cup \sigma\left(M(\mathrm{e}^{\mathrm{j}\omega})\mathrm{diag}\left\{\frac{\mathrm{e}^{-\mathrm{j}D_i\omega}}{\mathrm{e}^{\mathrm{j}\omega}-1},\ i \in \overline{S}\right\}\right) \\
&= \{0\} \cup \sigma\left(M(\mathrm{e}^{\mathrm{j}\omega})K^{1/2}\mathrm{diag}\left\{k_i\frac{\mathrm{e}^{-\mathrm{j}D_i\omega}}{\mathrm{e}^{\mathrm{j}\omega}-1},\ i \in \overline{S}\right\}K^{1/2}\right) \\
&= \{0\} \cup \sigma\left(K^{1/2}M(\mathrm{e}^{\mathrm{j}\omega})K^{1/2}\mathrm{diag}\left\{k_i\frac{\mathrm{e}^{-\mathrm{j}D_i\omega}}{\mathrm{e}^{\mathrm{j}\omega}-1},\ i \in \overline{S}\right\}\right).
\end{aligned}$$

By the definition of $M(z)$ given by (5.17) we have

$$\begin{aligned}
&\rho\left(K^{1/2}M(\mathrm{e}^{\mathrm{j}\omega})K^{1/2}\right) \\
&= \rho\left(M(\mathrm{e}^{\mathrm{j}\omega})K\right) \\
&= \rho\left(\kappa^{1/2}(WX^{\star -1} + X^{\star 1/2}R_f^{\mathrm{T}}(\mathrm{e}^{-\mathrm{j}\omega})P'R_f(\mathrm{e}^{\mathrm{j}\omega})X^{\star 1/2})\kappa^{1/2}K\right) \\
&= \rho\left(X^{\star 1/2}(WX^{\star -2} + R_f^{\mathrm{T}}(\mathrm{e}^{-\mathrm{j}\omega})P'R_f(\mathrm{e}^{\mathrm{j}\omega}))X^{\star 1/2}\kappa K\right) \\
&= \rho\left((WX^{\star -1} + R_f^{\mathrm{T}}(\mathrm{e}^{-\mathrm{j}\omega})P'R_f(\mathrm{e}^{\mathrm{j}\omega})X^{\star})\kappa K\right) \\
&= \rho\left(H(WX^{\star -1} + R_f^{\mathrm{T}}(\mathrm{e}^{-\mathrm{j}\omega})P'R_f(\mathrm{e}^{\mathrm{j}\omega})X^{\star})\kappa K H^{-1}\right).
\end{aligned}$$

Recall that $H = \mathrm{diag}\{h_i, h_i > 0,\ i \in \overline{S}\}$. By Corollary 2.31 of Gershgorin's disc lemma, the spectral radius of any matrix is bounded by its maximum absolute row sum. Therefore, by the definitions of all matrices (5.12)–(5.17) and the assumption of Theorem 5.2, we get

$$\begin{aligned}
&\rho\left(K^{1/2}M(\mathrm{e}^{\mathrm{j}\omega})K^{1/2}\right) \\
&\le \max_{i\in\overline{S}}\left\{\kappa_i\left(\sum_{l\in\overline{L}}R_{li}p_l + \sum_{l\in\overline{L}}\sum_{r\in\overline{S}}\left|\frac{h_i}{h_r}R_{li}\mathrm{e}^{\mathrm{j}\omega(d_{li}^{\rightarrow}-d_{lr}^{\rightarrow})}p_l'R_{lr}x_r^{\star}\right|\right)k_i^{-1}\right\} \\
&= \max_{i\in\overline{S}}\left\{\kappa_i\left(\sum_{l\in\overline{L}}R_{li}p_l + \sum_{l\in\overline{L}}R_{li}p_l'\sum_{r\in\overline{S}}\frac{h_i}{h_r}R_{lr}x_r^{\star}\right)\left(2\sin\frac{\pi}{2(2D_i+1)}\right)^{-1}\right\} \\
&= \max_{i\in\overline{S}}\left\{\kappa_i\left(\sum_{l\in L_i}p_l + \sum_{l\in L_i}p_l'\sum_{r\in S_l}\frac{h_i}{h_r}x_r^{\star}\right)\left(2\sin\frac{\pi}{2(2D_i+1)}\right)^{-1}\right\} \\
&< 1.
\end{aligned}$$

Then, by Vinnicombe's lemma (Lemma 2.34) we get

$$\begin{aligned}
&\lambda\left(M(\mathrm{e}^{\mathrm{j}\omega})\mathrm{diag}\left\{\frac{\mathrm{e}^{-\mathrm{j}D_i\omega}}{\mathrm{e}^{\mathrm{j}\omega}-1},\ i \in \overline{S}\right\}\right) \\
&\in \rho\left(K^{1/2}M(\mathrm{e}^{\mathrm{j}\omega})K^{1/2}\right)\mathrm{Co}\left(0 \cup \left\{k_i\frac{\mathrm{e}^{-\mathrm{j}D_i\omega}}{\mathrm{e}^{\mathrm{j}\omega}-1},\ i \in \overline{S}\right\}\right)
\end{aligned}$$

holds for all $\omega \in [-\pi, \pi]$.

Now, we can see the problem has been converted to the scalability condition for the first-order time-delayed system of the form

$$G(z) = k\frac{z^D}{z-1}.$$

This problem was solved in Chapter 3 (§3.3.2) by Theorem 3.16. Using this theorem we know that

$$\rho\left(K^{1/2}M(\mathrm{e}^{\mathrm{j}\omega})K^{1/2}\right)\mathrm{Co}\left(0\cup\left\{k_i\frac{\mathrm{e}^{-\mathrm{j}D_i\omega}}{\mathrm{e}^{\mathrm{j}\omega}-1},\ i\in\overline{S}\right\}\right)$$

does not contain the point $(-1, \mathrm{j}0)$ when $\rho\left(K^{1/2}M(\mathrm{e}^{\mathrm{j}\omega})K^{1/2}\right) < 1$. And hence, the eigenloci $\lambda\left(M(\mathrm{e}^{\mathrm{j}\omega})\mathrm{diag}\left\{\frac{\mathrm{e}^{-\mathrm{j}D_i\omega}}{\mathrm{e}^{\mathrm{j}\omega}-1},\ i\in\overline{S}\right\}\right)$ cannot enclose the point $(-1, \mathrm{j}0)$ for all $\omega\in[-\pi,\pi]$. The theorem is thus proved. □

Theorem 5.2 preserves the elegant property of Conjecture 5.1 being decentralized and locally implemented: each end system needs knowledge only of its own round-trip delay. Moreover, it enlarges the stability region of control gains and admissible communication delays.

From Theorem 5.2 we know that the stability region of control gain is given by $0 < \kappa_i < \kappa_i^\star(H)$ where

$$\kappa_i^\star(H) = \frac{2\sin\frac{\pi}{2(2D_i+1)}}{\sum_{l\in L_i} p_l + \sum_{l\in L_i} p_l' \sum_{r\in S_l}\frac{h_i}{h_r}x_r^\star},\quad i\in\overline{S}.$$

Consider H as a parameter one can draw a curved surface $\kappa(H)$ in the gain space. The stability region is covered by this curved surface and bounded by $\kappa_i > 0$. However, Conjecture 5.1 suggests that the stability region is a hyper-rectangle given by $0 < \kappa_i < \kappa_i^\star(I)$, which is inside the region given by Theorem 5.2.

Since $\sin\left(\frac{\pi}{2(2D_i+1)}\right)$ is strictly monotone when $D_i \geq 0$, fixing control gains κ_i, we get from Theorem 5.2 the bounds of the admissible values of delay constants as follows

$$\begin{aligned}0 \leq D_i &< D_i^\star(H)\\ &= \mathrm{Ing}\left(\frac{1}{2}\left(\frac{\pi}{2\arcsin\left(\kappa_i(\sum_{l\in L_i} p_l + \sum_{l\in L_i} p_l'\sum_{r\in N_l}\frac{h_i}{h_r}x_r^\star)/2\right)}-1\right)\right)\end{aligned}$$

where $\mathrm{Ing}(x)$ represents the function which maps x to the nearest integer towards infinity.

To illustrate the theoretic results we perform simulation as follows. We simulate a network with one link shared by three sources. The routing matrix is $R = [1\ 1\ 1]$. The congestion indication function is chosen as $p(y) = 1 - \mathrm{e}^y(1-y)$, where $y = \sum_{i=1}^{3} x_i(t - d_{\overrightarrow{1i}})$ is the network load. Setting $[w_1, w_2, w_3] = [0.005, 0.01, 0.015]$ we get the equilibrium of the transmission rate is $x^\star = [0.0255, 0.0509, 0.0763]^\mathrm{T}$ by (5.3). Then we can calculate p and p' at $x^\star$. We first fix the round-trip delay $D_i (i = 1, 2, 3)$ as 80(ms), 60(ms), and 70(ms). Then, scaling on the

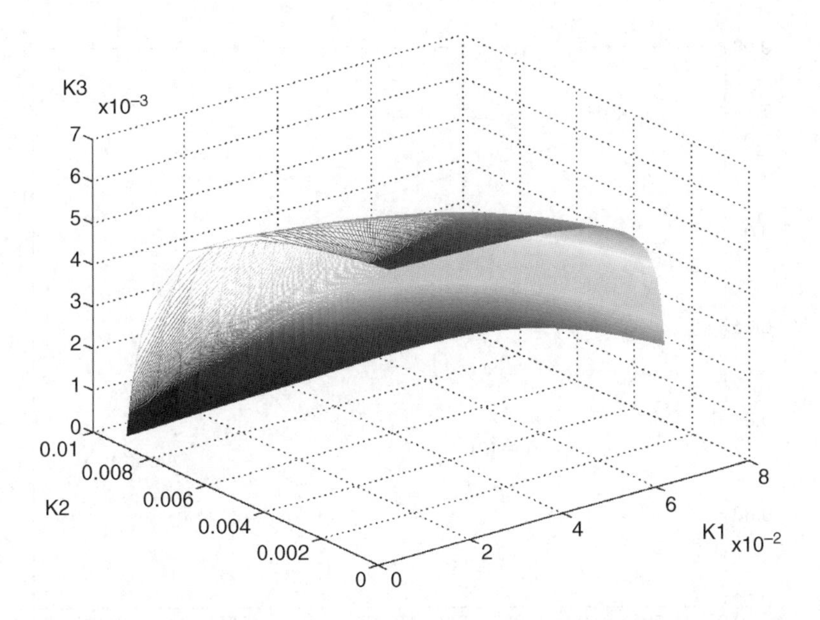

Figure 5.1 The curved surface of critical gains.

parameter matrix $H = \mathrm{diag}\{h_1, h_2, h_3\}$ (for simplicity of calculation, h_1 can always be taken as unity, i.e., $h_1 = 1$), we can obtain a curved surface of the critical gains, $\kappa^\star(H)$, shown in Figure 5.1. Setting $\kappa_3 = 0.0448$, we get a two-dimensional transversal of $\kappa^\star(H)$ as shown in Figure 5.2. In the figure, "Area(2)" is the stability region determined by Conjecture 5.1 while the entire area covered by the curve $\kappa^\star(H)$ ("Area(1)"∪"Area(2)"∪"Area(3)") is the stability region of control gains determined by Theorem 5.2.

Similarly, for a fixed $[\kappa_1, \kappa_2, \kappa_3] = [0.0527, 0.0588, 0.0448]$ we can also obtain the range of values of the admissible delays. Its two-dimensional transversal at $D_3 = 70$(ms) is shown in Figure 5.3, where "Area(2)" is the range of values of admissible delays determined by Conjecture 5.1, and "Area(1)" ∪ "Area(2)" ∪ "Area(3)" is the range of values of admissible delays determined by Theorem 5.2.

5.1.3 Scalable Stability Criterion for Continuous-Time Systems

The continuous-time version of system (5.4)–(5.5) is given by

$$\dot{x}_i(t) = \kappa_i \left(w_i - x_i(t - D_i) \sum_{l \in L_i} q_l(t - d_{li}^{\leftarrow}) \right)^+_{x_i}, \quad i \in \overline{S} \tag{5.20}$$

$$q_l(t) = p_l \left(\sum_{r \in S_l} x_r(t - d_{lr}^{\rightarrow}) \right), \quad l \in \overline{L}. \tag{5.21}$$

Then, an analog of Conjecture 5.1 for the continuous-time system (5.20)–(5.21) can be stated as follows (Vinnicombe 2000).

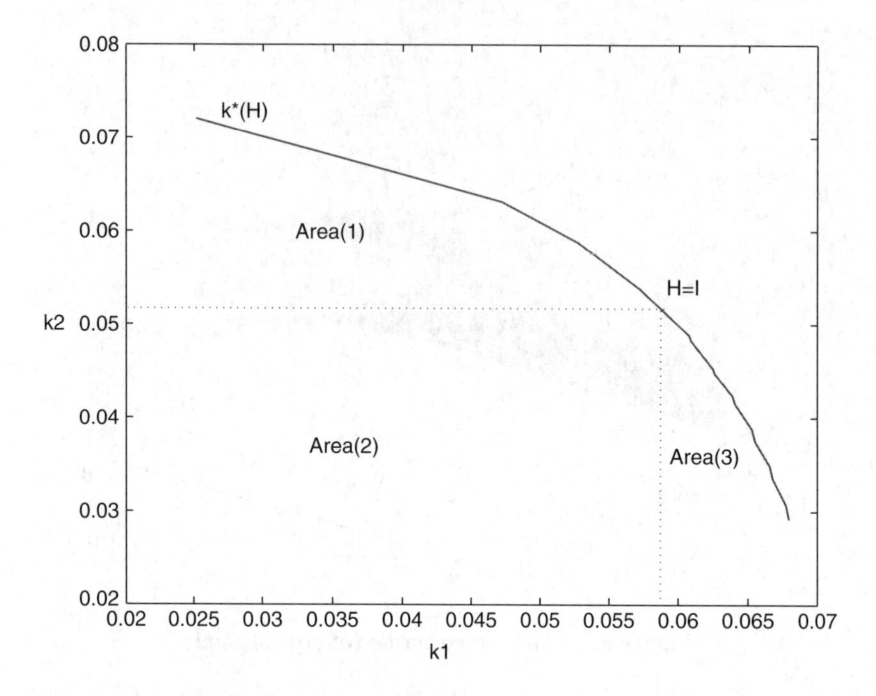

Figure 5.2 Stability region of control gains.

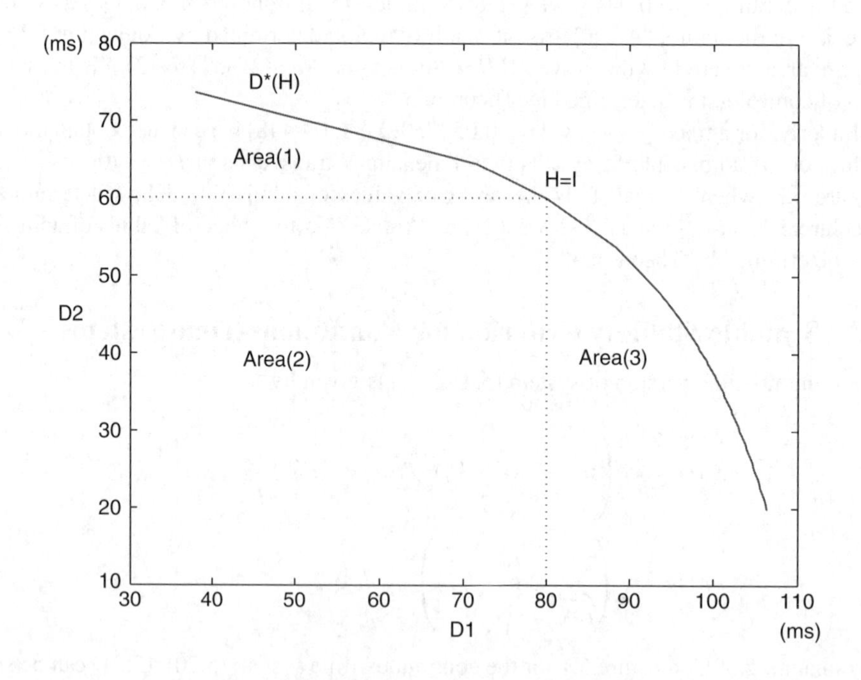

Figure 5.3 Range of admissible delays.

Conjecture 5.3 *Let $D_i, i \in \overline{S}$, be diverse round-trip delays. System (5.20)–(5.21) is locally stable at the equilibrium point $x^\star$, if*

$$\kappa_i D_i \left(\sum_{l \in L_i} p_l + \sum_{l \in L_i} p_l' \sum_{r \in S_l} x_r^\star \right) < \frac{\pi}{2} \tag{5.22}$$

holds for all $i \in \overline{S}$, where

$$p_l = p_l \left(\sum_{r \in S_l} x_r^\star \right) \tag{5.23}$$

and p_l' is the derivative of p_l evaluated at $y = \sum_{i \in S_l} x_i^\star$, i.e.,

$$p_l' = p_l' \left(\sum_{r \in S_l} x_r^\star \right). \tag{5.24}$$

Massoulie (2002) proved a weaker version of Conjecture 5.3, which says that system (5.20)–(5.21) with diverse round-trip delays is locally stable at the equilibrium point, if

$$\kappa_i D_i \left(\sum_{l \in L_i} p_l + \sum_{l \in L_i} p_l' \sum_{r \in S_l} x_r^\star \right) < 1. \tag{5.25}$$

Similar to the derivation for the discrete-time model, we can prove the following theorem which includes Conjecture 5.3 as a special result.

Theorem 5.4 *Let $D_i, i \in \overline{S}$, be diverse round-trip delays. System (5.20)–(5.21) is locally stable at the equilibrium point $x^\star$, if there exists a diagonal positive real number matrix $H = \mathrm{diag}\{h_i > 0,\ i \in \overline{S}\}$ such that*

$$\kappa_i D_i \left(\sum_{l \in L_i} p_l + \sum_{l \in L_i} p_l' \sum_{r \in S_l} \frac{h_i}{h_r} x_r^\star \right) < \frac{\pi}{2} \tag{5.26}$$

holds for all $i \in \overline{S}$.

Proof. Similar to the proof of Theorem 5.2 we can show that system (5.20)–(5.21) is locally stable at the equilibrium point if

$$(-1, \mathrm{j}0) \notin \rho\left(K^{1/2} M(\mathrm{j}\omega) K^{1/2}\right) \mathrm{Co}\left(0 \cup \left\{ k_i \frac{\mathrm{e}^{-\mathrm{j}D_i\omega}}{\mathrm{j}\omega},\ i \in \overline{S} \right\}\right) \tag{5.27}$$

holds for all $\omega \in (-\infty, \infty)$. So, the proof of the theorem is converted to the scalability condition for the first-order continuous-time system of the form

$$G(s) = k\frac{e^{-Ds}}{s}.$$

This problem has been solved in Chapter 3 (§3.3.1) by Theorem 3.10. Using this theorem we know that (5.27) holds for all $\omega \in (-\infty, \infty)$ if $\rho\left(K^{1/2}M(j\omega)K^{1/2}\right) < 1$, which is ensured by (5.26). The theorem is thus proved. □

Exercise 5.5 *Consider the system with dual algorithm (4.32)–(4.35). Prove that under diverse round-trip delays D_i, $i \in \overline{S}$, the system is locally stable about the equilibrium point, if there exists a diagonal positive real matrix $H = \text{diag}\{h_l, h_l > 0, l \in \overline{L}\}$ such that*

$$-\gamma_l \sum_{i \in S_l} R_{li} f_i' D_i \sum_{r \in L_i} \frac{h_l}{h_r} R_{ri} < \pi/2 \tag{5.28}$$

holds for all $l \in \overline{L}$, where f_i' is the derivative of the function $U_i'^{-1}(\cdot)$ at the equilibrium point.

Exercise 5.6 *Derive the expression of the equilibrium point, linearized the model and scalable local stability condition of the time-delayed primal algorithm given by (4.28)–(4.31).*

5.2 Stability of REM

By doing Exercise 5.5, one can see that the scalable stability results of the primal algorithm also exist for the dual algorithm, thanks to the symmetry between the primal algorithm and the dual algorithm. This may give the reader an impression that the Johari–Tan-like results may hold for other distributed congestion control algorithms. To give a more clear insight into this problem we study the stability of the second-order AQM control algorithm: REM. This control scheme was first proposed by Athuraliya, Low and Yin (2001) to overcome the drawback of the first-order dual algorithm that prices are proportional to link backlogs and thus the equilibrium can have large backlogs. Note that this algorithm is globally stable in the absence of communication delays (Paganini 2002). The local stability of the algorithm with one link or identical round-trip delay was studied in Yin and Low (2001, 2002). In this section we use the frequency-domain approach to analyze the stability of the REM algorithm in the presence of diverse round-trip delays.

5.2.1 Scalable Stability Criteria

The continuous-time version of the REM algorithm is given by (Paganini 2002)

$$\dot{b}_l(t) = (y_l(t) - c_l)^+_{b_l}, \tag{5.29}$$

$$\dot{p}_l(t) = \gamma_l\left(\alpha_l(b_l(t) - e_l) + y_l(t) - c_l\right)^+_{p_l}, \tag{5.30}$$

where $p_l(t) \in \mathbb{R}^+$ is the price (congestion measurement) at link l; $\gamma_l > 0$ the control gains; $y_l(t) \in \mathbb{R}^+$ the aggregate rate at link l; c_l the capacity of link l; $b_l(t) \in \mathbb{R}^+$ some internal variable for the price control, which often represents the queue length in the buffer of link l; e_l some desired value for the queue length at link l.

Note that unlike Paganini (2002) in which all the links share a common control gain, i.e., $\gamma_l = \gamma, \forall l \in \overline{L}$, here we assume that each link may have its own control gain. However, for the semi-homogeneousness of the system we assume that all the links have the same dynamics, and consequently, the control parameters α_l for the links in (4.23) take the same value, i.e., $\alpha_l = \alpha, \forall l \in \overline{L}$.

The transmitting rate at each source is generated by the following function:

$$x_i = U_i'^{-1}(q_i) \triangleq f_i(q_i(t)), \quad i \in \overline{S} \tag{5.31}$$

where $U_i'^{-1}(q_i)$ is the inverse of the derivative of $U_i(q_i)$, a utility function which is assumed to be differentiable and strictly concave. Hence, $f_i(q_i)$ is a strictly monotonically decreasing function of q_i.

For a given link $l \in \overline{L}$ and each source $i \in S_l$, there is a forward delay $d_{li}^{\rightarrow}$ of delivering packets from source i to link l, and a backward delay $d_{li}^{\leftarrow}$ of sending the feedback signal from link l to source i. Note that each route is subject to a round-trip delay D_i satisfying (4.27). Under propagation delays the following relationships are immediate:

$$y_l(t) = \sum_{i \in S_l} x_i(t - d_{li}^{\rightarrow}), \tag{5.32}$$

$$q_i(t) = \sum_{l \in L_i} p_l(t - d_{li}^{\leftarrow}). \tag{5.33}$$

Equations (5.29)–(5.33) constitute a dynamical feedback system which has a unique equilibrium $(b^\star, p^\star)$ satisfying

$$b_l^\star = e_l,$$

$$\sum_{i \in S_l} f_i\left(\sum_{l \in L_i} p_l^\star\right) = c_l.$$

For convenience, in the further development we denote

$$q_i^\star = \sum_{l \in L_i} p_l^\star,$$

$$y_l^\star = \sum_{i \in S_l} x_i^\star = \sum_{i \in S_l} f_i(q_i^\star),$$

as the aggregate price of all links used by source i and the aggregate rate of all sources which use link l at the equilibrium, respectively. The vectors $x^\star$, $q^\star \in \mathbb{R}^S$, $y^\star \in \mathbb{R}^L$ are defined by the above components across the set of sources (links).

We have assumed that the desired value of the queue length at each link is positive, i.e., $e_l > 0, \forall l \in \overline{L}$. Therefore, in the neighborhood of $(b^\star, p^\star)$, the system (5.29)–(5.30) is differentiable and simply governed by

$$\dot{b}_l(t) = y_l(t) - c_l, \tag{5.34}$$

$$\dot{p}_l(t) = \gamma(\alpha(b_l(t) - e_l) + y_l(t) - c_l). \tag{5.35}$$

For simplicity of statement, we call the system consisting of equations (5.31)–(5.35) as "system $\sum_{\text{rem}}$".

Define

$$\begin{aligned}\hat{b}_l(t) &= b_l(t) - b_l^\star,\\ \hat{p}_l(t) &= p_l(t) - p_l^\star,\\ \hat{q}_i(t) &= q_i(t) - q_i^\star,\\ \hat{y}_l(t) &= y_l(t) - y_l^\star,\\ \hat{x}_i(t) &= (x_i(t) - x_i^\star)/\sqrt{-f_i'},\end{aligned}$$

where f_i' is the derivative of the function $f_i(\cdot)$ at the equilibrium point. Note that $f_i' < 0$ because $f_i(\cdot)$ is a strictly monotonically decreasing function.

Then, we can linearize system $\sum_{\text{rem}}$ about the equilibrium point and get

$$\begin{cases}\dot{\hat{b}}_l(t) = \hat{y}_l(t),\\ \dot{\hat{p}}_l(t) = \gamma_l(\alpha\hat{b}_l(t) + \dot{\hat{b}}_l(t)),\\ \dot{\hat{x}}_i(t) = -\sqrt{-f_i'}\hat{q}_i(t),\\ \hat{y}_l(t) = \sum_{i\in S_l}\sqrt{-f_i'}\hat{x}_i(t - d_{li}^{\rightarrow}),\\ \hat{q}_i(t) = \sum_{l\in L_i}\hat{p}_l(t - d_{li}^{\leftarrow}).\end{cases} \tag{5.36}$$

Let $\hat{X}_i(s) = \mathcal{L}(\hat{x}_i(t))$, the Laplace transformation of $\hat{x}_i(t)$. The closed-loop system (5.36) can be represented in the Laplace domain as follows:

$$\hat{X}_i(s) = -\frac{\sqrt{-f_i'}(s+\alpha)}{s^2}\sum_{l\in L_i}\mathrm{e}^{-sd_{li}^{\leftarrow}}\gamma_l\sum_{r\in S_l}\mathrm{e}^{-sd_{lr}^{\rightarrow}}\sqrt{-f_r'}\hat{X}_r(s). \tag{5.37}$$

Denote

$$\hat{X}(s) = (\hat{X}_1(s), \ldots, \hat{X}_S(s))^T, \tag{5.38}$$

$$D(s) = \operatorname{diag}\left\{\frac{(s+\alpha)\mathrm{e}^{-D_i s}}{s^2},\ i \in \overline{S}\right\}, \tag{5.39}$$

$$\Gamma = \operatorname{diag}\{\gamma_l,\ l \in \overline{L}\}, \tag{5.40}$$

$$F = \operatorname{diag}\{-f_i',\ i \in \overline{S}\}, \tag{5.41}$$

where $D_i, i \in \overline{S}$ is the round-trip delay of the route associated with source i. Then, by using the relationship between the forward routing matrix and the backward routing matrix (4.72), equation (5.37) can be rewritten in the following matrix form:

$$X(s) = -D(s)F^{1/2}R_f^T(-s)\Gamma R_f(s)F^{1/2}X(s). \tag{5.42}$$

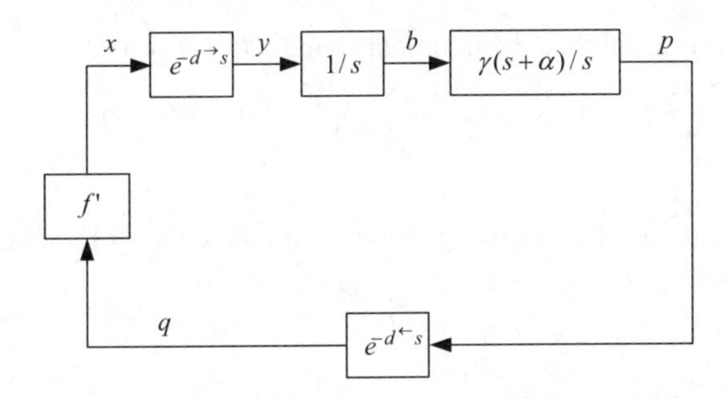

Figure 5.4 Closed-loop system of congestion control.

The characteristic equation of the above closed-loop system is

$$\det\left(I + D(s)F^{1/2}R_f^T(-s)\Gamma R_f(s)F^{1/2}\right) = 0. \tag{5.43}$$

Now, we see that system $\sum_{\text{rem}}$ is locally stable, if all the roots of the characteristic equation have negative real parts.

Before presenting our stability result for the general case, let us consider a special case when only one link and one source are involved in the network. In this case, the diagram of the closed-loop system (5.42) is shown in Figure 5.4.

Let $D = d^{\rightarrow} + d^{\leftarrow}$, and

$$W(s) = \frac{(s+\alpha)\mathrm{e}^{-Ds}}{s^2}. \tag{5.44}$$

Denote by G^M the gain margin of $W(s)$ which is defined by

$$G^M = 1/|W(\mathrm{j}\omega_c)| \tag{5.45}$$

where $\omega_c > 0$ is the frequency at which the phase of $W(\mathrm{j}\omega)$ is equal to π. ω_c is the minimal crossing frequency of $W(\mathrm{j}\omega)$ because the Nyquist plot of $W(j\omega)$ crosses the real axis of the complex plane for the first time at $\omega = \omega_c$, i.e., ω_c is the minimal frequency satisfying

$$\omega_c = \alpha\tan(\omega_c D). \tag{5.46}$$

According to the Nyquist criterion of stability (Theorem 2.20), we know that the closed-loop system shown in Figure 5.4 is asymptotically stable if and only if

$$-f'\gamma/G^M < 1. \tag{5.47}$$

Now, let us consider the general case when S sources and L links are involved in the network. For each $i \in \overline{S}$ we denote

$$A_i = D_i\alpha, \tag{5.48}$$

$$\omega_0(i) = \frac{\sqrt{\frac{3A_i - A_i^2 + \sqrt{(3A_i - A_i^2)^2 + 8A_i(1-A_i)}}{2}}}{D_i}. \tag{5.49}$$

Let $\hat{\imath} \in \overline{S}$ be the source which has the maximal round-trip delay, i.e.,

$$\hat{\imath} = \arg\max_{i \in \overline{S}} D_i. \tag{5.50}$$

We make the assumption on the control parameter and the delays of the network as follows.

Assumption 5.7

(A1) $A_i < 0.4495$, *for all* $i \in \overline{S}$;

(A2) $\omega_0(i) \leq D_{\hat{\imath}}^{-1} \arctan\left(\frac{\omega_0(i)}{\alpha}\right)$, *for all* $i \in \overline{S}$.

Now, we are in a position to present the following stability result.

Theorem 5.8 *Under Assumption 5.7, the system* $\sum_{\text{rem}}$ *is locally stable about the equilibrium point* $(b^\star, p^\star)$*, if there exists a diagonal positive real matrix* $H = \text{diag}\{h_l, h_l > 0, l \in \overline{L}\}$ *such that*

$$-\gamma_l \sum_{i \in S_l} R_{li} f_i'(G_i^M)^{-1} \sum_{r \in L_i} \frac{h_l}{h_r} R_{ri} < 1 \tag{5.51}$$

holds for all $l \in \overline{L}$, *where* G_i^M *is the gain margin of the transfer function*

$$W_i(s) = \frac{(s+\alpha)\mathrm{e}^{-D_i s}}{s^2}.$$

In particular, when H *is taken as the identity matrix, the stability condition reduces to*

$$-\gamma_l \sum_{i \in S_l} R_{li} f_i'(G_i^M)^{-1} \sum_{r \in L_i} R_{ri} < 1, \quad \forall l \in \overline{L}. \tag{5.52}$$

Proof. To prove the theorem let us return to equation (5.43). According to Theorem 2.20, all the roots of the characteristic equation (4.79) have negative real parts if and only if the eigenloci of $\text{diag}\left\{\frac{(s+\alpha)\mathrm{e}^{-D_i s}}{s^2}, i \in \overline{S}\right\} F^{1/2} R_f^T(-s) \Gamma R_f(s) F^{1/2}$, i.e.,

$$\lambda\left(\text{diag}\left\{\frac{(\mathrm{j}\omega+\alpha)\mathrm{e}^{-\mathrm{j}D_i\omega}}{(\mathrm{j}\omega)^2}, i \in \overline{S}\right\} F^{1/2} R_f^T(-\mathrm{j}\omega) \Gamma R_f(\mathrm{j}\omega) F^{1/2}\right),$$

do not enclose the point $(-1, \mathrm{j}0)$ for all $\omega \in [0, \infty)$, where $\lambda(\cdot)$ denotes the matrix eigenvalue.

Let us rewrite matrix

$$\text{diag}\left\{\frac{(s+\alpha)\mathrm{e}^{-D_i s}}{s^2}, i \in \overline{S}\right\} F^{1/2} R_f^T(-s) \Gamma R_f(s) F^{1/2}$$

as $G_0^{1/2}T(s)G_0^{1/2}Q_1(s)$, where

$$T(s) = \text{diag}\left\{G_i^M \frac{(s+\alpha)e^{-D_i s}}{s^2},\ i \in \overline{S}\right\}, \tag{5.53}$$

$$G_0 = \text{diag}\{1/G_i^M,\ i \in \overline{S}\}, \tag{5.54}$$

$$Q_1(s) = F^{1/2}R_f^T(-s)\Gamma R_f(s)F^{1/2}. \tag{5.55}$$

By the definition of $Q_1(s)$ we have

$$\begin{aligned}
&\rho\left(G_0^{1/2}Q_1(j\omega)G_0^{1/2}\right)\\
&= \rho\left(G_0^{1/2}F^{1/2}R_f^T(-s)\Gamma R_f(s)F^{1/2}G_0^{1/2}\right)\\
&= \rho\left(R_f^T(-s)\Gamma R_f(s)FG_0\right)\\
&= \rho\left(\Gamma R_f(s)FG_0R_f^T(-s)\right)\\
&= \rho\left(H\Gamma R_f(s)FG_0R_f^T(-s)H^{-1}\right).
\end{aligned}$$

Recall that $H = \text{diag}\{h_l, h_l > 0, l \in \overline{L}\}$. By Corollary 2.31 of Gershgorin's disc lemma, the spectral radius of any matrix is bounded by its maximum absolute column sum. Therefore, from (5.51) we get

$$\begin{aligned}
&\rho\left(G_0^{1/2}Q_1(j\omega)G_0^{1/2}\right)\\
&\le \max_{l\in\overline{L}}\left\{\left|\gamma_l\left(\sum_{i\in S_l} R_{li}e^{j\omega d_{li}^{\rightarrow}}(-f_i')(G_i^M)^{-1}\sum_{r\in L_i}\frac{h_l}{h_r}R_{ri}e^{-j\omega d_{ri}^{\rightarrow}}\right)\right|\right\}\\
&\le \max_{l\in\overline{L}}\left\{-\gamma_l\left(\sum_{i\in S_l} R_{li}f_i'(G_i^M)^{-1}\sum_{r\in L_i}\frac{h_l}{h_r}R_{ri}\right)\right\}\\
&< 1.
\end{aligned}$$

Note that $\{0\} \cup \sigma(G_0^{1/2}T(\mathrm{j}\omega)G_0^{1/2}Q_1(\mathrm{j}\omega)) = \{0\} \cup \sigma(T(\mathrm{j}\omega)G_0^{1/2}Q_1(\mathrm{j}\omega)G_0^{1/2})$. Therefore, it follows from Vinnicombe's lemma (Lemma 2.34) that

$$\begin{aligned}
&\lambda(G_0^{1/2}T(\mathrm{j}\omega)G_0^{1/2}Q_1(\mathrm{j}\omega))\\
&\in \rho\left(G_0^{1/2}Q_1(\mathrm{j}\omega)G_0^{1/2}\right)\text{Co}\left(0 \cup \left\{G_i^M\frac{(\mathrm{j}\omega+\alpha)e^{-\mathrm{j}D_i\omega}}{(\mathrm{j}\omega)^2},\ i \in \overline{S}\right\}\right)
\end{aligned}$$

holds for all $\omega \in [0, \infty)$.

Now, we see that the proof of the theorem has been converted to the problem of the scalability of the second-order time-delayed system of type II

$$G(s) = k\frac{(s+\alpha)e^{-Ds}}{s^2}. \tag{5.56}$$

The problem has been solved in Chapter 3 (§3.4.2) by Theorem 3.32. According to this theorem,

$$\rho\left(G_0^{1/2}Q_1(j\omega)G_0^{1/2}\right)\mathrm{Co}\left(0\cup\left\{G_i^M\frac{(j\omega+\alpha)e^{-jD_i\omega}}{(j\omega)^2},\ i\in\overline{S}\right\}\right)$$

does not contain the point $(-1, j0)$ when $\rho\left(G_0^{1/2}Q_1(j\omega)G_0^{1/2}\right) < 1$. And hence, the eigenloci $\lambda(G_0^{1/2}T(j\omega)G_0^{1/2}Q_1(j\omega))$ do not enclose the point $(-1, j0)$ for all $\omega\in[0,\infty)$. The theorem is thus proved. □

Remark 1. The clockwise property of the frequency response function of a time-delayed congestion control system plays a key role in obtaining scalable stability criteria such as Johari and Tan's conjecture as well as Theorem 5.2. Unfortunately, when a congestion control algorithm has a second-order dynamical law such as the REM algorithm, this clockwise property does not hold in the whole frequency band. Theorem 5.8, together with Theorem 3.32, indicates that if the network parameters satisfies Assumption 5.7, the clockwise property still holds in a certain frequency interval, and a scalable stability criteria can still be established for the REM algorithm.

Remark 2. The introduction of the diagonal matrix H can balance the control gains among link nodes and make the stability condition less conservative.

Remark 3. It can be shown that $(G_i^M)^{-1} = |G_i(j\omega_c)|$ increases while D_i increases. So, roughly speaking, $(G_i^M)^{-1}$ represents the size of the round-trip delay constant D_i. Therefore, the inequality (5.51) or (5.52) given by Theorem 5.8 can be regarded as a specification of the relationship between the control gain at a link and the weighted sum of the round-trip delay constants of the sources using the link. So the inequality (5.51) or (5.52) is useful for determining the control gain of each link if the round-trip delay constants are known.

The next theorem gives a dual stability condition for REM algorithm. The inequality provided by this theorem can be used for evaluating the maximal admissible round-trip delay constant of each source when all the control gains at links are already fixed.

Theorem 5.9 *Under Assumption 5.7, the system* $\sum_{\mathrm{rem}}$ *is locally stable about the equilibrium point* $(b^\star, p^\star)$, *if there exists a diagonal positive real matrix* $H = \mathrm{diag}\{h_i > 0, i\in\overline{S}\}$ *such that*

$$-f_i'(G_i^M)^{-1}\sum_{l\in L_i}R_{li}\gamma_l\sum_{r\in S_l}\frac{h_i}{h_r}R_{lr} < 1 \tag{5.57}$$

holds for all $i \in \overline{R}$. In particular, when H is taken as the identity matrix, the stability condition reduces to

$$-f_i'(G_i^M)^{-1}\sum_{l\in L_i} R_{li}\gamma_l \sum_{r\in S_l} R_{lr} < 1, \quad \forall i \in \overline{S}. \tag{5.58}$$

Proof. The spectral radius of any matrix is also bounded by its maximum absolute row sum. So, from (5.57) one can also get $\rho\left(G^{1/2}Q_1(\mathrm{j}\omega)G^{1/2}\right) < 1$. The rest of the proof is similar to the proof of Theorem 5.8. □

5.2.2 Dual Algorithm: the First-Order Limit Form of REM

From Theorem 5.8 or Theorem 5.9 it is easy to derive stability conditions for the first-order dual algorithm. If the control parameters α_l, $l \in \overline{L}$ in equation (4.23) are all set to be zero, the second-order dual algorithm reduces to the first-order one. In this case we note that $A_i = 0$, $\forall i \in \overline{S}$ and, therefore, Assumption 5.7 is automatically satisfied for any delay constants D_i. In this case, the open-loop transfer function for source i becomes

$$W_i(s) = \frac{(s+\alpha)\mathrm{e}^{-D_i s}}{s^2} = \frac{\mathrm{e}^{-D_i s}}{s}. \tag{5.59}$$

And it is easy to know that the gain margin for this transfer function is

$$G_i^M = \frac{\pi}{2D_i}.$$

Therefore, we get the following results just as corollaries of Theorem 5.8 and Theorem 5.9 for the first-order dual algorithm.

Theorem 5.10 *Suppose $\alpha = 0$. Then the system $\sum_{\text{rem}}$ is locally stable about the equilibrium point, if there exists a diagonal positive real matrix $H = \operatorname{diag}\{h_l, h_l > 0, l \in \overline{L}\}$ such that*

$$-\gamma_l \sum_{i\in S_l} R_{li} f_i' D_i \sum_{r\in L_i} \frac{h_l}{h_r} R_{ri} < \pi/2 \tag{5.60}$$

holds for all $l \in \overline{L}$. In particular, when H is taken as the identity matrix, the stability condition reduces to

$$-\gamma_l \sum_{i\in S_l} R_{li} f_i' D_i \sum_{r\in L_i} R_{ri} < \pi/2, \quad \forall l \in \overline{L}. \tag{5.61}$$

Theorem 5.11 *Suppose $\alpha = 0$. Then the system $\sum_{\text{rem}}$ is locally stable about the equilibrium point, if there exists a diagonal positive real matrix $H = \operatorname{diag}\{h_r, h_r > 0, r \in \bar{S}\}$ such that*

$$-f_i' D_i \sum_{l\in L_i} R_{li}\gamma_l \sum_{r\in S_l} \frac{h_i}{h_r} R_{lr} < \pi/2 \tag{5.62}$$

holds for all $i \in \overline{S}$. In particular, when H is taken as the identity matrix, the stability condition reduces to

$$-f_i' D_i \sum_{l \in L_i} R_{li} \gamma_l \sum_{r \in S_l} R_{lr} < \pi/2, \quad \forall i \in \overline{S}. \tag{5.63}$$

Remark. In the REM algorithm, there are mainly two kinds of parameters to be designed, namely, control gains γ_l and parameter α. Note that $1/\alpha$ can be viewed as the time constant of differentiation in the open-loop transfer function (5.56). While the relationship between the control gains and the round-trip delay constants is basically specified by condition (5.51) or (5.57), as explained in Remark 3 on Theorem 5.8, the relationship between the time constant of differentiation $1/\alpha$ and the round-trip delay constants is basically described by Assumption 5.7. The assumption implies that when the maximal round-trip delay in the network is large, the time constant of differentiation $1/\alpha$ should be also large (consequently, α should be small) so that delay effect can be overcome by the prediction effect of the differentiation operator. From (5.59) it is easy to see that when α is very small and approaches zero, the dynamics of REM algorithm is very similar to those of the first-order AQM algorithm, the stability of which is determined only by the scalable gain condition (5.60) or (5.62). However, this does not imply that the steady performance (equilibrium property) of the REM algorithm is also similar to that of the first-order AQM algorithm. Even in the case when α approaches zero (but not equal to zero), the REM algorithm still has the merit that the link backlogs are small at the equilibrium point.

5.2.3 Design of Parameters of REM

Now, we propose a simple scheme for designing the parameters of the REM algorithm with the purpose of guaranteeing the local asymptotic stability.

Let us illustrate the design procedure using a network with three links shared by four sources, which is shown in Figure 5.5. The routing matrix of this network is given by

$$R = \begin{bmatrix} 1 & 0 & 1 & 0 \\ 0 & 1 & 1 & 0 \\ 1 & 0 & 0 & 1 \end{bmatrix}.$$

Let the round-trip delays associated with the sources be $D_1 = 1.1(s)$, $D_2 = 1.0(s)$, $D_3 = 0.8(s)$, $D_4 = 0.7(s)$. According to (5.50), we know that $\hat{i} = 1$ in this network. Our purpose is to design the control parameter α and the control gains γ_l, $l = 1, 2, 3$, to guarantee the local asymptotic stability of the REM algorithm. Based on Theorem 5.8, the design procedure can be sketched as follows.

Step 1.
Draw a hyperbola defined by the equation $D\alpha = \bar{A} = 0.4495$ in the $\alpha - D$ plane as shown by Figure 5.6. By the condition (A1) of Assumption 5.7 we know that for each source $r \in \{1, 2, 3, 4\}$ the point (α, D_r) should be located beneath this hyperbola. Under the known

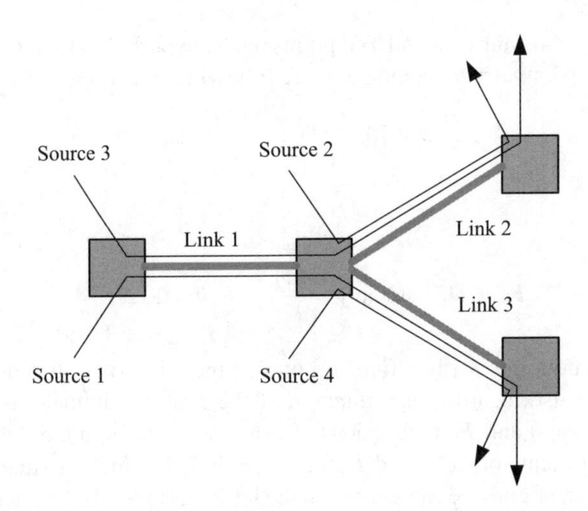

Figure 5.5 A network controlled by REM.

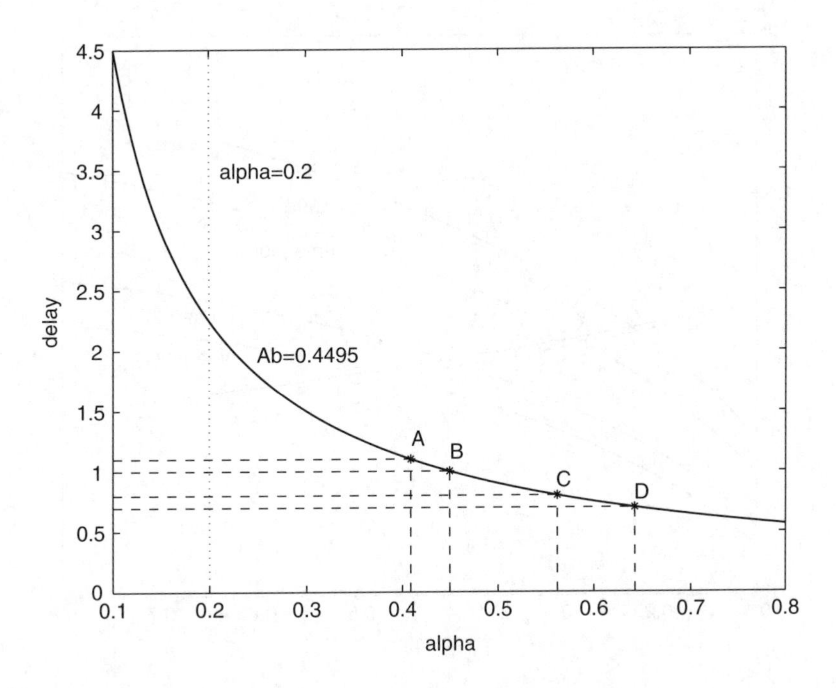

Figure 5.6 Design of parameters by Assumption 5.7 (A1). (Reproduced with permission from Tian Y.-P., "Stability analysis and design of the second-order congestion control for networks with heterogeneous delays," *IEEE/ACM Transactions on Networking*, **13**, 5, 1082–1093, 2005. © 2005 IEEE.)

round-trip delay constants, we find four points on the hyperbola, A, B, C, D, which suggest that the control parameter α should be in the following interval

$$\alpha \in [0, \bar{A}/D_1] \approx [0, 0.4]. \tag{5.64}$$

Step 2.
Denote

$$F_i = D_1^{-1} \arctan\left(\frac{\omega_0(i)}{\alpha}\right), \quad i = 1, 2, 3, 4.$$

By (5.49) we know that ω_0 is a function of A. Since the round-trip delay D_i is already given, it can be also regarded as a function of the control parameter α. So, we draw the curves of both $\omega_0(i)$ and F_i as functions of α as shown in Figure 5.7. Then, we can find the intersection points of $\omega_0(i)$ and F_i for $i = 1, 2, 3, 4$, which are marked as W, X, Y, Z respectively. In the figure we can see that in the left parts towards the intersection points the condition (A2) of Assumption 5.7 is satisfied. Therefore, another constraint on the control parameter can be determined graphically as

$$\alpha \in [0, 0.22]. \tag{5.65}$$

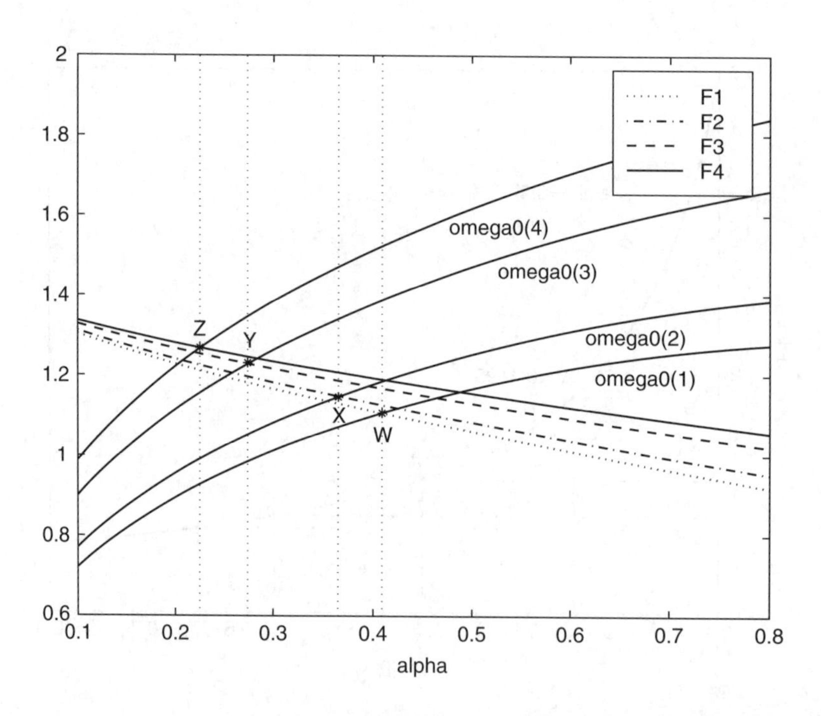

Figure 5.7 Design of parameters by Assumption 5.7 (A2). (Reproduced with permission from Tian Y.-P., "Stability analysis and design of the second-order congestion control for networks with heterogeneous delays," *IEEE/ACM Transactions on Networking*, **13**, 5, 1082–1093, 2005. © 2005 IEEE.)

Now, we can choose a control parameter in the intersection of the two intervals given by (5.64) and (5.65) as

$$\alpha = 0.2.$$

Step 3.
Draw the Nyquist plots of the time-delayed transfer functions

$$G_i(s) = \frac{(s+\alpha)e^{-D_i s}}{s^2}, \quad i = 1, 2, 3, 4$$

in Figure 5.8. Then, we can obtain

$$\begin{bmatrix} (G_1^M)^{-1} \\ (G_2^M)^{-1} \\ (G_3^M)^{-1} \\ (G_4^M)^{-1} \end{bmatrix} \approx \begin{bmatrix} 0.82 \\ 0.75 \\ 0.65 \\ 0.50 \end{bmatrix}.$$

Assume the link capacities are $l_1 = 2$(Mbps), $l_2 = 3$(Mbps), $l_3 = 2$(Mbps). The utility function of sources is chosen as $U_r(q_r) = w_r \log q_r$, where for each source the parameter w_r is set as $w_1 = 0.67$, $w_2 = 2.575$, $w_3 = 2.109$, $w_4 = 1.33$. Then, one can get source rates and aggregate prices at the equilibrium as $x_1^\star = 0.685(Mbps)$, $x_2^\star = 1.685(Mbps)$,

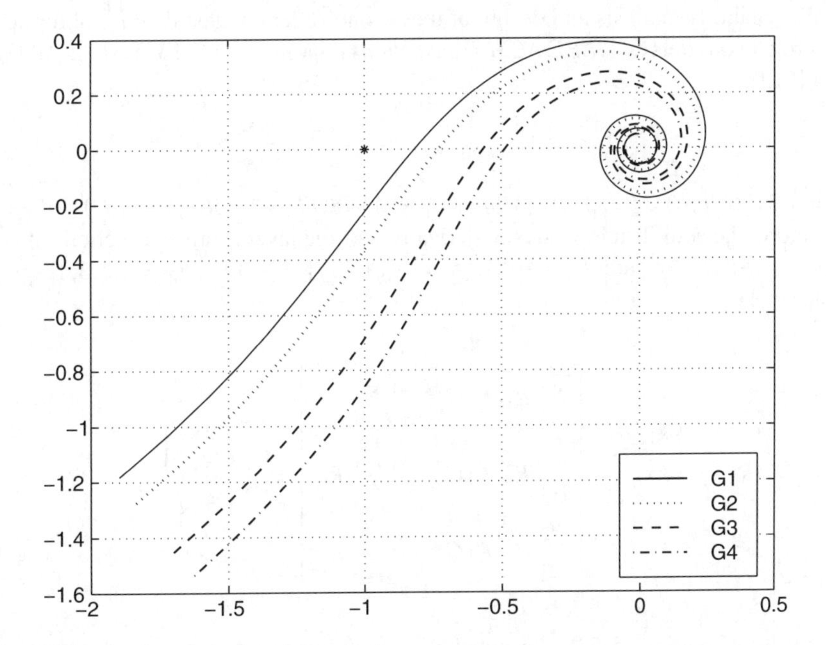

Figure 5.8 The Nyquist plots of round-trip dynamics. (Reproduced with permission from Tian Y.-P., "Stability analysis and design of the second-order congestion control for networks with heterogeneous delays," *IEEE/ACM Transactions on Networking*, **13**, 5, 1082–1093, 2005. © 2005 IEEE.)

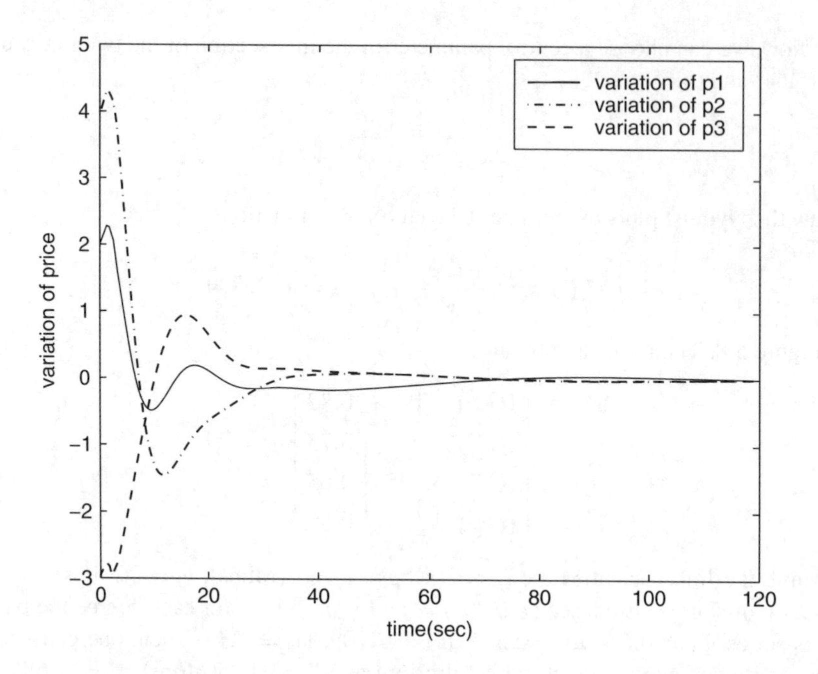

Figure 5.9 Variation of prices around the equilibrium. (Reproduced with permission from Tian Y.-P., "Stability analysis and design of the second-order congestion control for networks with heterogeneous delays," *IEEE/ACM Transactions on Networking*, **13**, 5, 1082–1093, 2005. © 2005 IEEE.)

$x_3^\star = 1.315(Mbps), x_4^\star = 1.315(Mbps), q_1^\star = 0.978, q_2^\star = 1.636, q_3^\star = 1.604, q_4^\star = 1.011$. Therefore, the equilibrium values of derivatives of the inverse utility functions of sources, $f_i'(q_i^\star) = -w_i/(q_i^\star)^2$, are $f_1' = -0.7$, $f_2' = -1.03$, $f_3' = -0.82$, $f_4' = -1.3$, respectively. Hence, we get

$$\begin{bmatrix} k_1 \\ k_2 \\ k_3 \end{bmatrix} = \begin{bmatrix} \sum_{i=1}^{4} R_{1i} f_i'(G_i^M)^{-1} \sum_{r=1}^{3} R_{ri} \\ \sum_{i=1}^{4} R_{2i} f_i'(G_i^M)^{-1} \sum_{r=1}^{3} R_{ri} \\ \sum_{i=1}^{4} R_{3i} f_i'(G_i^M)^{-1} \sum_{r=1}^{3} R_{ri} \end{bmatrix} = \begin{bmatrix} 2.214 \\ 1.839 \\ 1.798 \end{bmatrix}.$$

So, by the condition (5.52), we get the intervals of admissible control gains for links guaranteeing the local asymptotic stability of the REM algorithm as

$$\gamma_1 \in [0, k_1^{-1}] = [0, 0.45];$$

$$\gamma_2 \in [0, k_2^{-1}] = [0, 0.54];$$
$$\gamma_3 \in [0, k_3^{-1}] = [0, 0.55].$$

Figure 5.9 shows the simulation result of the variation of prices at three links under the choice of $\gamma_1 = \gamma_2 = \gamma_3 = 0.3$.

It is interesting to point out that, unlike the first-order control algorithm, condition (5.51) or condition (5.52) is not sufficient for the local stability of the second-order control algorithm. This can be shown by the following counter-example. We assume the round-trip delay D_4 in the above setting is changed from 0.7(s) to 0.02(s), and the parameters w_i are now set as $w_1 = 0.24$, $w_2 = 1.93$, $w_3 = 3.083$, $w_4 = 0.022$. All the other parameters including the control gains are the same as given previously. Then, one can get the source rates and the aggregate prices at the equilibrium as $x_1^\star = 0.410$(Mbps), $x_2^\star = 1.410$(Mbps), $x_3^\star = 1.590$(Mbps), $x_4^\star = 1.590$(Mbps), $q_1^\star = 0.585$, $q_2^\star = 1.369$, $q_3^\star = 1.939$, $q_4^\star = 0.014$. Therefore, the equilibrium values of derivatives of the inverse utility functions of the sources, $f_i'(q_i^\star) = -w_i/(q_i^\star)^2$, are $f_1' = -0.7$, $f_2' = -1.03$, $f_3' = -0.82$, $f_4' = -115$, respectively. It is easy to verify that condition (5.52) still holds for this new setting. But the variation of the price at link 3 goes to infinity as shown in Figure 5.10. The reason for this is perhaps that the condition (A2) of Assumption 5.7 does not hold any more for $\omega_0(4)$ in this case (see Figure 5.11).

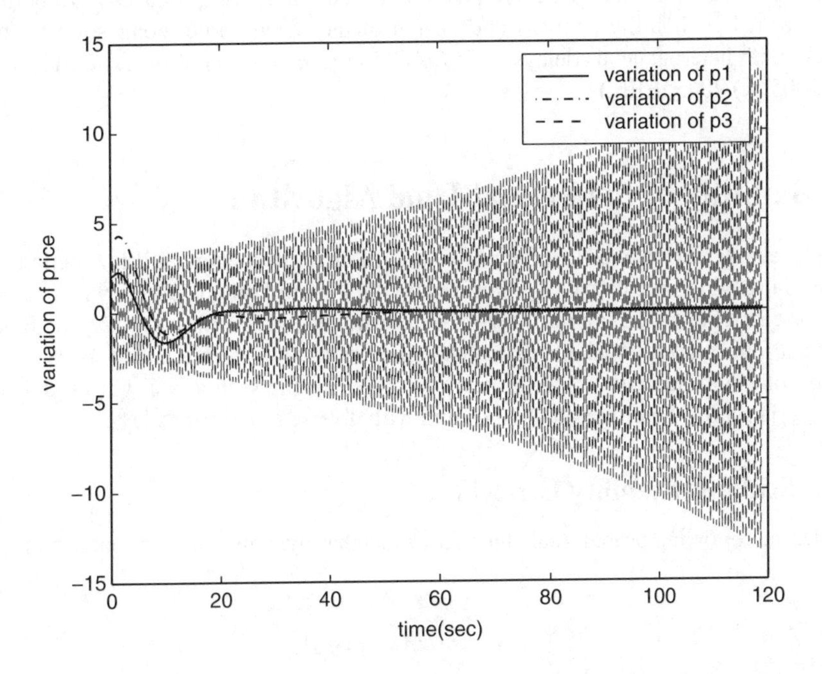

Figure 5.10 Counter-example: Variation of prices around the equilibrium. (Reproduced with permission from Tian Y.-P., "Stability analysis and design of the second-order congestion control for networks with heterogeneous delays," *IEEE/ACM Transactions on Networking*, **13**, 5, 1082–1093, 2005. © 2005 IEEE.)

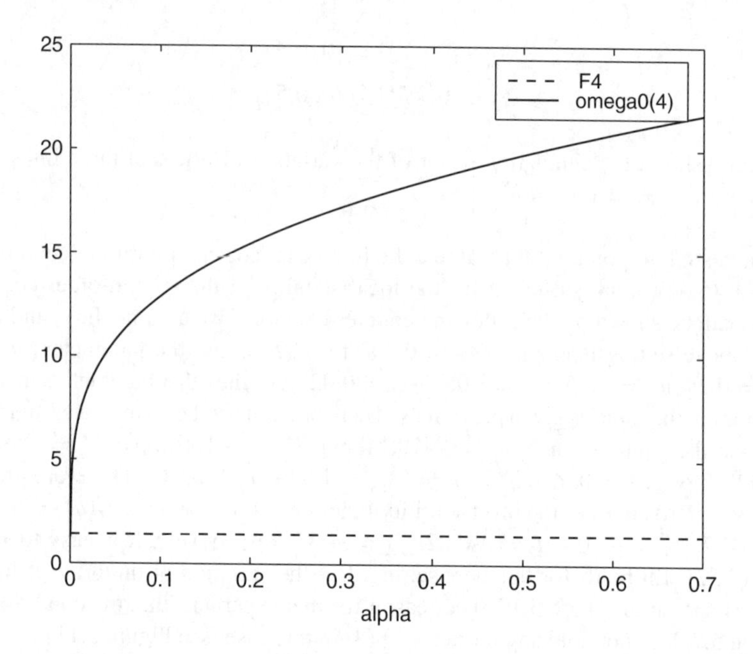

Figure 5.11 Counter-example: Prerequisite 2 does not hold. (Reproduced with permission from Tian Y.-P., "Stability analysis and design of the second-order congestion control for networks with heterogeneous delays," *IEEE/ACM Transactions on Networking*, **13**, 5, 1082–1093, 2005. © 2005 IEEE.)

5.3 Stability of the Primal-Dual Algorithm

In the last chapter we showed that the primal algorithm allows general utility functions, hence arbitrary fairness in rate allocation, but gives up tight control on utilization, whereas the dual algorithm can achieve very high utilization but is restricted to a specific class of utility functions. The primal-dual algorithm adopts dynamical adaptations both at links and at sources, and can achieve both high utilization and arbitrary fairness. In this section we will establish some stability criteria for the primal-dual algorithm with diverse round-trip delays.

5.3.1 Scalable Stability Criteria

Consider the following primal-dual algorithm which has dynamics at both source end and link end:

$$\dot{x}_i(t) = \left(k_i(U_i'(x_i) - q_i)\right)_{x_i}^+, \tag{5.66}$$

$$\dot{p}_l(t) = (\gamma_l(y_l(t) - \bar{c}_l))_{p_l}^+, \tag{5.67}$$

where $x_i(t) \in \mathbb{R}^+$, $k_i > 0$ are the transmission rate and control gain at source i, $i \in \overline{S}$, respectively; $p_l(t) \in \mathbb{R}^+$, $\gamma_l > 0$ are the price (congestion measurement) and control gain at link l,

$l \in \overline{L}$, respectively; $\bar{c}_l \le c_l$ is the desired aggregate equilibrium rate on link l; and $U_i'(x_i)$ is the derivative of $U_i(x_i)$, a utility function.

Similar to any other congestion control algorithm with propagation delays, we have equations which describe the interconnection of the sources and links:

$$y_l(t) = \sum_{i \in S_l} x_i(t - d_{li}^{\rightarrow}), \tag{5.68}$$

$$q_i(t) = \sum_{l \in L_i} p_l(t - d_{li}^{\leftarrow}), \tag{5.69}$$

where the forward delay $d_{li}^{\rightarrow}$ and backward $d_{li}^{\leftarrow}$ are subject to the well known assumption (4.27), which says all the links associated with the same route have the same round-trip delay.

Equations (5.66)–(5.69) constitute a dynamical feedback system with a unique equilibrium $(x^\star, p^\star)$ satisfying

$$\begin{aligned} \sum_{i \in S_l} x_i^\star &= \bar{c}_l, \\ \sum_{l \in L_i} p_l^\star &= U_i'(x_i^\star). \end{aligned} \tag{5.70}$$

For convenience in the following discussions, we denote

$$q_i^\star = \sum_{l \in L_i} p_l^\star,$$

$$y_l^\star = \sum_{i \in S_l} x_i^\star,$$

as the aggregate price of all links used by source i and the aggregate rate of all sources which use link l at the equilibrium, respectively. The vectors $x^\star, q^\star \in \mathbb{R}^S$, $y^\star, p^\star \in \mathbb{R}^L$ are defined by the above components across the set of sources or links in an obvious way.

Next, we study the local stability of the equilibrium $(x^\star, p^\star)$ in the presence of delays. In a neighborhood of $(x^\star, p^\star)$, the system (5.66)–(5.67) is differentiable and simply governed by

$$\dot{x}_i(t) = k_i(U_i'(x_i) - q_i), \; i \in \overline{S}, \tag{5.71}$$

$$\dot{p}_l(t) = \gamma_l(y_l(t) - \bar{c}_l), \; l \in \overline{L}. \tag{5.72}$$

For simplicity of statement, we call the system consisting of equations (5.68), (5.69), (5.71) and (5.72) "system $\sum_{\text{pd}}$".

Define

$$\begin{aligned} \hat{p}(t) &= p(t) - p^\star, \\ \hat{q}(t) &= q(t) - q^\star, \\ \hat{y}(t) &= y(t) - y^\star, \\ \hat{x}(t) &= x(t) - x^\star. \end{aligned}$$

Then, we can linearize system $\sum_{\text{pd}}$ about the equilibrium point and get

$$\begin{cases} \dot{\hat{x}}_i(t) = k_i(U_i''(x_i^\star)\hat{x}_i(t) - \hat{q}_i(t)), \\ \dot{\hat{p}}_l(t) = \gamma_l \hat{y}_l(t), \\ \hat{y}_l(t) = \sum_{i \in S_l} \hat{x}_i(t - d_{li}^{\rightarrow}), \\ \hat{q}_i(t) = \sum_{l \in L_i} \hat{p}_l(t - d_{li}^{\leftarrow}), \end{cases} \tag{5.73}$$

where $U_i''(x_i^\star) < 0$ is the derivative of $U_i'(x_i)$ at $x_i^\star$.

Remark. In last chapter, we considered a generalized form of the primal-dual algorithm where the control gains k_i in (5.66) a non-decreasing function of the transmission rate $k_i(x_i)$. In that case the linearized equations about the equilibrium point are

$$\begin{cases} \dot{\hat{x}}_i(t) = (k_i(x_i^\star)U_i''(x_i^\star) + k_i'(x_i^\star)U_i'(x_i^\star) - q_i^\star k_i'(x_i^\star))\hat{x}_i(t) - k_i(x_i^\star)\hat{q}_i(t), \\ \dot{\hat{p}}_l(t) = \gamma_l \hat{y}_l(t), \\ \hat{y}_l(t) = \sum_{i \in S_l} \hat{x}_i(t - d_{li}^{\rightarrow}), \\ \hat{q}_i(t) = \sum_{l \in L_i} \hat{p}_l(t - d_{li}^{\leftarrow}), \end{cases} \tag{5.74}$$

which have the same form as (5.73). So, the results given in this section also apply to the generalized form of the primal-dual algorithm.

Let $X_i(s) = \mathcal{L}(\hat{x}_i(t))$, the Laplace transformation of $\hat{x}_i(t)$. Then, the closed-loop system (5.73) can be represented in the Laplace domain as follows:

$$X_i(s) = -\frac{k_i}{s(s - k_i U_i''(x_i^\star))} \sum_{l \in L_i} \mathrm{e}^{-sd_{li}^{\leftarrow}} \gamma_l \sum_{r \in S_l} \mathrm{e}^{-sd_{lr}^{\rightarrow}} X_r(s). \tag{5.75}$$

Denote

$$X(s) = (X_1(s), \ldots, X_S(s))^{\mathrm{T}}, \tag{5.76}$$

$$D(s) = \mathrm{diag}\left\{ \frac{k_i \mathrm{e}^{-D_i s}}{s(s + \alpha_i)}, \quad i \in \overline{S} \right\}, \tag{5.77}$$

$$\Gamma = \mathrm{diag}\{\gamma_l, \quad l \in \overline{L}\}, \tag{5.78}$$

where

$$\alpha_i = -k_i U_i''(x_i^\star). \tag{5.79}$$

Then, by using the relationship between the time-delayed forward and backward routing matrices (4.72), equation (5.75) can be rewritten in the following matrix form:

$$X(s) = -D(s)R_f^{\mathrm{T}}(-s)\Gamma R_f(s)X(s), \tag{5.80}$$

where $R_f(s)$ is the time-delayed forward routing matrix. The characteristic equation of the above system is

$$\det\left(I + D(s)R_f^{\mathrm{T}}(-s)\Gamma R_f(s)\right) = 0. \tag{5.81}$$

Now, we see that system $\sum_{\mathrm{pd}}$ is locally stable, if all the roots of the characteristic equation have negative real parts.

Denote by G_i^M the gain margin of the transfer function

$$W_i(s) = \frac{\mathrm{e}^{-D_i s}}{s(s+\alpha_i)}, \tag{5.82}$$

which is defined by

$$G_i^M = 1/|W_i(\mathrm{j}\omega_c^i)|, \tag{5.83}$$

where $\omega_c^i > 0$ is the minimal frequency that satisfies the following equation:

$$D_i\omega_c^i = \frac{\pi}{2} - \arctan(\omega_c^i/\alpha_i). \tag{5.84}$$

Note that the locus of $W_i(\mathrm{j}\omega)$ crosses the real axis of the complex plane for the first time when $\omega = \omega_c^i$, which is the minimal crossing frequency of $W_i(\mathrm{j}\omega)$.

From (5.84) we have

$$\frac{\alpha_i}{\omega_c^i} = \tan(D_i\omega_c^i).$$

Using this equality, we get

$$\begin{aligned}(G_i^M)^{-1} &= \frac{1}{\omega_c^i\sqrt{(\omega_c^i)^2 + (\omega_c^i)^2\tan^2(D_i\omega_c^i)}}\\ &= \frac{D_i\sin(D_i\omega_c^i)}{\alpha_i D_i\omega_c^i}\\ &< \frac{D_i}{\alpha_i}.\end{aligned} \tag{5.85}$$

The following theorem gives some sufficient conditions for the asymptotic stability of the primal-dual algorithm.

Theorem 5.12 *Suppose that the following precondition holds for the primal-dual algorithm:*

$$(D_i\alpha_i - D_j\alpha_j)(\omega_c^i - \omega_c^j) \geq 0, \quad \forall i, j \in \overline{S},\ i \neq j. \tag{5.86}$$

Then system $\sum_{\rm pd}$ is locally stable about the equilibrium point $(x^\star, p^\star)$, if there exists a diagonal positive real matrix $H = \operatorname{diag}\{h_i > 0,\ i \in \overline{S}\}$ such that

$$k_i(G_i^M)^{-1}\sum_{l\in L_i} R_{li}\gamma_l \sum_{r\in S_l}\frac{h_i}{h_r}R_{lr} < 1, \quad \forall i \in \overline{S}; \tag{5.87}$$

or, if there exists a diagonal positive real matrix $H' = \operatorname{diag}\{h'_l > 0,\ l \in \overline{L}\}$ such that

$$\gamma_l \sum_{i\in S_l} R_{li}k_i(G_i^M)^{-1}\sum_{r\in L_i}\frac{h'_l}{h'_r}R_{ri} < 1, \quad \forall l \in \overline{L}. \tag{5.88}$$

Remark 1. Inequality (5.87) or (5.88) is a scalable stability condition, which does not require knowledge of the entire network. A precondition for the validity of such a scalable condition is that (5.86) holds. Therefore, equation (5.86) can be regarded as a scalability condition of the primal-dual algorithm.

Remark 2. As we have remarked before, the introduction of the diagonal matrix H (or H') can balance the control gains among source (or link) nodes and make the stability conditions less conservative. If H is taken as the identity matrix, condition (5.87) reduces to

$$k_i(G_i^M)^{-1}\sum_{l\in L_i} R_{li}\gamma_l \sum_{r\in S_l} R_{lr} < 1,\ \forall i \in \overline{S}. \tag{5.89}$$

Using (5.79) and (5.85), we can further get a more explicit sufficient condition than (5.87) as

$$\frac{D_i}{-U_i''(x_i^\star)}\sum_{l\in L_i} R_{li}\gamma_l \sum_{r\in S_l} R_{lr} < 1,\ \forall i \in \overline{S}. \tag{5.90}$$

Similarly, we can get an explicit sufficient condition, simpler than (5.88), as

$$\gamma_l \sum_{i\in S_l} R_{li}\left(-\frac{D_i}{U_i''(x_i^\star)}\right)\sum_{r\in L_i} R_{ri} < 1,\ \forall l \in \overline{L}. \tag{5.91}$$

A sufficient condition for the scalability condition (5.86), which is checked only by round-trip delays and the algorithm parameter α, is given by the following proposition.

Proposition 5.13 *Condition (5.86) holds if*

$$(D_i\alpha_i - D_j\alpha_j)(D_i - D_j) \le 0, \quad \forall i, j \in \overline{1, n},\ i \ne j. \tag{5.92}$$

Proof. Suppose (5.92) holds for $i, j = 1, 2$. Without loss of generality, we assume $D_1\alpha_1 - D_2\alpha_2 \le 0$. Then, (5.92) implies that $D_1 \ge D_2$ and $\alpha_1 \le \alpha_2$. To show that (5.86) holds, we only need to show $\omega_c^1 \le \omega_c^2$, or equivalently,

$$\arg W_2(\mathrm{j}\omega_c^1) \ge -\pi. \tag{5.93}$$

Indeed,

$$
\begin{aligned}
\arg W_2(\mathrm{j}\omega_c^1) &= -D_2\omega_c^1 - \arctan(\omega_c^1/\alpha_2) - \pi/2 \\
&= -D_2\omega_c^1 - \arctan(\omega_c^1/\alpha_2) - \pi/2 + D_1\omega_c^1 + \arctan(\omega_c^1/\alpha_1) - \pi/2 \\
&\quad \text{((5.84) is used at this stage)} \\
&= (D_1 - D_2)\omega_c^1 + \arctan(\omega_c^1/\alpha_1) - \arctan(\omega_c^1/\alpha_2) - \pi \\
&\geq -\pi.
\end{aligned}
$$

The last inequality holds because $\arctan(\omega_c^1/\alpha_1) - \arctan(\omega_c^1/\alpha_2) \geq 0$ when $\alpha_1 \leq \alpha_2$. □

Now, we give a simple example to illustrate the design of the parameter α and control gains in the primal-dual algorithm with the help of the scalability and stability conditions.

Example 5.14 *Design of parameters of the primal-dual algorithm.*

Consider the simple network with two links and three sources, which is illustrated by Figure 5.12. Source 2's route includes link 1 and link 2, source 1's route consists of link 1 and source 3's route consists of link 2. Obviously, the routing matrix of the network is

$$
R = \begin{bmatrix} 1 & 1 & 0 \\ 0 & 1 & 1 \end{bmatrix}.
$$

Let the round-trip delays associated with the sources be $D_1 = 1$(s), $D_2 = 0.7$(s), $D_3 = 0.4$(s). Assume that the link capacities are $c_1 = 3$(Mbps), $c_2 = 2$(Mbps), and the desired aggregate rates on links are equal to the link capacities, i.e., $\bar{c}_1 = c_1$, $\bar{c}_2 = c_2$. The utility function is $U_i(x_i) = w_i \log x_i$, where the parameter w_i is set to unity for each source. Then, from (5.70) it is easy to get the equilibrium point of the network as

$$
x^\star = [2.22, 0.78, 1.22]^{\mathrm{T}}, \quad p^\star = [0.45, 0.82]^{\mathrm{T}}.
$$

The control gains of sources and links can be designed as follows.

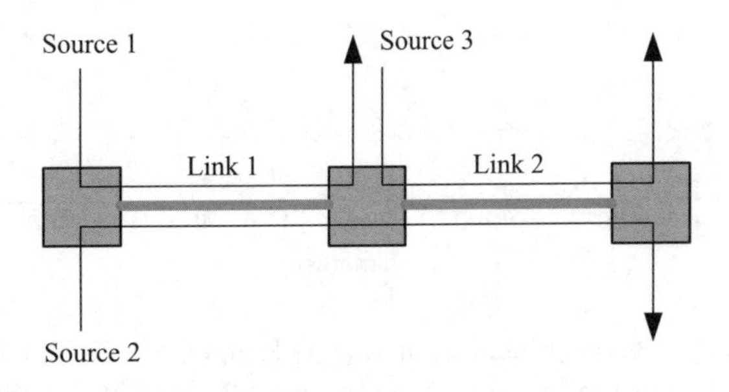

Figure 5.12 Network with two links and three sources.

(1) Choose the control gains of the sources k_i by the scalability condition (5.92).
Let $D_1\alpha_1 = 0.01$, $D_2\alpha_2 = 0.14$ and $A_3 = D_3\alpha_3 = 0.16$. Then, (5.92) is satisfied. Under this setting, we get $\alpha_1 = 0.01$, $\alpha_2 = 0.02$ and $\alpha_3 = 0.04$. So, from (5.79) it follows that

$$k_1 = 0.0493, \quad k_2 = 0.0122, \quad k_3 = 0.0595.$$

(2) Design the control gains of the links γ_l by the stability condition (5.91).
Under the parameters setting for the network, we have

$$\sum_{i \in S_1} R_{1i}\left(-\frac{D_i}{U_i''(x_i^\star)}\right)\sum_{r \in L_i} R_{ri} = 5.78,$$

$$\sum_{i \in S_2} R_{2i}\left(-\frac{D_i}{U_i''(x_i^\star)}\right)\sum_{r \in L_i} R_{ri} = 1.45.$$

Therefore, by (5.91) we know that the control gains should satisfy the following condition:

$$\gamma_1 \in (0,\ 5.78^{-1}), \quad \gamma_2 \in (0,\ 1.45^{-1}).$$

We take $\gamma_1 = 0.05$, $\gamma_2 = 0.2$. The simulation results (Figure 5.13 and Figure 5.14) show that the source rates and link prices converge to the equilibrium point.

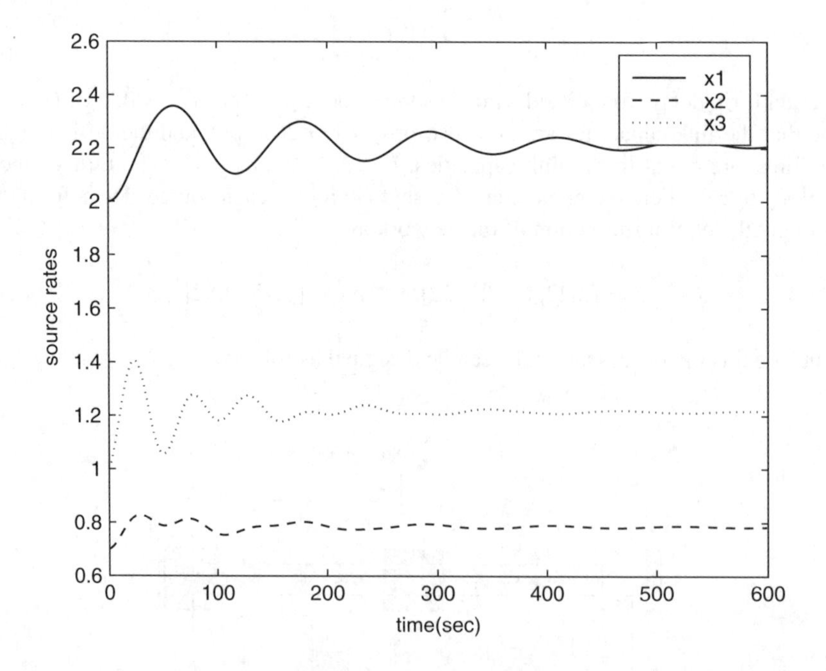

Figure 5.13 Source rates. (Reprinted from Tian Y.-P. and Chen G., "Stability of the primal-dual algorithm for congestion control," *International Journal of Control*, **79**, 6, 662–676, © 2006, by permission of Taylor & Francis Ltd, http://www.tandf.co.uk/journals.)

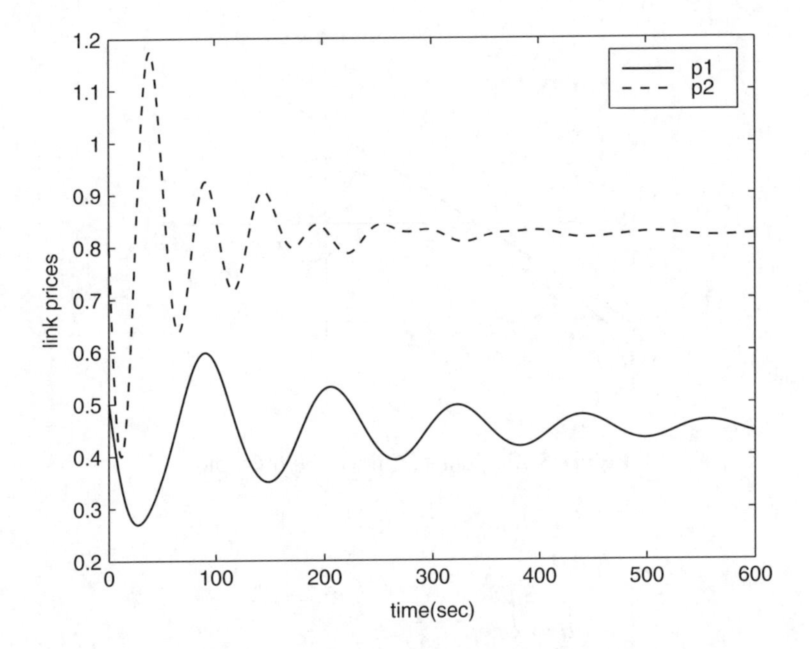

Figure 5.14 Link prices. (Reprinted from Tian Y.-P. and Chen G., "Stability of the primal-dual algorithm for congestion control," *International Journal of Control*, **79**, 6, 662–676, © 2006, by permission of Taylor & Francis Ltd, http://www.tandf.co.uk/journals.)

The above example shows that the scalability requirement and stability condition can usually be met by adjusting the control gains at the sources. When this is not the case, to maintain the local asymptotic stability we can strengthen condition (5.87) as described in the rest of this section.

Let

$$A_i = D_i \alpha_i. \tag{5.94}$$

Denote by $\breve{i}$ the source with minimal value of parameter A_i, i.e.,

$$\breve{i} = \arg\min_{i \in \bar{s}} A_i. \tag{5.95}$$

Let $T_c(\breve{i})$ be the slope of the tangent line to the Nyquist plot of $W_{\breve{i}}(\mathrm{j}\omega)$ at the intersection point C, as illustrated by Figure 5.15. It is easy to see that $T_c(\breve{i}) = |OE|/|EC|$. An analytic formula of $T_c(\breve{i})$ is given by

$$T_c(\breve{i}) = \left. \frac{\mathrm{Im} G'_{\breve{i}}(\omega)}{\mathrm{Re} G'_{\breve{i}}(\omega)} \right|_{\omega = \omega_c(\breve{i})}. \tag{5.96}$$

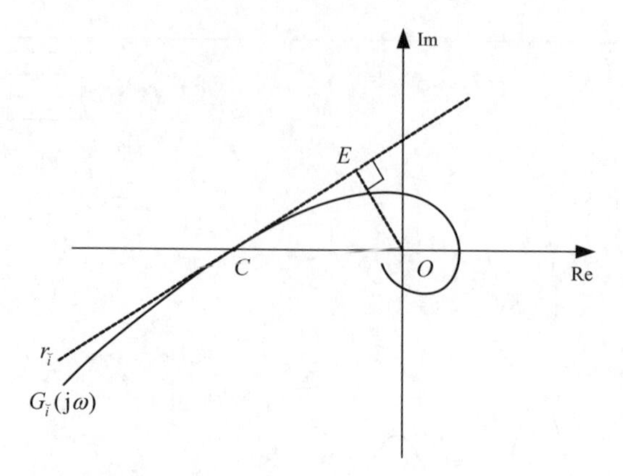

Figure 5.15 Tangent line of Nyquist plot.

Define

$$\mu_i = \begin{cases} 1, & \text{if } i = \check{i}; \\ \sqrt{\frac{(Tc(\check{i}))^2}{1+(Tc(\check{i}))^2}}, & \text{otherwise.} \end{cases} \tag{5.97}$$

Now, we are ready to give an alternative theorem for the local asymptotic stability of the primal-dual algorithm.

Theorem 5.15 *System* $\sum_{\text{pd}}$ *is locally stable about the equilibrium point* $(x^\star, p^\star)$, *if there exists a diagonal positive real matrix* $H = \text{diag}\{h_i > 0, \quad i \in \overline{S}\}$ *such that*

$$k_i(G_i^M)^{-1} \sum_{l \in L_i} R_{li}\gamma_l \sum_{r \in S_i} \frac{h_i}{h_r} R_{lr} < \mu_i, \quad \forall i \in \overline{S}; \tag{5.98}$$

or, if there exists a diagonal positive real matrix $H' = \text{diag}\{h'_l > 0, \quad l \in \overline{L}\}$ *such that*

$$\gamma_l \sum_{i \in S_l} R_{li} k_i (G_i^M \mu_i)^{-1} \sum_{r \in L_i} \frac{h'_l}{h'_r} R_{ri} < 1, \quad \forall l \in \overline{L}. \tag{5.99}$$

Remark. Using (5.79) and (5.85), we can get some more explicit sufficient conditions from (5.98) and (5.99) as

$$\frac{D_i}{-U_i''(x_i^\star)} \sum_{l \in L_i} R_{li}\gamma_l \sum_{r \in S_l} R_{lr} < \mu_i, \quad \forall i \in \overline{S}, \tag{5.100}$$

and

$$\gamma_l \sum_{i \in S_l} R_{li} \left(-\frac{D_i}{U_i''(x_i^\star)\mu_i} \right) \sum_{r \in L_i} R_{ri} < 1, \quad \forall l \in \overline{L}, \tag{5.101}$$

respectively. Compared to (5.90) or (5.91) given by Theorem 5.12, these conditions are sufficient for the stability of the primal-dual algorithm without preconditions. Condition (5.100) is almost scalable in the following sense: although the stability bound (μ_i in (5.100)) should be determined globally by considering all the routes in the network, actually the condition can be met by adjusting control parameters surrounding the sources so that the left-hand side of (5.100) is sufficiently small.

5.3.2 Proof of the Stability Criteria

Now, let us outline the procedure of the proofs of Theorems 5.12 and 5.15. As we show below, to close this proof procedure we need to use the geometric property of the frequency response function of the congestion control algorithm, which has been studied Chapter 3 in detail.

Proof of Theorem 5.12.

First, recall (5.81). According to the generalized Nyquist criterion (Theorem 2.20), all the roots of the characteristic equation (4.79) have negative real parts if and only if the eigenloci of

$$\mathrm{diag}\left\{\frac{k_i \mathrm{e}^{-D_i s}}{s(s+\alpha_i)}, i \in \bar{S}\right\} R_f^T(-s)\Gamma R_f(s),$$

i.e., $\lambda\left(\mathrm{diag}\left\{\frac{k_i \mathrm{e}^{-\mathrm{j}D_i\omega}}{(\mathrm{j}\omega)(\mathrm{j}\omega+\alpha_i)}, i \in \bar{S}\right\} R_f^{\mathrm{T}}(-\mathrm{j}\omega)\Gamma R_f(\mathrm{j}\omega)\right)$, do not enclose the point $(-1, \mathrm{j}0)$ for all $\omega \in [0, \infty)$.

Rewrite matrix

$$\mathrm{diag}\left\{\frac{k_i \mathrm{e}^{-D_i s}}{s(s+\alpha_i)}, i \in \bar{S}\right\} R_f^{\mathrm{T}}(-s)\Gamma R_f(s)$$

as $G^{1/2}T(s)G^{1/2}Q_1(s)$, where

$$T(s) = \mathrm{diag}\left\{G_i^M \frac{\mathrm{e}^{-D_i s}}{s(s+\alpha_i)}, i \in \bar{S}\right\}, \tag{5.102}$$

$$G = \mathrm{diag}\{k_i/G_i^M, i \in \bar{S}\}, \tag{5.103}$$

$$Q_1(s) = R_f^{\mathrm{T}}(-s)\Gamma R_f(s). \tag{5.104}$$

By the definition of $Q_1(s)$, we have

$$\begin{aligned}
&\rho\left(G^{1/2}Q_1(\mathrm{j}\omega)G^{1/2}\right)\\
&= \rho\left(G^{1/2}R_f^{\mathrm{T}}(-s)\Gamma R_f(s)G^{1/2}\right)\\
&= \rho\left(GR_f^{\mathrm{T}}(-s)\Gamma R_f(s)\right)\\
&= \rho\left(HGR_f^{\mathrm{T}}(-s)\Gamma R_f(s)H^{-1}\right),
\end{aligned}$$

where $H = \text{diag}\{h_i > 0, i \in \bar{S}\}$. It is well known that the spectral radius of a matrix is bounded by the maximum absolute value of its row sum. Therefore, by (5.87) we get

$$\begin{aligned}
&\rho\left(G^{1/2}Q_1(\mathrm{j}\omega)G^{1/2}\right)\\
&\leq \max_{i\in\bar{S}}\left\{\left|k_i(G_i^M)^{-1}\left(\sum_{l\in\overline{L}} R_{li}\mathrm{e}^{\mathrm{j}\omega d_1(l,i)}\gamma_l \sum_{r\in\bar{S}} \frac{h_i}{h_r} R_{lr}\mathrm{e}^{-\mathrm{j}\omega d_1(l,r)}\right)\right|\right\}\\
&\leq \max_{i\in\bar{S}}\left\{k_i(G_i^M)^{-1}\left(\sum_{l\in\overline{L}} R_{li}\gamma_l \sum_{r\in\bar{S}} \frac{h_i}{h_r} R_{lr}\right)\right\}\\
&< 1.
\end{aligned}$$

Since

$$\{0\} \cup \sigma(G^{1/2}T(\mathrm{j}\omega)G^{1/2}Q_1(\mathrm{j}\omega)) = \{0\} \cup \sigma(T(\mathrm{j}\omega)G^{1/2}Q_1(\mathrm{j}\omega)G^{1/2}),$$

it follows from Lemma 2.34 that

$$\begin{aligned}
&\lambda(G^{1/2}T(\mathrm{j}\omega)G^{1/2}Q_1(\mathrm{j}\omega))\\
&\in \rho\left(G^{1/2}Q_1(\mathrm{j}\omega)G^{1/2}\right)\text{Co}\left(0 \cup \left\{G_i^M \frac{\mathrm{e}^{-\mathrm{j}D_i\omega}}{\mathrm{j}\omega(\mathrm{j}\omega+\alpha_i)}, i \in \bar{S}\right\}\right)
\end{aligned}$$

holds for all $\omega \in [0, \infty)$.

Now, by Theorem 3.23, under the condition (5.86),

$$\rho\left(G^{1/2}Q_1(\mathrm{j}\omega)G^{1/2}\right)\text{Co}\left(0 \cup \left\{G_l^M \frac{\mathrm{e}^{-\mathrm{j}D_i\omega}}{\mathrm{j}\omega(\mathrm{j}\omega+\alpha_i)}, i \in \bar{S}\right\}\right)$$

does not contain the point $(-1, \mathrm{j}0)$ when $\rho\left(G^{1/2}Q_1(\mathrm{j}\omega)G^{1/2}\right) < 1$. Hence, the eigenloci

$$\lambda(G^{1/2}T(\mathrm{j}\omega)G^{1/2}Q_1(\mathrm{j}\omega))$$

will not enclose the point $(-1, \mathrm{j}0)$ for all $\omega \in [0, \infty)$. Thus, we have proved the sufficiency of (5.87) for the asymptotic stability of system $\sum_{\text{pd}}$ under precondition (5.86).

Note that the spectral radius of a matrix is also bounded by the maximum absolute value of its column sum. So, from (5.88) one also gets $\rho\left(G^{1/2}Q_1(\mathrm{j}\omega)G^{1/2}\right) < 1$. Therefore, the sufficiency of (5.88) for the asymptotic stability of system $\sum_{\text{pd}}$ under precondition (5.86) can be proved in a similar way. Theorem 5.12 is thus proved. □

Proof of Theorem 5.15.

To prove Theorem 5.15, we first need to modify (5.102) and (5.103) as follows:

$$T(s) = \text{diag}\left\{\mu_i G_i^M \frac{\mathrm{e}^{-D_i s}}{s(s+\alpha_i)}, i \in \bar{S}\right\}, \tag{5.105}$$

$$G = \text{diag}\{k_i/(G_i^M \mu_i), i \in \bar{S}\}. \tag{5.106}$$

Then, in a way similar to the proof of Theorem 5.12, by using Theorem 3.24 instead of Theorem 3.23, we can prove Theorem 5.15. □

5.4 Time-Delayed Feedback Control

This section briefly reviews various forms and stability results of time-delayed feedback control (TDFC).

5.4.1 Time-Delayed State as a Reference

We explain the basic idea of the TDFC as it was originally proposed for stabilizing unstable periodic orbits of chaotic systems.

Consider a controlled continuous-time system,

$$\dot{x}(t) = f(x(t), u(t)), \tag{5.107}$$

where $x \in \mathbb{R}^n$ is the system state, $f : \mathbb{R}^n \times \mathbb{R}^p \to \mathbb{R}^n$ is a nonlinear vector-valued function, and $u(t) \in \mathbb{R}^p$ is the control input to be designed.

If the uncontrolled system (5.107) with $u(t) = 0$ is chaotic, then there are an infinite number of unstable periodic orbits (UPOs) in its attractive region (the reader who is interested in chaotic dynamics is referred to the following general introductions: Moon 1987; Peitgen, Jürgens and Saupe 1993; Schuster 1984; Wiggins 1988). Let $\bar{x}_T(t)$ be an unstable T-periodic solution of the uncontrolled system, which satisfies

$$\dot{\bar{x}}_T(t) = f(\bar{x}_T, 0), \quad \text{and} \quad \bar{x}_T(t) = \bar{x}_T(t - T). \tag{5.108}$$

Since it is difficult to get analytical or even numerical expressions of the target orbit $\bar{x}_T(t)$ because of its instability, there is no explicit reference for the stabilization of $\bar{x}_T(t)$. Pyragas (1992) proposed the following TDFC to solve the problem:

$$u(t) = K(x(t) - x(t - T)), \tag{5.109}$$

where $K \in \mathbb{R}^{p \times n}$ is a constant gain matrix to be designed. The initial value of the control system is given as $u(t) = 0$, $x(t) = \phi(t)$, $\forall\, t \in [-T, 0]$. Obviously, when the system trajectory converges to a T-periodic orbit, the feedback control term in (5.109) vanishes automatically. Therefore, it guarantees that the reached orbit is indeed an inherent solution of the original uncontrolled system. In such a controller the time-delayed state $x(t - T)$ is used as a target reference.

By simulations, Pyragas (1992) showed that by choosing an appropriate control gain K, the target UPO could be stabilized. Many experiments have confirmed this assertion (see, e.g., Ott, Simmendinger and Hess 1996; Parmananda, Madrigal and de Ciencias 1999). However, the theoretic research on TDFC shows that it is quite difficult to analyze the stability property of TDFC and, hence, it is also difficult to give analytic and effective methods for selecting the control gain (Just *et al.* 1997). Moreover, the domains of the control parameters over which the stabilization can be achieved are usually very small and limited so that the method generally fails to track higher-periodic orbits. To overcome the last problem, the original TDFC was soon being extended to the multiple-delay setting. Listed below are some versions of the extended TDFC (Basso, Genesio and Tesi 1997; Nakajima and Ueda 1998; Socolar *et al.* 1994):

$$u(t) = \sum_{i=1}^{m} K_i[x(t) - x(t - iT)]; \tag{5.110}$$

$$u(t) = K[x(t) - \sum_{i=1}^{m} \mu_i x(t - iT)], \tag{5.111}$$

where $\sum_{i=1}^{m} \mu_i = 1$;

$$u(t) = K[x(t) - R_m \sum_{i=1}^{m} r_i x(t - iT)], \tag{5.112}$$

where $R_m = \left(\sum_{i=1}^{m} r_i\right)^{-1}$;

$$u(t) = K[x(t) - (1 - \rho) \sum_{i=1}^{\infty} \rho^{i-1} x(t - iT)], \tag{5.113}$$

where $0 \le \rho < 1$. It is easy to verify that (5.110), (5.111) and (5.112) are essentially equivalent to each other. Notice also that $\sum_{i=1}^{\infty}(1 - \rho)\rho^{i-1} = 1$ when $0 \le \rho < 1$. So, when $m = \infty$ and $\mu_i = (1 - \rho)\rho^{i-1}$, (5.111) reduces to (5.113).

When the measured signal is the system output $y \in \mathbb{R}^q$ instead of the system state, given by

$$y(t) = h(x(t)), \tag{5.114}$$

the TDFC based on the output is

$$u(t) = K(y(t) - y(t - T)). \tag{5.115}$$

It is well know that an output feedback is not always as powerful as a state feedback in stabilization. The following dynamical output TDFC can be used for overcoming the limitation of the output TDFC:

$$\begin{cases} u(t) = K_1(x_c(t) - x_c(t - T)), \\ \dot{x}_c(t) = f(x_c(t), u(t)) + K_2(h(x_c(t)) - y(t)), \end{cases} \tag{5.116}$$

where $x_c(t) \in \mathbb{R}^n$ is the state of the observer system, and K_1 and K_2 are some gain matrices to be designed. A separation principle is proved for the dynamical output TDFC.

Theorem 5.16 *(Tian and Chen 2001) The closed-loop system consisting of (5.107) and (5.116) is stable at the inherent T-periodic UPO if and only if the following two systems,*

$$\dot{x}(t) = f(x(t), K_1(x(t) - x(t - T))) \tag{5.117}$$

and

$$\dot{x}(t) = f(x(t), 0) + K_2(y(t) - h(\bar{x}_T(t))), \tag{5.118}$$

are both stable at the same UPO, where $y(t) = h(x(t))$ is the measured output of system (5.107).

The TDFC method can be easily extended to discrete-time systems. Suppose that for a nonlinear control system

$$x(k+1) = f(x(k), u(k)), \tag{5.119}$$

$\Gamma = \{x_1^\star, x_2^\star, \ldots x_T^\star\}$ is a T-periodic UPO of the uncontrolled system, i.e.,

$$x_1^\star = f(x_T^\star, 0), \text{ and } x_{i+1}^\star = f(x_i^\star, 0), \quad i = 1, 2, \ldots, T-1. \tag{5.120}$$

Then, the TDFC and extended TDFC for stabilizing Γ become

$$u(k) = K(x(k) - x(k-T)) \tag{5.121}$$

and

$$u(k) = \sum_{i=1}^{m} K_i(x(k) - x(k-iT)), \tag{5.122}$$

respectively.

5.4.2 TDFC for Stabilization of an Unknown Equilibrium

Let $x^\star$ be an unstable fixed point (UFP) of the discrete-time system (5.119) without control, i.e.,

$$x^\star = f(x^\star, 0). \tag{5.123}$$

Since a UFP can be considered as a period one UPO, it may be stabilized by the one-step TDFC

$$u(k) = K(x(k) - x(k-1)). \tag{5.124}$$

The extended TDFC for stabilizing UFPs is given

$$u(k) = \sum_{i=1}^{m} K_i(x(k) - x(k-i)). \tag{5.125}$$

Compared to conventional stabilization strategies, TDFC has an advantage that it does not need the value of the equilibrium point to be stabilized.

An equilibrium point of a continuous-time system can be considered as a trivial period orbit with arbitrary period. So the TDFC for stabilizing the equilibrium point of system (5.107) can be given as

$$u = K(x(t) - x(t-\tau)), \tag{5.126}$$

where $\tau > 0$ is called self-delay. Here, The self-delay τ is usually taken as some small constant in stabilization of equilibrium points (Kokame *et al.* 2001).

5.4.3 Limitation of TDFC in Stabilization

Ushio (1996) studied the stabilization of UFPs of discrete-time systems, and found that the TDFC is subject to a substantial limitation, which is now referred to as the *odd number limitation*. Later, the limitation of TDFC for continuous-time systems was also derived (Just *et al.* 1998; Nakajima 1997). After that, similar results were obtained for some other cases (Hino, Yamamoto and Ushio 2002; Kokame *et al.* 2001).

Assume that the linearized system around the UFP $x^\star$ is

$$\hat{x}(k+1) = A\hat{x}(k) + Bu(k), \tag{5.127}$$

where $A = (\partial/\partial x)f(x^\star, 0)$, $B = (\partial/\partial u)f(x^\star, 0)$, $\hat{x}(k) = x(k) - x^\star$.

The odd number limitation for stabilizing UFPs of discrete-time systems is given by the following theorem.

Theorem 5.17 *(Ushio 1996) If the Jacobian matrix A of system (5.119), evaluated at the target UFP $x^\star$, has an odd number of real eigenvalues that are greater than unity, then system (5.119) cannot be stabilized by the TDFC (5.124) at $x^\star$ with any choice of the constant feedback gain matrix K.*

Actually, if the real matrix A has an odd number of real eigenvalues that are greater than unity, then the following inequality holds:

$$\det(I - A) < 0.$$

It should be noted that the TDFC method also fails to stabilize UFPs in the case of $\det(I - A) = 0$ (Yamamoto and Ushio 2003). So, the odd number limitation can be further modified as follows:

$$(A, B) \text{ is stabilizable via TDFC} \Rightarrow \det(I - A) > 0. \tag{5.128}$$

The proof of (5.128) is quite easy. Indeed, if the feedback control is taken as $u = K[x(k) - x(k-1)]$, then the linearized system of the closed-loop system is

$$\hat{x}(k+1) = A\hat{x}(k) + BK[\hat{x}(k) - \hat{x}(k-1)]. \tag{5.129}$$

It is obvious that the characteristic polynomial is

$$d(z) = \det[z^2 I - z(A + BK) + BK]. \tag{5.130}$$

Since the controller $u = K[x(k) - x(k-1)] = K[\hat{x}(k) - \hat{x}(k-1)]$ should make the closed-loop system (5.129) stable, all the roots of (5.130) lie in the unit circle, which implies $d(1) > 0$ by the Jury stability criterion. Meanwhile, from (5.130) we can see that $d(1) = \det(I - A)$. Therefore, (5.128) is proved.

Now, we consider the stabilization of a UPO. Assume that for nonlinear system (5.119) the linearized periodic time-varying system around the UPO Γ is

$$\hat{x}(k+1) = A(k)\hat{x}(k) + B(k)u(k), \tag{5.131}$$

where $A(k) = (\partial/\partial x)f(x_k^\star, 0)$, $B(k) = (\partial/\partial u)f(x_k^\star, 0)$, $\hat{x}(k) = x(k) - x_k^\star$. Write the state transition matrix for $A(k)$ as

$$\Phi_A(i, j) = \begin{cases} A(i-1)\cdots A(j+1)A(j), & i > j \\ I_n, & i = j. \end{cases} \tag{5.132}$$

The odd number limitation for stabilizing UPOs of discrete-time systems is then given by the following theorem.

Theorem 5.18 *(Hino, Yamamoto and Ushio 2002) If the state transition matrix $\Phi_A(T, 0)$ has an odd number of real eigenvalues that are greater than unity, then system (5.119) cannot be stabilized by the TDFC (5.121) at Γ with any choice of the constant feedback gain matrix K.*

Actually, this limitation can also be generalized to the following form:

$$(A(k), B(k)) \text{ is stabilizable via TDFC} \Rightarrow \det(I - \Phi_A(T, 0)) > 0. \tag{5.133}$$

For the continuous-time system (5.107), assume that the linearized system around the equilibrium point $x^\star$ is

$$\dot{\hat{x}}(t) = A\hat{x}(t) + Bu(t), \tag{5.134}$$

where $A = (\partial/\partial x)f(x^\star, 0)$, $B = (\partial/\partial u)f(x^\star, 0)$, $\hat{x}(t) = x(k) - x^\star$.

The odd number limitation for stabilizing equilibrium points of continuous-time systems is stated as follows.

Theorem 5.19 *(Kokame et al. 2001) If the Jacobian matrix A of system (5.107), evaluated at the target equilibrium $x^\star$, has an odd number of real eigenvalues that are greater than zero, then the system (5.107) cannot be stabilized by the TDFC (5.126) at $x^\star$ with any choices of the constant feedback gain matrix K and positive number τ.*

Similarly, the generalized form of this limitation is

$$(A, B) \text{ is stabilizable via TDFC} \Rightarrow \det(-A) > 0. \tag{5.135}$$

The assertion (5.135) can be easily proved as follows. If the feedback control is taken as $u = K[x(t) - x(t-\tau)]$, then the linearized system of the closed-loop system is

$$\dot{\hat{x}}(t) = A\hat{x}(t) + BK[\hat{x}(t) - \hat{x}(t-\tau)]. \tag{5.136}$$

It is obvious that the characteristic polynomial is

$$d(s) = \det[sI - A - (1 - e^{-s\tau})BK]. \tag{5.137}$$

Since the controller $u = K[x(t) - x(t-\tau)] = K[\hat{x}(t) - \hat{x}(t-\tau)]$ should make the closed-loop system (5.136) stable, all the roots of (5.137) lie in the open left-half plane, which implies $d(0) > 0$. Moreover, from (5.137) we can see $d(0) = \det(-A)$. Therefore, (5.135) is proved.

In the case of stabilizing a T-periodic UPO, the linearized system around the target UPO $\bar{x}_T(t)$ is

$$\dot{\hat{x}}(t) = A(t)\hat{x}(t) + B(t)u(t), \tag{5.138}$$

where $A(t) = (\partial/\partial x)f(\bar{x}_T, 0)$, $B(t) = (\partial/\partial u)f(\bar{x}_T, 0)$ are both T-periodic, and $x(t) = x(t) - \bar{x}_T(t)$.

Using the Floquet theory for periodic systems, Nakajima (1997) proved the odd number limitation for stabilizing UPOs of continuous-time systems.

Theorem 5.20 *(Nakajima 1997) If the linear variational equation (5.138) about the target hyperbolic UPO has an odd number of real characteristic multipliers greater than unity, then the UPO can not be stabilized by the TDFC with any choice of the constant feedback gain matrix K.*

It follows from Theorem 5.16 that such a limitation also holds for the observer-based dynamic output TDFC method.

The odd number limitation discussed above describes a necessary condition for the stabilizability via TDFC or extended TDFC. In the discrete-time case, Ushio (1996) obtained a necessary and sufficient condition for the first-order and second-order controllable systems by using Jury's stability test. Up to now, necessary and sufficient conditions of stabilizability of TDFC are obtained only for single-input systems.

Consider an n-order nonlinear discrete-time system described by (5.119), where $u(k) \in R$ is the control input, $x(k) \in \mathbb{R}^n$ is the state, and $f : \mathbb{R}^n \times R \to \mathbb{R}^n$ is a smooth mapping. Assume that $x^\star$ is an unstable fixed point of the open-loop system, i.e., $x^\star = f(x^\star, 0)$. Write $A = (\partial/\partial x)f(x^\star, 0)$ and $b = (\partial/\partial u)f(x^\star, 0)$. Then, TDFC of (5.124) reduces to a scalar form

$$u(k) = p(x(k) - x(k-1)), \tag{5.139}$$

where $p \in \mathbb{R}^{1\times n}$.

A necessary and sufficient condition for the stabilizability via TDFC is given as follows.

Theorem 5.21 *(Zhu and Tian 2005) Assume that (A, b) is controllable. There exists a TDFC in the form of (5.139) such that the closed-loop system of (5.119) is stable if and only if*

$$0 < \det(I_n - A) < 2^{n+1}. \tag{5.140}$$

Note that for $n = 1$ and $n = 2$, condition (5.140) is exactly the same as the results obtained by Ushio (1996).

Suppose that (A, b) is not controllable. Without loss of generality, let (A, b) be in the form of the following controllability decomposition:

$$A = \begin{bmatrix} A_1 & A_3 \\ 0 & A_2 \end{bmatrix}, \quad b = \begin{bmatrix} b_1 \\ 0 \end{bmatrix}.$$

Then, an extension of Theorem 5.21 can be given as follows.

Theorem 5.22 *(Zhu and Tian 2005) The following statements are equivalent:*

1) There is a TDFC in the form of (5.139) such that the closed-loop system of (5.119) is stable;

2) Matrix A_2 is stable and

$$0 < \det(I_{n_1} - A_1) < 2^{n_1+1}; \tag{5.141}$$

3) System (A, b) is stabilizable and

$$0 < \det(I_n - A) < 2^{n_1+1} \prod_{i=1}^{n-n_1} (1 - \theta_i); \tag{5.142}$$

where n_1 is the dimension of the controllable subspace of system (A, b), and θ_i $(i = 1, 2, \ldots, n - n_1)$ are the uncontrollable poles of system (A, b).

Consider the extended TDFC

$$u(k) = \sum_{i=1}^{n} p_i \left[x_i(k) - \sum_{j=1}^{m} \lambda_{ij} x_i(k - m - 1 + j) \right], \tag{5.143}$$

where $\lambda_{ij} \in R$, satisfying

$$\sum_{j=1}^{m} \lambda_{ij} = 1, \; i = 1, \ldots, n, \tag{5.144}$$

and p_i, $i = 1, \ldots, n$, are control gains to be designed.

Then, a full characterization of the limitation of the extended TDFC is given as follows.

Theorem 5.23 *(Tian and Zhu 2004) Assume that (A, b) is controllable. There exists an extended TDFC in the form of (5.143) such that the closed-loop system of (5.119) is stable if and only if*

$$0 < \det(I_n - A) < 2^{n+m}. \tag{5.145}$$

This result shows that the upper bound in the above-described condition can be enlarged by increasing the number of delays in the feedback. In other words, a system which cannot be stabilized by the conventional TDFC may still be stabilized by the extended TDFC, in contrast to the early claim that extended TDFC has no advantage over the conventional TDFC in overcoming the odd number limitation (Nakajima and Ueda 1998).

Kokame *et al.* (2001) considered the stabilization of equilibrium points for single-input continuous-time systems by applying TDFC, and showed that if the odd number limitation is excluded, then the UFP can be stabilized by TDFC.

Theorem 5.24 *(Kokame et al. 2001) Assume that the linearized system of (5.107) about the target UFP, denoted $(A, \; b)$, is controllable and* $\det(-A) > 0$. *Then, there exists a pair of K and τ such the closed loop of (5.107) with TDFC (5.126) is locally asymptotically stable.*

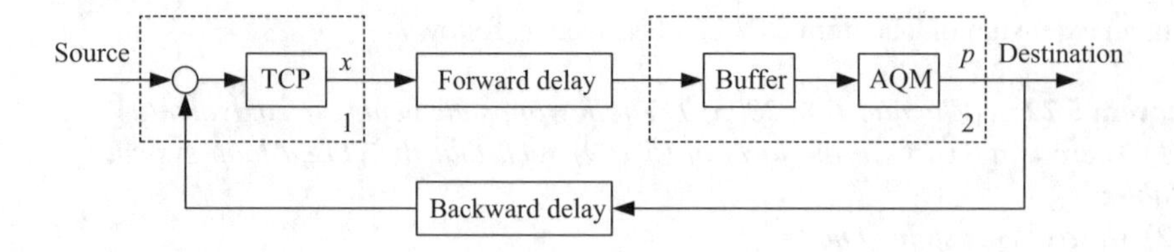

Figure 5.16 TCP-AQM as a closed loop control system.

5.5 Stabilization of Congestion Control Systems by Time-Delayed Feedback Control

5.5.1 Introduction of TDFC into Distributed Congestion Control Systems

The basic idea of the congestion control is to adjust the sending rate of the source or decide the length of the queue in the buffer according to a certain algorithm in order to avoid congestion. Therefore, congestion control algorithms are usually classified into two types. One is implemented in TCP, which adjusts the data transmission windows at source nodes based on the congestion signal. The other is the AQM control strategy, which calculates the packet loss probability to adjust the queue length by detecting the queue in the buffer and judging the congestion circumstance in the network communication.

A network with congestion control algorithms (TCP and AQM) can be viewed as a closed-loop control system as shown in Figure 5.16. The feedback signal to the source is the congestion information from the link. The dashed frame 1 and the dashed frame 2 in the diagram correspond to the TCP algorithm at the source and the AQM strategy at the link, respectively, where x denotes the packets sending rate of the user and p denotes the congestion indication.

Such a time-delayed feedback system may involve very complicated dynamics such as Hopf bifurcation (Li *et al.* 2004; Yang and Tian 2005) and chaos (Chen, Wang and Han 2004; Jiang, Wang and Xi 2004) when the stability conditions of a equilibrium point are destroyed. Moreover, the equilibrium point of a huge network varies with the change of the number of users and is not available for individual sources or links. Therefore, it is a natural idea to apply the TDFC method to the congestion control for enhancing the stability of the control algorithms.

Figure 5.17 gives a kind of combination of the TDFC with currently used congestion control algorithms. In this framework the TDFC becomes active only if oscillations are detected in

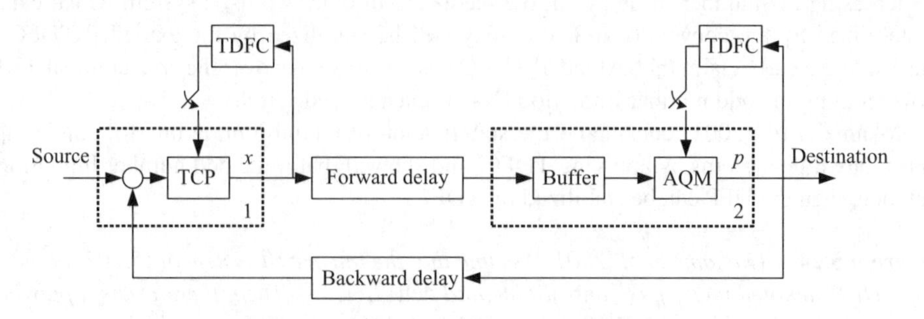

Figure 5.17 Introduction of TDFC into congestion control.

the network. The application of TDFC to the congestion control algorithm has the following advantages: (1) it does not need the steady state of the congestion algorithm; (2) it does not change the original algorithm and its steady state; (3) it can obtain various types of dynamical performance by adjusting the gain and the delay time of the TDFC.

5.5.2 Stabilizability under TDFC

Since a congestion control system is usually a time-delayed system, we first consider the following linear time-delayed system

$$\dot{x}(t) = \sum_{i=0}^{n_d} A_i x(t - D_i) + Bu(t) \tag{5.146}$$

where $A_i \in \mathbb{R}^{n\times n}$ and $B \in \mathbb{R}^{n\times p}$ are system matrices, $x(t) \in \mathbb{R}^n$ is the state, $u(t) \in \mathbb{R}^p$ is the control input, $D_i \in \mathbb{R}^+ (i \in \overline{1, n_d})$ are delay constants. The following lemma shows the limitation of TDFC in the stabilization of such systems.

Lemma 5.25 *Time-delayed system (5.146) can not be stabilized by the TDFC (5.126) if*

$$\det\left(-\sum_{i=0}^{n_d} A_i\right) \le 0.$$

Proof. The closed-loop form of system (5.146) with TDFC (5.126) is given by

$$\dot{x}(t) = \sum_{i=0}^{n_d} A_i x(t - D_i) + BK(x(t) - x(t - \tau)), \tag{5.147}$$

where $K \in \mathbb{R}^{n\times n}$ and $\tau > 0$. The characteristic function is

$$f(s) = \det\left(sI - \sum_{i=0}^{n_d} A_i \mathrm{e}^{-D_i s} - BK(1 - \mathrm{e}^{-\tau s})\right).$$

By assumption, $f(0) = \det(-\sum_{i=0}^{n_d} A_i) \le 0$. On the other hand, for the real positive s, $f(s)$ tends to infinity as s increases. From continuity, $f(s)$ has to vanish for some non-negative real number s. Thus, the system (5.146) can not be stabilized by the TDFC. □

From Lemma 5.25, we know that applying TDFC in congestion control algorithms with delays may also encounter some limitation similar to the odd number limitation.

Now, let us consider application of TDFC of the form

$$u_i(t) = -\frac{h_i}{\tau_i}(x_i(t) - x_i(t - \tau_i)) \tag{5.148}$$

to Kelly's primal algorithm in continuous-time form, i.e.,

$$\dot{x}_i(t) = \kappa_i \left(w_i - x_i(t - D_i) \sum_{l \in L_i} q_l(t - d_{li}^{\leftarrow}) \right)_{x_i}^{+} + u_i(t), \quad i \in \overline{S} \tag{5.149}$$

$$q_l(t) = p_l \left(\sum_{r \in S_l} x_r(t - d_{lr}^{\rightarrow}) \right), \quad l \in \overline{L}. \tag{5.150}$$

where $h_i > 0$, $\tau_i > 0$ are some parameters of the TDFC to be designed.

Define

$$\hat{x}_i(t) = \frac{x_i(t) - x_i^{\star}}{\sqrt{\kappa_i x_i^{\star}}}, \quad i = \overline{S},$$

where $x^{\star} = [x_1^{\star}, \ldots, x_S^{\star}]^{\mathrm{T}}$ is the unique equilibrium point of the system. Linearizing the system (5.149)–(5.150) with (5.148) about the equilibrium yields

$$\begin{aligned} \dot{\hat{x}}_i = & -\kappa_i w_i x_i^{\star -1} \hat{x}_i(t - D_i) - \sqrt{\kappa_i x_i^{\star}} \sum_{l \in L_i} R_{li} p_l' \sum_{r \in S_l} R_{lr} \sqrt{\kappa_r x_r^{\star}} \hat{x}_r(t - d_{lr}^{\rightarrow} - d_{li}^{\leftarrow}) \\ & - h_i(\hat{x}_i(t) - \hat{x}_i(t - \tau)), \end{aligned} \tag{5.151}$$

where p_l' is the derivative of p_l evaluated at $y = \sum_{i \in S_l} x_i^{\star}$, i.e., $p_l' = p_l'(\sum_{i \in S_l} x_i^{\star})$. Assuming the initial state is zero, we take the Laplace transform and obtain

$$\begin{aligned} s\hat{X}_i(s) = & - \left(\kappa_i w_i x_i^{\star -1} \hat{X}_i(s) + \sqrt{\kappa_i x_i^{\star}} \sum_{l \in L_i} R_{li} p_l' \mathrm{e}^{d_{li}^{\rightarrow} s} \sum_{r \in S_l} R_{lr} \sqrt{\kappa_r x_r^{\star}} \mathrm{e}^{-d_{lr}^{\rightarrow} s} \hat{X}_r(s) \right) \mathrm{e}^{-D_i s} \\ & - h_i(1 - \mathrm{e}^{-\tau s}) \hat{X}_i(s), \end{aligned} \tag{5.152}$$

where $\hat{X}_i(s) = \mathcal{L}(\hat{x}_i(t))$ is the Laplace transform of $\hat{x}_i(t)$. Define the following matrices:

$$\kappa = \mathrm{diag}\{\kappa_i,\ i \in \overline{S}\}, \tag{5.153}$$

$$X^{\star} = \mathrm{diag}\{x_i^{\star},\ i \in \overline{S}\}, \tag{5.154}$$

$$W = \mathrm{diag}\{w_i,\ i \in \overline{S}\}, \tag{5.155}$$

$$P' = \mathrm{diag}\{p_l',\ l \in \overline{L}\}, \tag{5.156}$$

$$R_f(s) = \{R_{li} \mathrm{e}^{-s d_{li}^{\rightarrow}},\ i \in \overline{S},\ l \in \overline{L}\}, \tag{5.157}$$

$$M(s) = \kappa^{1/2} W X^{\star -1} \kappa^{1/2} + \kappa^{1/2} X^{\star 1/2} R_f^{\mathrm{T}}(-s) P' R_f(s) X^{\star 1/2} \kappa^{1/2}. \tag{5.158}$$

Then, (5.152) can be written in the matrix form as

$$s\hat{X}(s) = \left(-M(s) \mathrm{diag}\left\{ \mathrm{e}^{-sD_i}, i \in \overline{S} \right\} - \mathrm{diag}\left\{ \frac{h_i}{\tau_i}(1 - \mathrm{e}^{-s\tau_i}) \right\} \right) \hat{X}(s) \tag{5.159}$$

where $\hat{X}(s) = [\hat{X}_1(s), \ldots, \hat{X}_S(s)]^{\mathrm{T}}$.

For $s = 0$, we have

$$-M(0)\mathrm{diag}\{\mathrm{e}^{-0\times D_i}\} = -\kappa^{1/2}WX^{\star-1}\kappa^{1/2} - \kappa^{1/2}X^{\star 1/2}R^{\mathrm{T}}P'RX^{\star 1/2}\kappa^{1/2},$$

where $R = R_f(0)$ is the routing matrix. Because $\kappa_i, x_i^\star, w_i, p_l'$ are all positive in the primal algorithm, $-M(0)\mathrm{diag}\{\mathrm{e}^{-0\times D_i}\}$ is a negatively definite matrix. By Lemma 5.25, we find that system (5.151) does not have the odd number limitation. This encourages us to apply the TDFC to the primal algorithm.

Theorem 5.26 *There exist h_i and τ_i for $i \in \overline{S}$ such that the primal algorithm (5.149)–(5.150) with TDFC given by (5.148) is locally stable at the equilibrium point.*

This theorem suggests that when the TDFC method is combined with the primal algorithm proposed by Kelly, Maulloo and Tan (1998), for any round-trip delays the network can be stabilized at the prescribed equilibrium point which is unknown for each source node, and thus the stability is much enhanced. In Chapter 1 we have shown some numerical examples of applying TDFC to eliminating oscillations in congestion control system.

Before giving a complete proof of Theorem 5.26, we need to make some preparation. First, let us prove the following lemma.

Lemma 5.27 *The quasi-polynomial*

$$f(s) = s + h(1 - \mathrm{e}^{-\tau s})$$

has no RHP roots for any $h > 0$ and $\tau > 0$.

Proof. Let

$$F(s) = 1 + h\frac{1 - \mathrm{e}^{-\tau s}}{s}.$$

Then, $f(s)$ has no RHP roots if and only if $F(s)$ has no RHP zeros. By Nyquist the stability criterion, $F(s)$ has no RHP zeros if and only if the Nyquist plot of

$$G(\mathrm{j}\omega) := h\frac{1 - \mathrm{e}^{-\mathrm{j}\tau\omega}}{\mathrm{j}\omega}$$

does not enclose the point $(-1, \mathrm{j}0)$ in the complex plane. So,

$$\mathrm{Re}(G(\mathrm{j}\omega)) = \frac{h\sin\tau\omega}{\omega},$$

$$\mathrm{Im}(G(\mathrm{j}\omega)) = -\frac{h(1 - \cos\tau\omega)}{\omega}.$$

It is easy to get

$$\lim_{\omega\to 0^+}\mathrm{Im}(G(\mathrm{j}\omega)) = 0, \qquad \lim_{\omega\to 0^+}\mathrm{Re}(G(\mathrm{j}\omega)) = h\tau.$$

Now, let

$$\mathrm{Im}(G(\mathrm{j}\omega)) = 0,$$

then we have $\tau\omega = 2k\pi (k \neq 0)$ which implies

$$\mathrm{Re}(G(\mathrm{j}\omega)) = 0.$$

Therefore, for any $h > 0$ the Nyquist plot of $G(\mathrm{j}\omega)$ intersects the real axis of the complex plane only at the origin. □

Now, let us investigate the role of TDFC for stabilization of a first-order time-delayed system. The following lemma shows that an analog of the odd number limitation holds for the first-order time-delayed system.

Lemma 5.28 *The first-order time-delayed system*

$$\dot{x} = -ax(t - D) + u, \tag{5.160}$$

with any delay constant $D > 0$ can be asymptotically stabilized by TDFC (5.148) if and only if $a > 0$.

Proof. *(Necessity)* Since system (5.160) is a special case of system (5.146) with $n = 1$ and $n_d = 1$, the necessity part is provided by Lemma 5.25.

(Sufficiency) The closed-loop form of system (5.160) with TDFC (5.148) is given by

$$\dot{x}(t) = -ax(t - D) - \frac{h}{\tau}(x(t) - x(t - \tau)) \tag{5.161}$$

where $a > 0$, $D > 0$. The characteristic equation of the above system is

$$s + a\mathrm{e}^{-Ds} + \frac{h}{\tau}(1 - \mathrm{e}^{-s\tau}) = 0. \tag{5.162}$$

We will prove that there exists h and τ such that all the roots of this equation lie inside the LHP.

Let

$$F_1(s) = \frac{a\mathrm{e}^{-Ds}}{s + \frac{h}{\tau}(1 - \mathrm{e}^{-s\tau})}.$$

Then, the characteristic equation can be written as $1 + F_1(s) = 0$, $s \neq 0$. We adopt the Nyquist stability criterion to continue the proof. Consider the frequency property function of $F_1(s)$ given by

$$\begin{aligned} F_1(\mathrm{j}\omega) &= \frac{a\mathrm{e}^{-\mathrm{j}D\omega}}{\mathrm{j}\omega + \frac{h}{\tau}(1 - \mathrm{e}^{-\mathrm{j}\omega\tau})} \\ &= \frac{a\mathrm{e}^{-\mathrm{j}D\omega}}{\mathrm{j}\omega + \frac{h}{\tau}((1 - \cos\omega\tau) + \mathrm{j}\sin\omega\tau)}, \quad \omega \in [0, \infty) \end{aligned} \tag{5.163}$$

As shown by Lemma 5.27, $F_1(s)$ has no RPH poles. So, by the Nyquist stability criterion, all the roots of (5.162) lie inside the LHP if and only if the Nyquist plot of $F_1(\mathrm{j}\omega)$ does not enclose the point $(-1, \mathrm{j}0)$ in the complex plane. Let ω_{c1} denote the frequency at which $F_1(\mathrm{j}\omega)$ crosses the real axis for the first time. It suffices to prove that $|F_1(\mathrm{j}\omega)| < 1$ for $\omega \in [\omega_{c1}, \infty)$.

We choose $\tau < D$. Thus,

$$\arg(F_1(\mathrm{j}\omega)) = -D\omega - \arctan \frac{\omega + \frac{h}{\tau}\sin\omega\tau}{\frac{h}{\tau}(1-\cos\omega\tau)}$$

for $\omega \in (0, \frac{\pi}{D})$, where $\arg(\cdot)$ denotes the phase. By noticing that

$$\arg(F_1(\mathrm{j}\omega)) \in (-\pi, 0), \quad \forall \omega \in \left(0, \frac{\pi}{2D}\right),$$

and

$$\arg\left(F_1\left(\mathrm{j}\frac{\pi}{D}\right)\right) \in \left(-\frac{3}{2}\pi, -\pi\right),$$

we know that

$$\omega_{c1} \in \left(\frac{\pi}{2D}, \frac{\pi}{D}\right)$$

from the continuous property of $\arg(F_1(\mathrm{j}\omega))$. So we have

$$\frac{a}{\omega(1+h^\star)} < 1, \quad \forall \omega \in [\omega_{c1}, \infty),$$

where

$$h^\star = \frac{2aD}{\pi} - 1. \tag{5.164}$$

Therefore,

$$\begin{aligned} |F_1(\mathrm{j}\omega)| &= \frac{a}{\frac{1}{\tau}\sqrt{h^2(1-\cos\omega\tau)^2 + (\omega\tau + h\sin\omega\tau)^2}} \\ &= \frac{1+h^\star}{\sqrt{h^2(1-\cos\omega\tau)^2/(\omega\tau)^2 + (1+h\sin\omega\tau/(\omega\tau))^2}} \frac{a}{\omega(1+h^\star)} \qquad (5.165) \\ &< \frac{1+h^\star}{\sqrt{h^2(1-\cos\omega\tau)^2/(\omega\tau)^2 + (1+h\sin\omega\tau/(\omega\tau))^2}} \qquad (5.166) \\ &\triangleq F_2(\omega\tau) \end{aligned}$$

Choose h such that $\frac{|h^\star|}{h} < 1$. There exists a positive constant $q_1 < \pi/2$ such that $\frac{\sin q}{q} \geq \frac{|h^\star|}{h}$ for $q \in (0, q_1]$. Thus, $F_2(\omega\tau) < 1$ for $\omega\tau \in (0, q_1]$, i.e. $\omega \in (0, q_1/\tau]$. Obviously, there exists a positive constant $\xi < D$ such that $\omega_{c1} < q_1/\tau$, and hence, $|F_1(\mathrm{j}\omega)|_{\omega\in[\omega_{c1}, q_1/\tau]} < 1$ with $\tau \in (0, \xi)$.

In the following, we prove that by choosing τ we can ensure $|F_1(\mathrm{j}\omega)| < 1$ for $\omega \in (q_1/\tau, \infty)$ with $h > |h^\star|$. It is equivalent to ensure that $G(\omega) > a^2$ for $\omega \in (q_1/\tau, \infty)$ with

$$G(\omega) = \frac{1}{\tau^2}[h^2(1-\cos\omega\tau)^2 + (\omega\tau + h\sin\omega\tau)^2]. \tag{5.167}$$

For $h > |h^\star|$, where $h^\star$ is given by (5.164), we consider the following two cases with different values of h.

Case 1: $h \leq 1.5\pi$.
From (5.167) it is easy to see

$$G(\omega) > \frac{1}{\tau^2}[h^2(1 - \cos q_1)^2]$$

for $\omega\tau \in [q_1, 2\pi - q_1]$, and hence, there exists a positive constant ξ_{11} such that

$$G(\omega) > \frac{1}{\tau^2}[h^2(1 - \cos q_1)^2] > a^2$$

for $\tau \in (0, \xi_{11})$ and

$$G(\omega) \geq \frac{1}{\tau^2}(\omega\tau + h\sin\omega\tau)^2 \geq \frac{1}{\tau^2}(\omega\tau - h)^2$$

for $\omega\tau \in (2\pi - q_1, \infty)$. Thus, there also exists a positive constant ξ_{12} such that

$$G(\omega) \geq \frac{1}{\tau^2}(\omega\tau - h)^2 > a^2$$

for $\tau \in (0, \xi_{12})$. Therefore, by choosing

$$\tau \in (0, \xi_1 := \min\{\xi_{11}, \xi_{12}\}),$$

we can ensure $G(\omega) > a^2$, i.e.,

$$|F_1(\mathrm{j}\omega)| < 1$$

for $\omega \in (q_1/\tau, \infty)$.

Case 2: $h > 1.5\pi$.
Define

$$\chi(g) = g + h\sin g.$$

Obviously, for $g > 0$, the equation $\chi(g) = 0$ has solutions when $h > 1.5\pi$. Take these solutions of $\chi(g) = 0$ as a sequence $\{g(n), n = 1, \ldots\}$ with certain length. Rank the elements of $\{g(n), n = 1, \ldots\}$ increasingly. Then, we get

$$(2n-1)\pi < g(2n-1) < g(2n) < 2n\pi.$$

To consider the minimums of $G(\omega)$, we compute the derivative of $G(\omega)$ with respect to ω as follows

$$\begin{aligned} f(\omega\tau) &:= \frac{\partial G(\omega)}{\partial\omega} \\ &= 2\frac{1}{\tau}(h^2\sin\omega\tau(1 - \cos\omega\tau) + (\omega\tau + h\sin\omega\tau)(1 + h\cos\omega\tau)). \end{aligned} \tag{5.168}$$

Obviously,

$$f(\omega\tau) > 0, \quad \forall \omega\tau \in \left[q_1, \arccos\left(-\frac{1}{h}\right)\right],$$

and

$$f(\omega\tau) < 0, \quad \forall \omega\tau \in \left[\pi, \min\left\{2\pi - \arccos\left(-\frac{1}{h}\right), g(1)\right\}\right].$$

Thus, $G(\omega)$ may have a minimum at $\omega\tau = \theta_{21} \in \left(\arccos\left(-\frac{1}{h}\right), \pi\right)$, and

$$G(\omega)_{\omega\tau=\theta_{21}} = \frac{1}{\tau^2}[h^2(1-\cos\theta_{21})^2 + (\theta_{21} + h\sin\theta_{21})^2].$$

Therefore, we can choose a positive constant ξ_{21} to make $G(\omega)_{\omega\tau=\theta_{21}} > a^2$ with $\tau \in (0, \xi_{21})$. Then, $f(\omega\tau) > 0$ for $\omega\tau \in [2\pi, 2\pi + \arccos(-\frac{1}{h})]$. Hence, $G(\omega)$ has a minimum at $\omega\tau = \theta_{22} \in (\min\{2\pi - \arccos(-\frac{1}{h}), g(1)\}, 2\pi)$, and

$$G(\omega)_{\omega\tau=\theta_{22}} = \frac{1}{\tau^2}[h^2(1-\cos\theta_{22})^2 + (\theta_{22} + h\sin\theta_{22})^2].$$

Therefore, there also exists a positive constant ξ_{22} such that $G(\omega)_{\omega\tau=\theta_{22}} > a^2$ with $\tau \in (0, \xi_{22})$. Similarly, when $\omega\tau > 2\pi$, we can also ensure that $G(\omega) > a^2$ at the minimum of $G(\omega)$ by choosing τ. Therefore, we can choose a positive constant ξ_2 finally to make $G(\omega) > a^2$, i.e. $|F_1(j\omega)| < 1$ for $\omega \in (q_1/\tau, \infty)$ with $\tau \in (0, \xi_2)$.

As discussed above, there exist $K = h/\tau$ with $h > |h^\star|$, and $\tau \in (0, \min(\xi, \xi_1))$ for case 1 or $\tau \in (0, \min(\xi, \xi_2))$ for case 2 such that system (5.160) is stable. Lemma 5.28 is thus proved. □

Now, we provide the proof of Theorem 5.26 as follows.

Proof of Theorem 5.26

From (5.159) we get the characteristic equation of the closed-loop system with TDFC as follows:

$$I + M(s)\mathrm{diag}\left\{\frac{e^{-sD_i}}{s + \frac{h_i}{\tau_i}(1 - e^{-s\tau_i})}\right\} = 0.$$

Since the equation

$$s + \frac{h_i}{\tau_i}(1 - e^{-s\tau_i}) = 0$$

has no RHP roots for any $i \in \overline{S}$ according to Lemma 5.27, from the general Nyquist stability criterion (Theorem 2.20) we know that the system (5.159) is asymptotically stable, if and only if the eigenloci of $M(\mathrm{j}\omega)\mathrm{diag}\{\frac{\mathrm{e}^{-\mathrm{j}\omega D_i}}{\mathrm{j}\omega+\frac{h_i}{\tau_i}(1-\mathrm{e}^{-\mathrm{j}\omega\tau_i})}\}$, i.e.

$$\lambda\left(M(\mathrm{j}\omega)\mathrm{diag}\left\{\frac{\mathrm{e}^{-\mathrm{j}\omega D_i}}{\mathrm{j}\omega+\frac{h_i}{\tau_i}(1-\mathrm{e}^{-\mathrm{j}\omega\tau_i})}\right\}\right),$$

do not enclose the point $(-1, \mathrm{j}0)$ for $\omega \in [0, \infty)$. Denote

$$\tilde{M}(\mathrm{j}\omega) = \left(\mathrm{diag}\left\{\frac{2D_i}{(1+h_i^\star)\pi}\right\}\right)^{1/2} M(\mathrm{j}\omega)\left(\mathrm{diag}\left\{\frac{2D_i}{(1+h_i^\star)\pi}\right\}\right)^{1/2}, \tag{5.169}$$

where

$$h_i^\star = \frac{2k_i D_i}{\pi}\left(\sum_{l\in L_i} p_l + \sum_{l\in L_i} p_l' \sum_{r\in N_l} x_r^\star\right) - 1, \tag{5.170}$$

and

$$p_l = p_l\left(\sum_{i\in N_l} x_i^\star\right). \tag{5.171}$$

Then, we have

$$\begin{aligned}
&\{0\}\cup\lambda\left(M(\mathrm{j}\omega)\mathrm{diag}\left\{\frac{\mathrm{e}^{-\mathrm{j}\omega D_i}}{\mathrm{j}\omega+\frac{h_i}{\tau_i}(1-\mathrm{e}^{-\mathrm{j}\omega\tau_i})}\right\}\right)\\
&=\{0\}\cup\lambda\left(\tilde{M}(\mathrm{j}\omega)\mathrm{diag}\left\{\frac{\pi}{2D_i}\frac{\mathrm{e}^{-\mathrm{j}\omega D_i}}{\mathrm{j}\omega}\frac{\mathrm{j}(1+h_i^\star)\omega}{\mathrm{j}\omega+\frac{h_i}{\tau_i}(1-\mathrm{e}^{-\mathrm{j}\omega\tau_i})}\right\}\right).
\end{aligned}$$

Hence, from Lemma 2.34 it follows that

$$\begin{aligned}
&\lambda\left(M(\mathrm{j}\omega)\mathrm{diag}\left\{\frac{\mathrm{e}^{-\mathrm{j}\omega D_i}}{\mathrm{j}\omega+\frac{h_i}{\tau_i}(1-\mathrm{e}^{-\mathrm{j}\omega\tau_i})}\right\}\right)\\
&\in\rho\left(\tilde{M}(\mathrm{j}\omega)\right)\mathrm{Co}\left(0\cup\left\{\frac{\pi}{2D_i}\frac{\mathrm{e}^{-\mathrm{j}\omega D_i}}{\mathrm{j}\omega}\frac{\mathrm{j}(1+h_i^\star)\omega}{\mathrm{j}\omega+\frac{h_i}{\tau_i}(1-\mathrm{e}^{-\mathrm{j}\omega\tau_i})}\right\}\right).
\end{aligned}$$

Using Corollary 2.31 of Gershgorin's disc lemma and (5.169)–(5.171) as well, we have

$$\begin{aligned}
\rho\left(\tilde{M}(\mathrm{j}\omega)\right) &\le \max_{i\in\overline{S}}\left\{\kappa_i\left(\sum_{l\in\overline{L}} R_{li}p_l + \sum_{l\in\overline{L}}\sum_{r\in\overline{S}}|R_{li}\mathrm{e}^{\mathrm{j}\omega(\overrightarrow{d}_{li}-\overrightarrow{d}_{lr})}p_l' R_{lr}x_r^\star|\right)\frac{2D_i}{(1+h_i^\star)\pi}\right\}\\
&= \max_{i\in\overline{S}}\left\{\kappa_i\left(\sum_{l\in L_i} p_l + \sum_{l\in L_i} p_l'\sum_{r\in N_l} x_r^\star\right)\frac{2D_i}{(1+h_i^\star)\pi}\right\}\\
&= 1
\end{aligned}$$

Define

$$F_{1i}(\mathrm{j}\omega) = \frac{\pi}{2D_i}\frac{\mathrm{e}^{-\mathrm{j}\omega D_i}}{\mathrm{j}\omega}$$

and

$$F_{2i}(\mathrm{j}\omega) = \frac{\pi}{2D_i}\frac{\mathrm{e}^{-\mathrm{j}\omega D_i}}{\mathrm{j}\omega}\frac{\mathrm{j}(1+h_i^\star)\omega}{\mathrm{j}\omega + \frac{h_i}{\tau_i}(1-\mathrm{e}^{-\mathrm{j}\omega\tau_i})}.$$

Then, we just need to prove that there exist h_i and τ_i such that

$$(-1, \mathrm{j}0) \notin \mathrm{Co}(0 \cup \{F_{2i}(\mathrm{j}\omega), i \in \overline{S}\})$$

for all $\omega \in (0, \infty)$.

Choose h_i such that $\frac{|h_i^\star|}{h_i} < 1$. There always exists a positive constant $q_i < \pi/2$ such that $\frac{\sin q}{q} \geq \frac{|h_i^\star|}{h_i}$ for $q \in (0, q_i]$, and

$$\left|\frac{\mathrm{j}(1+h_i^\star)\omega}{\mathrm{j}\omega + \frac{h_i}{\tau_i}(1-\mathrm{e}^{-\mathrm{j}\omega\tau_i})}\right| = \frac{1+h_i^\star}{\sqrt{(1+h_i\frac{\sin\omega\tau_i}{\omega\tau_i})^2 + h_i^2\frac{(1-\cos\omega\tau_i)^2}{(\omega\tau_i)^2}}}$$
$$< 1, \quad \omega\tau_i \in (0, q_i] \tag{5.172}$$

From the proof of Lemma 5.28 we know that $F_{2i}(\mathrm{j}\omega)$ crosses the real axis at $\omega_{ic1} \in (\frac{\pi}{2D_i}, \frac{\pi}{D_i})$ for the first time with $0 < \tau_i < D_i$. Thus, there always exist $\{\varsigma_{i1} > 0, i \in \overline{S}\}$ such that $\min\{q_i/\tau_i, i \in \overline{S}\} > \max\{\omega_{ic1}, i \in \overline{S}\}$ with $\tau_i \in (0, \varsigma_{i1})$.

When $\omega \in (0, \min\{q_i/\tau_i, i \in \overline{S}\}]$, we have

$$|F_{1i}(\mathrm{j}\omega)| > |F_{2i}(\mathrm{j}\omega)| \tag{5.173}$$

from (5.172) and the fact that

$$\begin{aligned}\arg(F_{1i}(\mathrm{j}\omega)) &= \left(-\omega D_i - \frac{\pi}{2}\right)\\ &< \arg(F_{2i}(\mathrm{j}\omega))\\ &= -\omega D_i - \arctan\left(\frac{\omega + \frac{h_i}{\tau_i}\sin\omega\tau_i}{\frac{h_i}{\tau_i}(1-\cos\omega\tau_i)}\right).\end{aligned} \tag{5.174}$$

According to the proof of Theorem 3.10, all the Nyquist plots of $F_{1i}(\mathrm{j}\omega)$, $i \in \overline{S}$ have the same shape, and cross the real axis for the first time at $(-1, \mathrm{j}0)$. Let $r(-1, \mathrm{j}0)$ denote the tangent line to $F_{1i}(\mathrm{j}\omega)$, $i \in \overline{S}$ at $(-1, \mathrm{j}0)$. Then, by Lemma 3.9, the complex plane is divided into two parts by $r(-1, \mathrm{j}0)$, and the Nyquist plot of $F_{1i}(\mathrm{j}\omega)$ is in one side of $r(-1, \mathrm{j}0)$ (see Figure 5.18). Because of (5.173) and (5.174) as well as the fact that $|F_{1i}(\mathrm{j}\omega)| = \frac{\pi}{2D_i\omega}$ decreases as the phase $\arg(F_{1i}(\mathrm{j}\omega))$ decreases, $|F_{2i}(\mathrm{j}\omega)|$ is less than $|F_{1i}(\mathrm{j}\omega)|$ at the same phase. Therefore, for $\omega \in (0, \min\{q_i/\tau_i, i \in \overline{S}\}]$, the Nyquist plot of $F_{2i}(\mathrm{j}\omega)$ is on the same side of $r(-1, \mathrm{j}0)$ as that of $F_{1i}(\mathrm{j}\omega)$, and it doesn't have any intersection with $r(-1, \mathrm{j}0)$ (Figure 5.18). Therefore,

$$(-1, \mathrm{j}0) \notin \kappa\mathrm{Co}(0 \cup F_{2i}(\mathrm{j}\omega),\ i \in \overline{S})$$

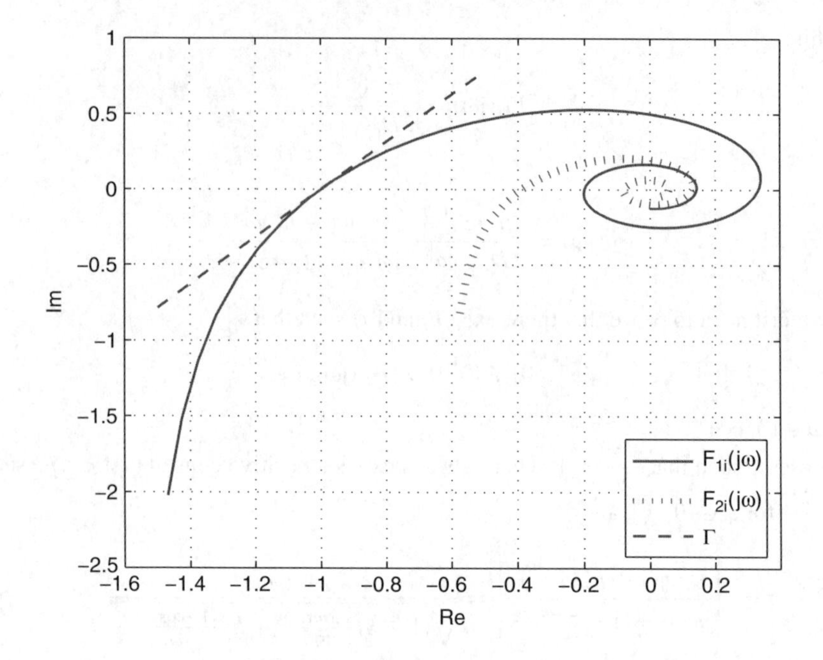

Figure 5.18 Nyquist plots.

for all $\omega \in (0, \min\{q_i/\tau_i, i \in \overline{S}\}]$, $0 < \kappa \leq 1$. By Lemma 2.34, we come to the conclusion that

$$(-1, \mathrm{j}0) \notin \lambda\left(M(\mathrm{j}\omega)\mathrm{diag}\left\{\frac{\mathrm{e}^{-\mathrm{j}\omega D_i}}{\mathrm{j}\omega + \frac{h_i}{\tau_i}(1 - \mathrm{e}^{-\mathrm{j}\omega\tau_i})}\right\}\right)$$

for all $\omega \in (0, \min\{q_i/\tau_i, i \in \overline{S}\}]$ with $\tau_i \in (0, \varsigma_{i1})$.

When $\omega \in (\min\{q_i/\tau_i, i \in \overline{S}\}, \infty)$, analogous to the proof of Lemma 5.28, there always exist $\{\varsigma_{i2} > 0, i \in \overline{S}\}$ to make $|F_{2i}(\mathrm{j}\omega)| < 1$ for $\omega \in (\min\{q_i/\tau_i, i \in \overline{S}\}, \infty)$ with $\tau_i \in (0, \varsigma_{i2})$. (For illustration we show the general shape of the Nyquist plot of $F_{2i}(\mathrm{j}\omega)$, $i = 1, 2, 3$ for $\omega \in (\min\{q_i/\tau_i, i \in \overline{S}\}, \infty)$ in Figure 5.19.) Hence,

$$(-1, 0) \notin \kappa\mathrm{Co}(0 \cup F_{2i}(\mathrm{j}\omega), i \in \overline{S})$$

for all $\omega \in (\min\{q_i, i \in \overline{S}\}, \infty)$, $0 < \kappa \leq 1$. By Lemma 2.34, we come to the conclusion that

$$(-1, \mathrm{j}0) \notin \lambda\left(M(\mathrm{j}\omega)\mathrm{diag}\left\{\frac{\mathrm{e}^{-\mathrm{j}\omega D_i}}{\mathrm{j}\omega + \frac{h_i}{\tau_i}(1 - \mathrm{e}^{-\mathrm{j}\omega\tau_i})}\right\}\right)$$

for all $\omega \in (\min\{q_i/\tau_i, i \in \overline{S}\}, \infty)$ with $\tau_i \in (0, \varsigma_{i2})$.

From the above discussion, there exist $H_i = h_i/\tau_i$ with

$$h_i > |h_i^\star|$$

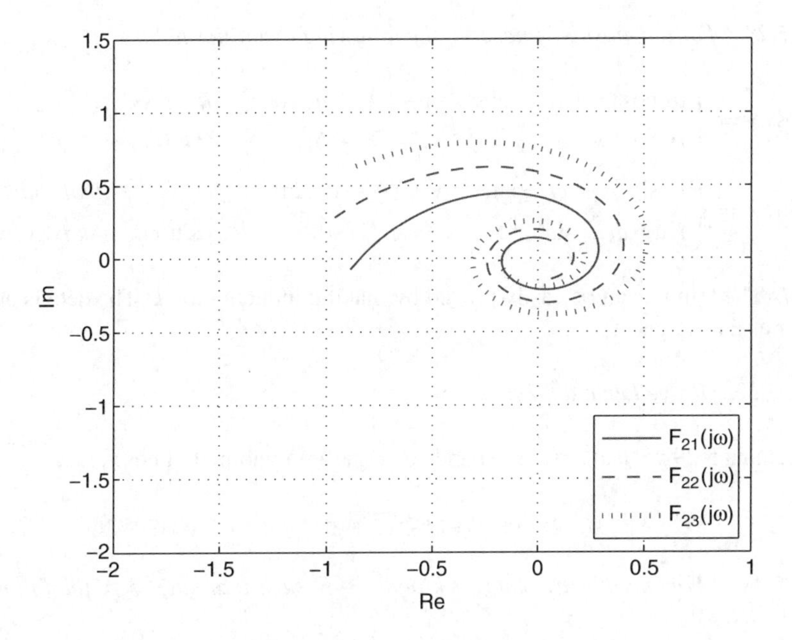

Figure 5.19 High frequency property of Nyquist plots.

and

$$\tau_i \in (0, \varsigma_i = \min\{\varsigma_{i1}, \varsigma_{i2}\})$$

such that

$$(-1, \mathrm{j}0) \notin \lambda\left(M(\mathrm{j}\omega)\mathrm{diag}\left\{\frac{\mathrm{e}^{-\mathrm{j}\omega D_i}}{\mathrm{j}\omega + \frac{h_i}{\tau_i}(1 - \mathrm{e}^{-\mathrm{j}\omega\tau_i})}\right\}\right)$$

for $\omega \in (-\infty, \infty)$. Therefore, the system (5.159) is stable. Theorem 5.26 is proved. □

5.5.3 Design of TDFC with Commensurate Self-Delays

Theorem 5.26 proves the existence of TDFC that stabilizes Kelly's primal algorithm with any round-trip delays. However, it does not provide an explicit procedure for the design of the control parameters (self-delays τ_i and gains h_i) of TDFC. In this subsection we present a design scheme by considering the case when self-delays and round-trip delays are commensurate. Note that delays $\tau_1, \ldots, \tau_m$ are said to be commensurate if τ_i/τ_j is rational for all $i, j \in \overline{1, m}, i \neq j$ (Gu, Kharitonov and Chen 2003). In this case we can always find a τ such that $\tau_i = k_i\tau$ for all $i \in \overline{1, m}$, where k_i are integers.

Let us first introduce some preliminary lemmas.

Lemma 5.29 *For any positive integer k, the following equalities hold:*

$$\cos kx = \begin{cases} a_k \cos^k x + a_{k-2} \cos^{k-2} x + \cdots + a_1 \cos x, & \text{if } k \text{ is odd,} \\ b_k \cos^k x + b_{k-2} \cos^{k-2} x + \cdots + b_1, & \text{if } k \text{ is even;} \end{cases}$$

$$\sin kx = \begin{cases} c_k \sin^k x + c_{k-2} \sin^{k-2} x + \cdots + c_1 \sin x, & \text{if } k \text{ is odd,} \\ \sin x(d_{k-1} \sin^{k-1} x + d_{k-3} \sin^{k-3} x + \cdots + d_1 \sin x), & \text{if } k \text{ is even.} \end{cases}$$

Proof. The lemma can be easily proved by making induction on k. The details are left to the reader as an exercise. □

Exercise 5.30 *Prove Lemma 5.29.*

By Lemma 5.29 we can always expand $\cos kx$ as a polynomial of $\cos x$, i.e.,

$$\cos kx = e_k \cos^k x + e_{k-2} \cos^{k-2} x + \cdots + e_1 \triangleq v_k(\cos x). \tag{5.175}$$

Lemma 5.31 *For any positive integer k and $x \in \mathbb{R}$ such that $\sin x \neq 0$, the following inequality holds*

$$\frac{|\sin kx|}{|\sin x|} \leq k;$$

moreover, the equality holds if and only if $k = 1$.

Proof. Obviously, the equality holds if $k = 1$. Therefore, the rest task of the proof is to prove that

$$\frac{|\sin kx|}{|\sin x|} < k$$

for any integer $k > 1$.

Since the function $g(x) = \frac{|\sin kx|}{|\sin x|}$ is π-periodic and symmetric about $x = -\pi/2$, i.e.,

$$g(x) = g(x + \pi), \quad g(x) = g(\pi - x),$$

we just need to show the equality is true for $x \in (0, \pi/2]$.

When $0 < x \leq \frac{\pi}{2k}$, we have $\sin x > 0$, $\sin kx > 0$. Let $f(x) = \sin kx - k \sin x$. Then, in the given interval we have $f'(x) = k(\cos kx - \cos x) < 0$ for $k > 1$. Therefore, $f(x) < f(0) = 0$.

When $\frac{\pi}{2k} < x \leq \frac{\pi}{2}$, we have

$$g(x) \leq \frac{1}{\sin x} < \frac{1}{\sin \frac{\pi}{2k}} \leq k.$$

At the last step of the above derivation we used the well-known inequality $\sin \frac{\pi}{2} x \geq x$ for any $x \in (0, 1]$. □

Lemma 5.32 *Suppose that $a > 0$, $D \in \mathbb{R}$, $\tau = D/n$, where n is any positive integer. Then,*

$$\mathrm{Re}\left(\frac{a\mathrm{e}^{-\mathrm{j}D\omega}}{\mathrm{j}\omega + \hbar(1 - \mathrm{e}^{-\mathrm{j}\tau\omega})}\right) > -1 \tag{5.176}$$

holds for all $\omega \neq 0$, if

$$\hbar > \max\left\{\frac{a}{2}(n|e_n| + (n-1)|e_{n-1}| + \cdots + |e_1|), \frac{(2n-1)a}{2}\right\}, \tag{5.177}$$

where $e_1, \ldots, e_n$ are coefficients of $v_n(\cos x)$ defined by (5.175).

Proof. For convenience we denote $\xi = 1/n$ and

$$g(\mathrm{j}\omega) = \frac{a\mathrm{e}^{-\mathrm{j}D\omega}}{\mathrm{j}\omega + \hbar(1 - \mathrm{e}^{-\mathrm{j}\tau\omega})}.$$

Due to the symmetry of $g(\mathrm{j}\omega)$ we just need to prove (5.176) for $\omega > 0$. Since

$$\mathrm{Re}(g(\mathrm{j}\omega)) = a\frac{\hbar\cos D\omega - \omega\sin D\omega - \hbar\cos D(1-\xi)\omega}{2\hbar^2 + \omega^2 - 2\hbar^2\cos D\xi\omega + 2\omega\hbar\sin D\xi\omega},$$

$\mathrm{Re}(g(\mathrm{j}\omega)) > -1$ if and only if

$$\begin{aligned} &a(\hbar\cos D\omega - \omega\sin D\omega - \hbar\cos D(1-\xi)\omega) \\ &\quad + 2\hbar^2 + \omega^2 - 2\hbar^2\cos D\xi\omega + 2\omega\hbar\sin D\xi\omega > 0. \end{aligned} \tag{5.178}$$

Denote $x = D\omega$. Then, (5.178) holds for all $\omega \neq 0$ and $D \in \mathbb{R}$ if and only if

$$\begin{aligned} f(x, D) &:= (2\hbar^2 - 2\hbar^2\cos\xi x + a\hbar\cos x - a\hbar\cos(1-\xi)x)D^2 \\ &\quad + (2x\hbar\sin\xi x - ax\sin x)D + x^2 > 0 \end{aligned} \tag{5.179}$$

for $x \neq 0$ and $D \in \mathbb{R}$.

We first consider the case when

$$2\hbar^2 - 2\hbar^2\cos\xi x + a\hbar\cos x - a\hbar\cos(1-\xi)x = 0. \tag{5.180}$$

Straightforward calculation yields

$$\begin{aligned} &2\hbar^2 - 2\hbar^2\cos\xi x + a\hbar\cos x - a\hbar\cos(1-\xi)x \\ &= 4\hbar^2\sin^2\frac{\xi}{2}x - 2a\hbar\sin\frac{\xi}{2}x\sin\frac{2-\xi}{2}x \qquad (5.181) \\ &= 2\hbar\sin^2\frac{\xi}{2}x\left(2\hbar - a\frac{\sin\frac{2-\xi}{2}x}{\sin\frac{\xi}{2}x}\right) \\ &= 2\hbar\sin^2\frac{\xi}{2}x\left(2\hbar - a\frac{\sin k\frac{x}{k+1}}{\sin\frac{x}{k+1}}\right), \qquad (5.182) \end{aligned}$$

where $k = 2n - 1$. By Lemma 5.31 and (5.177), we know that

$$2\hbar - a\frac{\sin k\frac{x}{k+1}}{\sin\frac{x}{k+1}} > 0. \tag{5.183}$$

So, (5.180) holds if and only if

$$\sin\frac{\xi}{2}x = 0.$$

This implies that $\sin\xi x = \sin x = 0$, and $2x\hbar\sin\xi x - ax\sin x = 0$, which further implies that $f(x, D) = x^2$ in this case. Thus, $f(x, D) > 0$ for all $x \neq 0$ and $D \in \mathbb{R}$.

Now, we consider the case when

$$2\hbar^2 - 2\hbar^2\cos\xi x + a\hbar\cos x - a\hbar\cos(1-\xi)x \neq 0. \tag{5.184}$$

By (5.182) and (5.183), we have $\sin\frac{\xi}{2}x \neq 0$, and

$$2\hbar^2 - 2\hbar^2\cos\xi x + a\hbar\cos x - a\hbar\cos(1-\xi)x > 0.$$

So, in this case $f(x, D)$ can be considered as a quadratic polynomial of D with a positive coefficient of the second order term. Then, $f(x, D) < 0$ if and only if the discriminant

$$\Delta = \beta^2 - 4\alpha\zeta < 0, \tag{5.185}$$

where α, β, ζ are the coefficients of the quadratic polynomial $f(x, D)$. Straightforward calculation shows that

$$\Delta = -x^2(2\hbar\cos\xi x - a\cos x - (2\hbar + a))(2\hbar\cos\xi x - a\cos x - (2\hbar - a)).$$

So, (5.185) holds if

$$2\hbar\cos\xi x - a\cos x < 2\hbar - a. \tag{5.186}$$

By using Lemma 5.29 we have

$$\begin{aligned}
2\hbar\cos\xi x - a\cos x &= 2\hbar\cos\frac{x}{n} - a\cos x \\
&= 2\hbar\cos\frac{x}{n} - av_n\left(\cos\frac{x}{n}\right) \\
&\triangleq u(z),
\end{aligned}$$

where $z = \cos\frac{x}{n}$. Obviously, $-1 \leq z < 1$ because $\sin\frac{\xi}{2}x \neq 0$. It follows from (5.177) that

$$\begin{aligned}
|v'(z)| &= |ne_nz^{n-1} + (n-1)e_{n-1}z^{n-2} + \cdots + e_1| \\
&\leq n|e_n| + (n-1)|e_{n-1}| + \cdots + |e_1| \\
&< \frac{2h}{a},
\end{aligned}$$

which implies that

$$u'(z) > 0.$$

When $z = \cos\frac{x}{n} = 1$, we have $\cos x = 1$, and thus, $u(1) = 2\hbar - a$. So, $u(z) < u(1) = 2\hbar - a$ for all $-1 \leq z < 1$. Therefore, (5.185) is indeed true. The proof of the lemma is thus completed. □

Now, we can present the following result which provides an explicit procedure for the design of the parameters of the TDFC.

Theorem 5.33 *Consider the primal algorithm (5.149)–(5.150) with TDFC given by*

$$u_i(t) = -\hbar_i(x_i(t) - x_i(t - \tau_i)). \tag{5.187}$$

Take

$$\hbar_i = \frac{\tau_i}{n_i}, \tag{5.188}$$

where n_i, $i \in \overline{S}$, are any positive integers. Then, the closed-loop system is locally stable at the equilibrium point, if

$$\hbar > \max\left\{\frac{a_i}{2}(n_i|e_{n_i}| + (n_i - 1)|e_{n_i-1}| + \cdots + |e_1|), \frac{(2n_i - 1)a_i}{2}\right\},$$

where

$$a_i = \kappa_i\left(\sum_{l\in L_i} p_l + \sum_{l\in L_i} p' \sum_{r\in S_l} x_r^\star\right), \tag{5.189}$$

and $e_1, \ldots, e_{n_i}$ are coefficients of $v_{n_i}(\cos x)$ defined by (5.175).

Proof. Similar to the proof of Theorem 5.26, we get the characteristic equation of the linearized closed-loop system as

$$I + M(s)\mathrm{diag}\left\{\frac{\mathrm{e}^{-sD_i}}{s + \hbar_i(1 - \mathrm{e}^{-s\tau_i})}\right\} = 0.$$

Since the equation

$$s + \hbar_i(1 - \mathrm{e}^{-s\tau_i}) = 0$$

has no RHP roots for any $i \in \overline{S}$ according to Lemma 5.27, from the general Nyquist stability criterion (Theorem 2.20) we know that the system (5.159) is asymptotically stable, if and only if the eigenloci of $M(\mathrm{j}\omega)\mathrm{diag}\{\frac{\mathrm{e}^{-\mathrm{j}\omega D_i}}{\mathrm{j}\omega+\hbar_i(1-\mathrm{e}^{-\mathrm{j}\omega\tau_i})}\}$, i.e.

$$\lambda\left(M(\mathrm{j}\omega)\mathrm{diag}\left\{\frac{\mathrm{e}^{-\mathrm{j}\omega D_i}}{\mathrm{j}\omega+\hbar_i(1-\mathrm{e}^{-\mathrm{j}\omega\tau_i})}\right\}\right),$$

do not enclose the point $(-1, \mathrm{j}0)$ for $\omega \in [0, \infty)$, where $M(s)$ is defined by (5.158).

Denote

$$\tilde{M}(\mathrm{j}\omega) = \mathrm{diag}\left\{\frac{1}{\sqrt{a_i}},\ i \in \overline{S}\right\} M(\mathrm{j}\omega)\mathrm{diag}\left\{\frac{1}{\sqrt{a_i}},\ i \in \overline{S}\right\}. \tag{5.190}$$

Then, we have

$$\begin{aligned}&\{0\} \cup \lambda\left(M(\mathrm{j}\omega)\mathrm{diag}\left\{\frac{\mathrm{e}^{-\mathrm{j}\omega D_i}}{\mathrm{j}\omega+\hbar_i(1-\mathrm{e}^{-\mathrm{j}\omega\tau_i})}\right\}\right)\\ &= \{0\} \cup \lambda\left(\tilde{M}(\mathrm{j}\omega)\mathrm{diag}\left\{\frac{\mathrm{e}^{-\mathrm{j}\omega D_i}}{\mathrm{j}\omega+\hbar_i(1-\mathrm{e}^{-\mathrm{j}\omega\tau_i})}\right\}\right).\end{aligned}$$

Hence, from Lemma 2.34 it follows that

$$\begin{aligned}&\lambda\left(M(\mathrm{j}\omega)\mathrm{diag}\left\{\frac{\mathrm{e}^{-\mathrm{j}\omega D_i}}{\mathrm{j}\omega+\hbar_i(1-\mathrm{e}^{-\mathrm{j}\omega\tau_i})}\right\}\right)\\ &\in \rho\left(\tilde{M}(\mathrm{j}\omega)\right)\mathrm{Co}\left(0 \cup \left\{\frac{a_i\mathrm{e}^{-\mathrm{j}\omega D_i}}{\mathrm{j}\omega+\hbar_i(1-\mathrm{e}^{-\mathrm{j}\omega\tau_i})}\right\}\right).\end{aligned}$$

Using Corollary 2.31 of Gershgorin's disc lemma, (5.190) and (5.189), we have $\rho\left(\tilde{M}(\mathrm{j}\omega)\right) \leq 1$. By Lemma 5.32 we know

$$\mathrm{Re}\left(\frac{a_i\mathrm{e}^{-\mathrm{j}\omega D_i}}{\mathrm{j}\omega+\hbar_i(1-\mathrm{e}^{-\mathrm{j}\omega\tau_i})}\right) > -1,$$

and hence,

$$(-1, \mathrm{j}0) \notin \mathrm{Co}\left(0 \cup \left\{\frac{a_i\mathrm{e}^{-\mathrm{j}\omega D_i}}{\mathrm{j}\omega+\hbar_i(1-\mathrm{e}^{-\mathrm{j}\omega\tau_i})}\right\}\right).$$

Theorem 5.33 is thus proved. □

Example 5.34 *Design of TDFC for stabilizing the primal algorithm with diverse round-trip delays.*

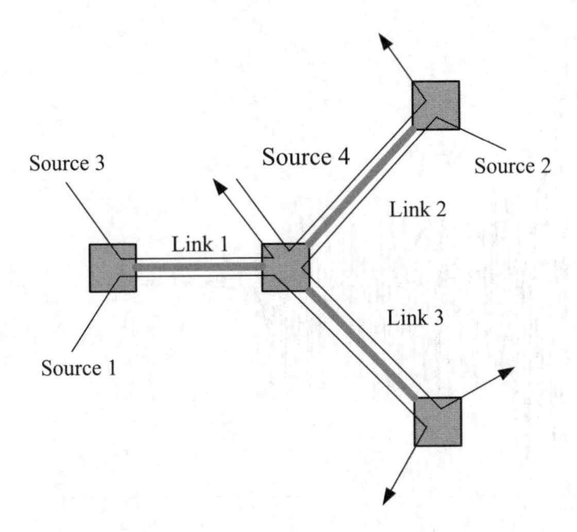

Figure 5.20 A network controlled by TDFC.

Consider a network with three links shared by four sources, which is illustrated by Figure 5.20. Obviously, the routing matrix of this network is

$$R = \begin{bmatrix} 1 & 0 & 1 & 0 \\ 0 & 1 & 0 & 1 \\ 1 & 1 & 0 & 0 \end{bmatrix}.$$

The congestion control algorithm is given by

$$\dot{x}_i(t) = \kappa_i \left(w_i - x_i(t - D_i) \sum_{l \in L_i} q_l(t - d_{li}^{\leftarrow}) \right) - \hbar_i(x_i(t) - x_i(t - \tau_i)), \quad i \in \overline{1, 4},$$

$$q_l(t) = p_l \left(\sum_{r \in S_l} x_r(t - d_{lr}^{\rightarrow}) \right), \quad l \in \overline{1, 3},$$

where $p_l(y) = 1 - (1 - y)e^y$, $\forall l \in \overline{1, 3}$. Let $w_1 = w_2 = 0.0019$ and $w_3 = w_4 = 0.0009$. Then, we get the equilibrium as $x_i^\star = 0.075$(Mbps), $\forall i \in \overline{1, 4}$.

Take $D_1 = D_4 = 500$s, $D_2 = D_3 = 100$s; and $\kappa_1 = 0.0611$, $\kappa_2 = 0.3053$, $\kappa_3 = 0.6107$, $\kappa_4 = 0.1221$. Then, oscillation is observed in the network. Now, we take two sets of parameters of TDFC as $\tau_i = D_i$, $\hbar_i = a_i$ and $\tau_i = D_i/2$, $\hbar_i = 3a_i$, respectively, where a_i is determined by (5.189). The TDFC is put on after $t \geq 10^4$s. The simulation results for the two case are shown in Figure 5.21 and Figure 5.22, respectively. We see that in both cases the network is stabilized by TDFC. Now, we take $\tau = kD_i$, where k is set as $0.1, 0.2, 0.3, \ldots, 1.4$, respectively; and $\hbar_i = a_i$. Then, we have found the network is also stabilized. The experiments tell us that the TDFC designed by Theorem 5.33 is robust against non-commensurate self-delays.

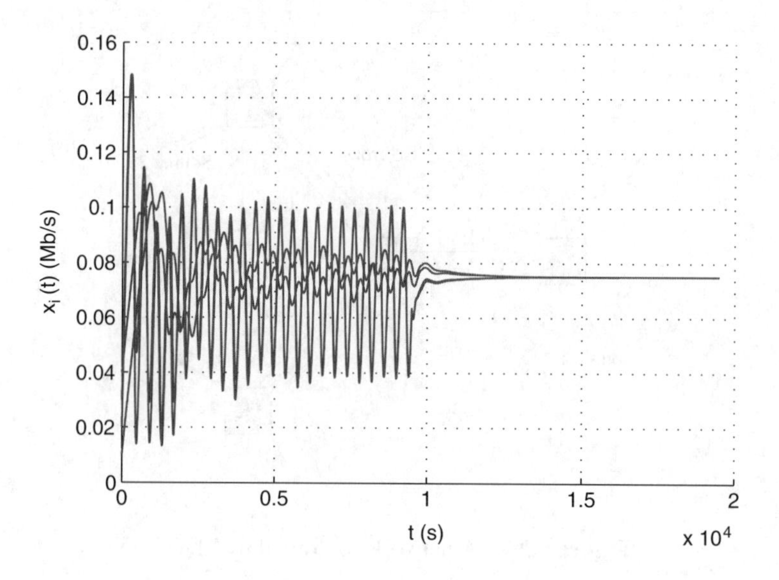

Figure 5.21 Source rates under TDFC.

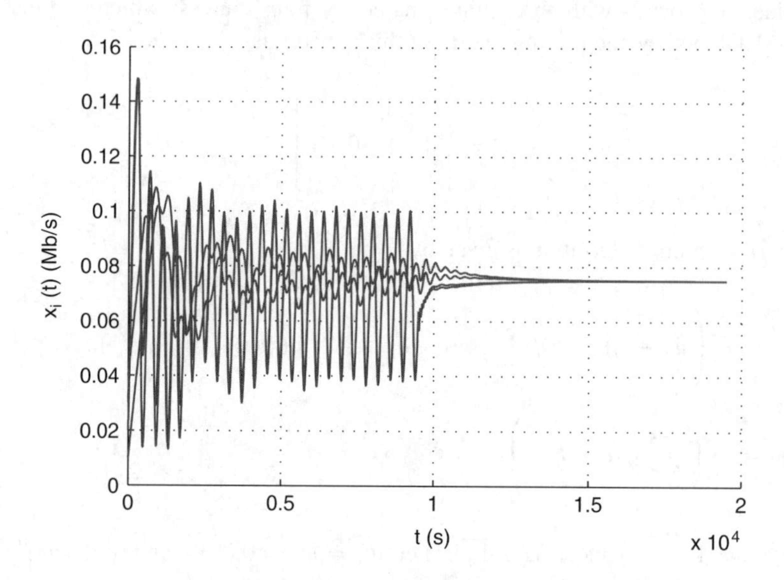

Figure 5.22 Sources rate under TDFC.

5.6 Notes and References

5.6.1 Stability of Congestion Control with Propagation Delays

As one of most important topics in congestion control, the stability issue has drawn much attention (Kelly 2003; Liu *et al.* 2003). The primal algorithm and the dual algorithm have been proved globally asymptotically stable in the absence of delay (Kelly, Maulloo and Tan 1998; Low and Lapsley 1999; Paganini 2002). Johari and Tan (2001) investigated the primal algorithm with propagation delays. They gave a nice decomposition of the transfer function matrix of the feedback congestion control system into a product of a diagonal and a Hermitian matrix, and used this to derive some sufficient conditions for the local stability of networks with the same round-trip delays for different TCP connections. For a more general case of networks with diverse round-trip delays, they proposed a conjecture (Conjecture 5.1) on the local stability of the algorithm. A weaker version of the continuous-time analogue of the conjecture (Conjecture 5.3) was proved by Massoulie (2002). To prove Conjecture 5.3 Vinnicombe (2000) proposed an elegant lemma (Lemma 2.34) which relates the eigenloci of the product of a Hermitian matrix and a diagonal matrix to the product of the spectral radius of the Hermitian matrix and the convex hull of the entries of the diagonal matrix. Using this lemma and the generalized Nyquist criterion (Desoer and Wang 1980), Vinnicombe illustrated the correctness of Conjecture 5.3. Tian and Yang (2004b) proved a more general stability criterion (Theorem 5.4) which includes Conjecture 5.3 as a special case. However, the scheme of Vinnicombe (2000) does not show the correctness of the original conjecture of Johari and Tan (Conjecture 5.1). Using the clockwise property of Nyquist plot of the time-delayed Tian and Yang (2004a) gave a rigorous proof of the Conjecture 5.1 via proving a more general stability criterion (Theorem 5.2). Theorem 5.2 preserves the elegance of Conjecture 5.1 being decentralized and locally implemented and enlarges the stability region of control gains and admissible communication delays. Tian (2005) and Tian and Chen (2006) analyzed the REM algorithm and the primal-dual algorithm respectively, and found that the clockwise property plus some other nice geometric properties of the dynamics of congestion control algorithms in a certain frequency interval determines the the scalability of these congestion control algorithms.

5.6.2 Time-Delayed Feedback Control

The original TDFC proposed by Pyragas (1992) was extended to multiple-delay systems (Socolar *et al.* 1994), observer-based version (Tian and Chen 2001) and a predictor-based version (Nakajima 2002). Ushio (1996) first studied the stability of the TDFC for discrete-time systems. The multiple-delay TDFC (Konishi, Ishii and Kokame 1999), observer-based TDFC (Konishi and Kokame 1998) and the predictor-based TDFC (Ushio and Yamamoto 1999) for discrete-time systems were also proposed in parallel with continuous-time systems. Finally, the TDFC was even extended to systems described by partial differential equations (Bleich and Socolar 1996; Harrington and Socolar 2001) and hybrid systems (Tian, Yu and Chua 2004).

In stability analysis for discrete-time systems under TDFC, Ushio (1996) proved that the TDFC has an inherent limitation, known as "odd number limitation". Ushio and Yamamoto (1998) gave a linear matrix inequality condition for designing the gain matrix of TDFC for a discrete-time system. Ushio and Yamamoto (1998) also showed that the odd number limitation can be overcome by using a nonlinear time-delayed feedback control method. Konishi and Kokame (1998), Ushio and Yamamoto (1999), and Yamamoto, Hino and Ushio (2001) then showed that the odd number limitation for discrete-time systems can also be overcome by other methods such as observer-based TDFC, predictor-based TDFC and dynamical TDFC.

For continuous-time systems, Nakajima (1997) proved that the odd number limitation holds for general higher-dimensional continuous-time systems under TDFC. Nakajima and Ueda (1998) further showed that the odd number limitation also holds for TDFC with multiple-delays. Tian and Chen (2001) proved that the odd number limitation exists in the case of observer-based dynamical output TDFC. Lemma 5.25 which was first given by Liu and Tian (2008) shows that the odd number limitation exists in systems with multiple delays.

However, the odd number limitation just gives a necessary condition of the stabilizability under TDFC. Kokame *et al.* (2001) proved that for a continuous-time system with single input, this condition is also sufficient for the existence of a stabilizing TDFC. Lemma 5.28 extends this result to scalar time-delayed systems. But this conclusion does not apply to discrete-time systems. Zhu and Tian (2005) found a new upper bound and gives a necessary and sufficient condition of the existence of a stabilizing TDFC for discrete-time systems with single input. Tian and Zhu (2004) gives a full characterization on stabilizability single input discrete-time systems under extended TDFC.

Chen, Wang and Han (2004) and Jiang, Wang and Xi (2004) used TDFC to adjust some parameters of the RED algorithm, an AQM algorithm in the Internet congestion control (Floyd and Jacobson 1993). Liu and Tian (2008) introduced the TDFC as a stabilizer in distributed congestion control systems. Theorem 5.26 which is due to Liu and Tian (2008) shows that the primal algorithm with any diverse round-trip delays can always be stabilized by distributed TDFC. Theorem 5.33 is a new result which provides an explicit design procedure of TDFC for distributed time-delayed systems.

References

Athuraliya S, Low SH and Yin Q (2001). REM: Active Queue Management. *IEEE Network*, 15, 48–53.

Basso M, Genesio R, and Tesi A (1997). Stabilizing periodic orbits of forced systems via generalized Pyragas controllers. *IEEE Transactions on Circuits and Systems–I*, 44(10), 1023–1027.

Bleich ME and Socolar JES (1996). Controlling spatiotemporal dynamics with time-delay feedback. *Physics Review E*, 54, R17–R20.

Chen L, Wang X and Han Z (2004). Control chaos in internet congestion control model. *Chaos, Solitons & Fractals*, 21, 81–91.

Desoer CA and Wang Y-T (1980). On the generalized Nyquist Stability Criterion. *IEEE Transactions on Automatic Control*, 25(2), 187–196.

Floyd S and Jacobson V (1993). Random early detection gateways for congestion avoidance. *IEEE/ACM Transactions on Networking*, 1, 397–413.

Gu K, Kharitonov VL and Chen J (2003). *Stability of Time-delayed Systems*. Birkhäuser, Boston.

Harrington I and Socolar JES (2001). Limitation on stabilizing plane waves via time-delayed feedback. *Physics Review E*, 64(5), 056206(6).

Hino T, Yamamoto S and Ushio T (2002). Stabilization of unstable periodic orbits of chaotic discrete-time systems using prediction-based feedback control. *International Journal Bifurcation and Chaos*, 12(2), 439–446.

Jiang K, Wang X and Xi Y (2004). A robust RED algorithm based on time-delayed feedback control. *Proceeding of the 5th Asian Control Conference* 2, 708–713.

Johari R and Tan D (2001). End to end congestion control for the internet: delays and stability. *IEEE/ACM Transactions on Networking*, 9, 818–832.

Just W, Bernard T, Ostheimer M, Reibold E and Benner H (1997). Mechanism of time-delayed feedback control. *Physics Review Letters*, 78(2), 203–206.

Just W, Reckwerth D, Mockel J, Reibold E and Benner H (1998). Delayed feedback control of periodic orbits in autonomous systems. *Physics Review Letters*, 81(3), 562–565.

Kelly FP, Maulloo A and Tan D (1998). Rate control for communication networks: shadow prices proportional fairness and stability. *Journal of the Operational Research Society*, 49, 237–252.

Kelly FP (2003). Fairness and stability of end-to-end congestion control. *European Journal of Control*, 9, 149–165.

Kokame H, Hirata K, Konishi K and Mori T (2001). Difference feedback can stabilize uncertain steady states. *IEEE Transactions on Automatic Control*, 46, 1908–1913.

Konishi K and Kokame H (1998). Obsever-based delayed-feedback control for discrete-time chaotic systems. *Physics Letters A*, 248(5-6), 359–368.

Konishi K, Ishii M and Kokame H (1999). Stability of extended delayed-feedback control for discrete-time chaotic system. *IEEE Transactions on Circuits and Systems–I*, 46(10), 1285–1288.

Li C, Chen G, Liao X and Yu J (2004). Hopf bifurcation in an Internet congestion control model. *Chaos, Solitons & Fractals* 6, 853–62.

Liu C-L and Tian Y-P (2008). Eliminating oscillations in the Internet by time-delayed feedback control. *Chaos, Solitons and Fractals*, 35, 878–887

Liu S, Başar T and Srikant R (2003). Controlling the Internet: A survey and some new results. *Proceedings of IEEE Conference on Decision and Control*, Maui, Hawaii.

Low SH and Lapsley DE (1999). Optimization flow control–I: basic algorithm and convergence. *IEEE/ACM Transactions on Networking*, 7, 861–874.

Massoulie L (2002). Stability of distributed congestion control with heterogeneous feedback delays. *IEEE Transactions on Automatic Control*, 47, 895–902.

Moon FC (1987). *Chaotic vibrations*. John Wiley, New York.

Nakajima H (1997). On analytical properties of delayed feedback control of chaos. *Physics Letters A*, 232, 207–210.

Nakajima H and Ueda Y (1998). Limitation of generalized delayed feedback control. *Physica D*, 111, 143–150.

Nakajima H (2002). Delayed feedback control with state predictor for continuous-time chaotic system. *International Journal of Bifurcation and Chaos*, 12(5), 1067–1077.

Ott E, Simmendinger C and Hess O (1996). Controlling delayed-induced chaotic behavior of a semiconductor laser with optical feedback. *Physics Letters A*, 216, 97–105.

Paganini F (2002). A global stability result in network flow control. *Systems & Control Letters*, 40, 165–172.

Parmananda P, Madrigal R and de Ciencias MRF (1999). Stabilization of unstable steady states and periodic orbits in an electrochemical system using delayed-feedback control. *Physics Review E*, 59(5), 5266–5271.

Peitgen H-O, Jürgens H and Saupe D (1993). *Chaos and Fractal, new frontiers of science*. Springer-Verlag.

Pyragas K (1992). Continuous control of chaos by self-controlling feedback. *Physics Letters A*, 170, 421–428.

Schuster H (1984). *Deterministic Chaos*. Physik Verlag, Weinheim.

Socolar JES, Sukow DW and Gauthier DJ (1994). Stabilizing unstable periodic orbits in fast dynamics, *Physics Review E*, 50(5), 3245–3248.

Tian Y-P and Chen G (2001). A separation principle for dynamical delayed output feedback control of chaos. *Physics Letters A* 284, 31–42.

Tian Y-P, Yu X and Chua LO (2004). Time-delayed impulsive control of hybrid chaotic systems. *International Journal of Bifurcation and Chaos*, 14(3), 1091–1104.

Tian Y-P and Yang H-Y (2004a). Stability of the Internet congestion control with diverse delays. *Automatica*, 40, 1533–1541.

Tian Y-P and Yang H-Y (2004b). Stability of distributed congestion control with diverse communication delays. *Proceedings of the World Congress on Intelligent Control and Automation* 2, 1438–1442.

Tian Y-P and Zhu J (2004). Full characterization on limitation of generalized delayed feedback control for discrete-time systems. *Physica D* 198, 248–257.

Tian Y-P (2005). Stability analysis and design of the second-order congestion control for networks with heterogeneous delays. *IEEE/ACM Transactions on Networking*, 13, 1082–1093.

Tian Y-P and Chen G (2006). Stability of the primal-dual algorithm for congestion control. *International Journal of Control*, 79, 662–676.

Ushio T (1996). Limitation of delayed feedback control in nonlinear discrete-time systems. *IEEE Transactions on Circuits and Systems–I*, 43, 815–816.

Ushio T and Yamamoto S (1998). Delayed feedback with nonlinear estimation in chaotic discrete-time system. *Physics Letters A*, 247, 112–118.

Ushio T and Yamamoto S (1999). Prediction-based control of chaos. *Physics Letters A*, 264, 30–35.

Vinnicombe G (2000). On the stability of end-to-end congestion control for the Internet. *Technical report CUED/F-INFENG/TR.*, No.398, 2000.

Wiggins S (1988). *Global Bifurcations and Chaos.* Analytical Methods, Springer-Verlag.

Yamamoto S, Hino T and Ushio T (2001). Dynamical delayed feedback controllers for chaotic discrete-time systems. *IEEE Transactions on Circuits and Systems–I*, 48(6), 785–789.

Yamamoto S and Ushio T (2003). Odd number limitation in delayed feedback control. *Chaos Control: Theory and Application*, edited by G. Chen and X. Yu, Lecture Notes in Control and Information Sciences, 292, 71–87, Springer, Berlin.

Yang H-Y and Tian Y-P (2005). Hopf bifurcation in REM algorithm with communication delay. *Chaos, Solitons & Fractals*, 25, 1093–1105.

Yin Q and Low SH (2001). Convergence of REM flow control at a single link. *IEEE Communication Letters*, 5, 119–121.

Yin Q and Low SH (2002). On stability of REM algorithm with uniform delay. *GLOBECOM '02, IEEE*, 3, 2649–2653.

Zhu J and Tian Y-P (2005). Necessary and sufficient conditions for stabilizability of discrete-time systems via delayed feedback control. *Physics Letters A* 343, 95–107.

6

Consensus in Homogeneous Multi-Agent Systems

Like seeks to like, and (birds) of the same note respond to one another – this is a rule of Heaven.

—Zhuang Zi (368–286 BC), 'The Old Fisherman'

Cooperation is a central topic in the design of a distributed control system. One of the most convenient and efficient approaches to undertaking a cooperative task for the agents of a distributed control system is to achieve some agreement (consensus). This chapter introduces basic notions in consensus problems, such as consensus protocol, the existence of consensus solutions and consentability. A unified approach to treating the consensus problem as a stability problem is presented. The consensus problem for systems with homogeneous agents of the first order, second order and nth order is investigated.

6.1 Introduction to Consensus Problem

In this section we introduction some basic notions in consensus problems by considering a simple multi-agent system: a system with integrator agents over an ideal network, i.e., a network without communication delays.

6.1.1 Integrator Agent System

Consider a multi-agent system with a fixed topology digraph $G = (V, E, A)$. Each agent can be considered as a node in the digraph, and an information flow between two agents can be regarded as a directed path between two nodes. Each dynamic agent is described by

$$\dot{x}_i(t) = u_i(t), \tag{6.1}$$

where $i \in \overline{1, n}$ and $x_i \in \mathbb{R}$, and $u_i \in \mathbb{R}$. The control u_i adopts the following *consensus protocol*

$$u_i(t) = \gamma \sum_{j \in N_i} a_{ij}(x_j(t) - x_i(t)), \tag{6.2}$$

where $\gamma > 0$ is the control gain.

Frequency-Domain Analysis and Design of Distributed Control Systems, First Edition. Yu-Ping Tian.

We say a *consensus solution* (or *consensus* in short) is asymptotically achieved in the system if

$$|x_i(t) - x_j(t)| \to 0 \text{ as } t \to \infty, \quad \forall i, j \in \overline{1, n}, \quad i \neq j. \tag{6.3}$$

6.1.2 Existence of Consensus Solution

Let $x = [x_1, x_2, \ldots, x_n]^{\mathrm{T}}$. With the control protocol (6.2), the closed-loop system can be written in the matrix form as

$$\dot{x} = -\gamma L x, \tag{6.4}$$

where $L = \{l_{ij}\}$ is the Laplacian matrix of digraph G.

By Theorem 1.9, $LC = 0$ for any $C \in \operatorname{span}(\mathbf{1}_n)$. This implies that the system (6.1) with protocol (6.2) (namely, system (6.4)) indeed contains a consensus solution $x_1(t) = x_2(t) = \cdots = x_n(t) = c$, where $c \in \mathbb{R}$ is a constant determined by the initial condition of the system. Actually, all the consensus solutions $x(t) = c\mathbf{1}_n$, $c \in \mathbb{R}$ form a continuum of equilibria of the system, denoted by X_e.

6.1.3 Consensus as a Stability Problem

To check if the system asymptotically achieves a consensus solution in X_e, let us introduce the error variables

$$e_i = x_{i+1} - x_1, \quad i \in \overline{1, (n-1)}. \tag{6.5}$$

Denote $e = [e_1, \ldots, e_{n-1}]^{\mathrm{T}}$. Then, we get

$$\dot{e} = -\gamma \tilde{L} e, \tag{6.6}$$

where

$$\tilde{L} = \begin{bmatrix} l_{22} - l_{12} & \cdots & l_{2n} - l_{1n} \\ \vdots & \vdots & \vdots \\ l_{n2} - l_{12} & \cdots & l_{nn} - l_{1n} \end{bmatrix}. \tag{6.7}$$

Obviously, system (6.4) reaches the consensus asymptotically if and only if system (6.6) is asymptotically stable.

Proposition 6.1 *Let the eigenvalues of L be denoted as* $0, \mu_2, \ldots, \mu_n$. *Then, the eigenvalues of* $\tilde{L}$ *defined by (6.7) are* $\mu_2, \ldots, \mu_n$.

Proof. By choosing a nonsingular transformation matrix

$$P = \begin{bmatrix} 1 & 0 \\ -\mathbf{1}_{n-1} & I_{n-1} \end{bmatrix}, \tag{6.8}$$

we have $PLP^{-1} = \begin{bmatrix} 0 & \phi \\ 0 & \tilde{L} \end{bmatrix}$, where $\phi = [l_{12}, l_{13}, \ldots, l_{1n}]$. This implies that the eigenvalues of $\tilde{L}$ are exactly the remaining $n-1$ eigenvalues of L excluding zero. □

Therefore, by Theorem 1.9, all the eigenvalues of $\tilde{L}$ have positive real parts if and only if the digraph G contains a globally reachable node. Thus, the following theorem is obtained.

Theorem 6.2 *System (6.1) with protocol (6.2) reaches a consensus solution asymptotically if and only if the interconnection topology digraph G contains a globally reachable node.*

6.1.4 Discrete-Time Systems

Consider the discrete-time system

$$x_i(k+1) = x_i(k) + u_i(k) \tag{6.9}$$

with consensus protocol

$$u_i(k) = \gamma \sum_{j \in N_i} a_{ij}(x_j(k) - x_i(k)). \tag{6.10}$$

In a similar way to that shown in Section 6.1.3, the consensus problem for the discrete-time system can be converted to the stability problem for the following system

$$e(k+1) = e(k) - \gamma \tilde{L} e(k).$$

Then, one can easily obtain the consensus condition for the system which is summarized in the following theorem.

Theorem 6.3 *System (6.9) with protocol (6.10) reaches a consensus solution asymptotically if and only if the interconnection topology digraph $G(V, E, A)$ has a globally reachable node and the following condition holds:*

$$\gamma \mathrm{Re}(\mu_i) < 2, \quad \forall i \in \overline{2, n}, \tag{6.11}$$

where μ_i's are the eigenvalues of the Laplacian matrix $L(G)$ except for zero.

Remark. By Corollary 2.31 of Gershgorin's disc lemma, it is easy to get a sufficient but scalable condition for (6.11) as

$$\gamma \sum_{j \in N_i} a_{ij} < 1, \quad \forall i \in \overline{1, n}. \tag{6.12}$$

6.1.5 Consentability

For a multi-agent system, if there exists a consensus protocol such that a consensus can be reached asymptotically, then we say the system is *consentable*. Theorem 6.2 (Theorem 6.3) tells us that there always exists a control gain γ such that system (6.1) with protocol (6.4) (system (6.9) with protocol (6.10)) can reach a consensus solution provided the interconnection topology digraph G contains a globally reachable node. So, for the system with integrator agents over ideal networks the connectivity of the graph serves as the sole condition of consentability.

Exercise 6.4 *Prove that Theorem 6.2 and Theorem 6.3 are true for the case when the states in (6.1) and (6.9) are p-dimensional vectors, i.e.,* $x_i \in \mathbb{R}^p, \forall i \in \overline{1, n}$.

(Hint: Reformulate the consensus problem as a stability problem, i.e.,

$$\dot{e} = -\gamma \tilde{L} \otimes I_p e$$

for the continuous-time system, and

$$e(k+1) = e(k) - \gamma \tilde{L} \otimes I_p e(k)$$

for the discrete-time system, where $\otimes$ denotes the Kronecker product, $e = [e_1^{\mathrm{T}}, \ldots, e_{n-1}^{\mathrm{T}}]^{\mathrm{T}}$, and $e_i = x_{i+1} - x_1, \;\; i \in \overline{1, (n-1)}$.)

6.2 Second-Order Agent System

6.2.1 Consensus and Stability

Now, we consider dynamic agents with second-order dynamics

$$\begin{aligned} \dot{\xi}_i(t) &= \zeta_i(t) \\ \dot{\zeta}_i(t) &= -\beta \xi_i(t) - \alpha \zeta_i(t) + u_i(t), \quad i \in \overline{1, n} \end{aligned} \tag{6.13}$$

where $\xi_i(t) \in \mathbb{R}$ and $\zeta_i(t) \in \mathbb{R}$ are the states (for many physical systems they represent the position and velocity); $u_i(t) \in \mathbb{R}$ is the control input; $\alpha \in \mathbb{R}$ and $\beta \in \mathbb{R}$ are some constant. Obviously, when $\alpha = \beta = 0$, the agent (6.13) reduces to the so-called double-integrator which has been widely studied in references. Through such a general second-order agent different consensus dynamics including periodic dynamics and positive exponential dynamics can be realized by choosing different consensus gains. The control u_i adopts the following consensus protocol

$$u_i = \sum_{j \in N_i} a_{ij} [\gamma_0 (\xi_j - \xi_i) + \gamma_1 (\zeta_j - \zeta_i)], \tag{6.14}$$

where N_i is the set of neighbors of agent i.

Remark. For simplicity of symbolization and statement we let $\xi_i(t) \in \mathbb{R}$ and $\zeta_i(t) \in \mathbb{R}$ in (6.13). However, by using the Kronecker product technique all the discussion in this section can be easily extended to the case when $\xi_i(t) \in \mathbb{R}^p$ and $\zeta_i(t) \in \mathbb{R}^p$, and the results still hold.

Let $\xi = [\xi_1, \xi_2, \ldots, \xi_n]^{\mathrm{T}}$ and $\zeta = [\zeta_1, \zeta_2, \ldots, \zeta_n]^{\mathrm{T}}$. With the control protocol (6.14), the closed-loop system of can be written in the matrix form as

$$\begin{bmatrix} \dot{\xi} \\ \dot{\zeta} \end{bmatrix} = \Gamma \begin{bmatrix} \xi \\ \zeta \end{bmatrix}, \tag{6.15}$$

where

$$\Gamma = \begin{bmatrix} 0_{n\times n} & I_n \\ -\beta I_n - \gamma_0 L & -\alpha I_n - \gamma_1 L \end{bmatrix}. \tag{6.16}$$

It is said that a consensus is asymptotically achieved in system (6.15) if $|\xi_i - \xi_j| \to 0$ and $|\zeta_i - \zeta_j| \to 0$ as $t \to \infty$.

To analyze the stability property of system (6.15), we first research the eigenvalues of Γ. Denote the eigenvalues of L by $\mu_1 = 0, \mu_2, \ldots, \mu_n$. Simple computation shows that

$$\begin{aligned} \det(\lambda I_{2n} - \Gamma) &= \det \begin{bmatrix} \lambda I_n & -I_n \\ \beta I_n + \gamma_0 L & \lambda I_n + \alpha I_n + \gamma_1 L \end{bmatrix} \\ &= \det \begin{bmatrix} 0 & -I_n \\ \beta I_n + \gamma_0 L + \lambda(\lambda I_n + \alpha I_n + \gamma_1 L) & \lambda I_n + \alpha I_n + \gamma_1 L \end{bmatrix} \\ &= \det[\lambda^2 I_n + (\alpha I_n + \gamma_1 L)\lambda + (\beta I_n + \gamma_0 L)]. \end{aligned} \tag{6.17}$$

Writing L in upper triangular Jordan blocks with $\mu_i (i \in \overline{1, n})$ as diagonal elements, from (6.17) we get

$$\det(\lambda I_{2n} - \Gamma) = \prod_{i=1}^{n} [\lambda^2 + (\alpha + \gamma_1 \mu_i)\lambda + (\beta + \gamma_0 \mu_i)]. \tag{6.18}$$

Thus for each μ_i, there exist two eigenvalues of Γ, denoted by λ_{i1} and λ_{i2} respectively. Since $\mu_1 = 0$, we obtain two eigenvalues λ_{11} and λ_{12} of Γ:

$$\lambda_{11,\,12} = \frac{-\alpha \pm \sqrt{\alpha^2 - 4\beta}}{2}, \tag{6.19}$$

which are determined only by parameters of the agent. Obviously, $\lambda_{11,12} = 0$ if and only if $\alpha = 0$, $\beta = 0$. Now, we show that the eigenvalues of Γ excluding $\lambda_{11,12}$ determine if the consensus is achieved in system (6.15), and $\lambda_{11,12}$ determine the dynamics of the achieved consensus solution.

Theorem 6.5 *System (6.15) reaches a consensus asymptotically if and only if*

$$\mathrm{Re}(\lambda_{ij}) < 0, \; i = 2, 3, \ldots, n; \; j = 1, 2. \tag{6.20}$$

And the dynamics of the consensus solution are determined by the equation

$$\begin{bmatrix} \dot{\xi}_1 \\ \dot{\zeta}_1 \end{bmatrix} = \begin{bmatrix} 0 & 1 \\ -\beta & -\alpha \end{bmatrix} \begin{bmatrix} \xi_1 \\ \zeta_1 \end{bmatrix}. \tag{6.21}$$

Proof. Let

$$\begin{aligned} \tilde{\xi} &= [\xi_2, \ldots, \xi_n]^{\mathrm{T}}, \\ \tilde{\zeta} &= [\zeta_2, \ldots, \zeta_n]^{\mathrm{T}}, \\ x(t) &= \tilde{\xi} - \xi_1 \mathbf{1}_{n-1}, \\ y(t) &= \tilde{\zeta} - \zeta_1 \mathbf{1}_{n-1}. \end{aligned}$$

Then, the consensus is achieved asymptotically if and only if $x(t) \to 0$ and $y(t) \to 0$ as $t \to \infty$. Let the nonsingular transformation matrix P be given by (6.8). Then, we have

$$\begin{bmatrix} \xi_1 \\ x \\ \zeta_1 \\ y \end{bmatrix} = \begin{bmatrix} P & 0 \\ 0 & P \end{bmatrix} \begin{bmatrix} \xi \\ \zeta \end{bmatrix}, \tag{6.22}$$

and

$$PLP^{-1} = \begin{bmatrix} 0 & \phi \\ 0 & \tilde{L} \end{bmatrix} \tag{6.23}$$

where $\tilde{L}$ was given by (6.7).

With the linear transformation (6.22), the closed-loop system (6.15) is transformed to

$$\begin{bmatrix} \dot{\xi}_1 \\ \dot{x} \\ \dot{\zeta}_1 \\ \dot{y} \end{bmatrix} = \begin{bmatrix} 0 & 0 & 1 & 0 \\ 0 & 0 & 0 & I_{n-1} \\ -\beta & -\gamma_1 \phi & -\alpha & -\gamma_0 \phi \\ 0 & -\beta - \gamma_1 \tilde{L} & 0 & -\alpha - \gamma_0 \tilde{L} \end{bmatrix} \begin{bmatrix} \xi_1 \\ x \\ \zeta_1 \\ y \end{bmatrix},$$

or equivalently,

$$\begin{bmatrix} \dot{\xi}_1 \\ \dot{\zeta}_1 \\ \dot{x} \\ \dot{y} \end{bmatrix} = \begin{bmatrix} 0 & 1 & 0 & 0 \\ -\beta & -\alpha & -\gamma_1 \phi & -\gamma_0 \phi \\ 0 & 0 & 0 & I_{n-1} \\ 0 & 0 & -\beta - \gamma_1 \tilde{L} & -\alpha - \gamma_0 \tilde{L} \end{bmatrix} \begin{bmatrix} \xi_1 \\ \zeta_1 \\ x \\ y \end{bmatrix}. \tag{6.24}$$

Hence, the consensus is achieved if and only if all the eigenvalues of

$$\tilde{\Gamma} := \begin{bmatrix} 0 & I_{n-1} \\ -\beta - \gamma_1 \tilde{L} & -\alpha - \gamma_0 \tilde{L} \end{bmatrix} \tag{6.25}$$

lie in the left-half complex plane. Since we have shown that the eigenvalues of $\tilde{L}$ are the remaining $n - 1$ eigenvalues of L excluding zero, i.e., $\mu_2, \ldots, \mu_n$ (Proposition 6.1), it follows from (6.18) that the eigenvalues of $\tilde{\Gamma}$ are exactly the remaining $2(n - 1)$ eigenvalues of Γ excluding $\lambda_{11,12} = 0$. Hence, system (6.15) reaches a consensus asymptotically if and only if $\mathrm{Re}(\lambda_{ij}) < 0, i = 2, 3, \ldots, n;\ j = 1, 2$.

When $x(t) \to 0$, $y(t) \to 0$, from (6.24) it is clear that the dynamics of the consensus solution is determined by equation (6.21), whose eigenvalues are given by (6.19). The theorem is thus proved. □

Obviously, if $\alpha > 0$ and $\beta > 0$, then the consensus solution is asymptotically stable. An asymptotically stable solution is called *trivial consensus solution* in this book. For a trivial consensus solution, cooperation of agents is not necessary because it can be easily achieved by taking zero consensus gains, i.e., $\gamma_0 = \gamma_1 = 0$. Therefore, to achieve a non-trivial consensus we make the following assumption for system (6.13).

Assumption 6.6 *In system (6.13) $\alpha \leq 0$ or $\beta \leq 0$.*

6.2.2 Consensus and Consentability Condition

Now, we study how agent parameters, protocol parameters and the Laplacian matrix determine the eigenvalues of system (6.15). According to (6.18), we need to analyze the stability of a class of quadratic polynomials with complex coefficients.

Lemma 6.7 *Let $\mu \in \mathbb{C}$. The two roots of the polynomial*

$$f_\mu(\lambda) = \lambda^2 + (\alpha + \gamma_1\mu)\lambda + \beta + \gamma_0\mu \tag{6.26}$$

lie in the open left-half complex plane if and only if

$$\alpha > p\gamma_1, \tag{6.27}$$

$$\beta > \frac{\gamma_0^2 q^2}{(\alpha - p\gamma_1)^2} - \frac{\gamma_0\gamma_1 q^2}{\alpha - p\gamma_1} + p\gamma_0, \tag{6.28}$$

where $p = \mathrm{Re}(-\mu)$ and $q = \mathrm{Im}(-\mu)$.

Proof. Denote

$$f_{\mu^*}(\lambda) = \lambda^2 + (\alpha + \gamma_1\mu^*)\lambda + \beta + \gamma_0\mu^*,$$

where μ^* is the conjugate of μ. It is easy to see that $f_\mu(\lambda_0^*) = (f_{\mu^*}(\lambda_0))^*$ for any complex number λ_0, which implies that λ_0^* is a root of $f_\mu(\lambda)$ if and only if λ_0 is a root of $f_{\mu^*}(\lambda)$. Therefore, the complex coefficient polynomial $f(\lambda)$ has no RHP roots if and only if the real coefficient polynomial $f_\mu(\lambda)f_{\mu^*}(\lambda)$ has no RHP roots.

Let $a = \alpha - p\gamma_1$, $b = \beta - p\gamma_0$, $c = q\gamma_1$ and $d = q\gamma_0$. With simple calculations, we have

$$f_\mu(\lambda)f_{\mu*}(\lambda) = \lambda^4 + a_1\lambda^3 + a_2\lambda^2 + a_3\lambda + a_4, \tag{6.29}$$

where

$$a_1 = 2a, \quad a_2 = a^2 + 2b + c^2,$$

$$a_3 = 2ab + 2cd, \quad a_4 = b^2 + d^2.$$

By Hurwitz stability criterion, all the roots have negative real parts if and only if

$$\Delta_1 = 2a > 0, \tag{6.30}$$

$$\Delta_2 = \begin{vmatrix} a_1 & 1 \\ a_3 & a_2 \end{vmatrix} = 2(a^3 + ab + ac^2 - cd) > 0, \tag{6.31}$$

$$\Delta_3 = \begin{vmatrix} a_1 & 1 & 0 \\ a_3 & a_2 & a_1 \\ 0 & a_4 & a_3 \end{vmatrix} = 4(a^2+c^2)[a^2b+(ac-d)d] > 0, \tag{6.32}$$

$$\Delta_4 = (b^2 + d^2)\Delta_3 > 0. \tag{6.33}$$

Obviously, (6.30)–(6.33) are equivalent to

$$a > 0, \tag{6.34}$$

$$b > \frac{cd}{a} - a^2 - c^2, \tag{6.35}$$

$$b > \frac{d^2}{a^2} - \frac{cd}{a}, \tag{6.36}$$

$$b^2 + d^2 \neq 0. \tag{6.37}$$

In the following, we will show that (6.35) and (6.37) can be removed. With simple computation, one has

$$\left(\frac{d^2}{a^2} - \frac{cd}{a}\right) - \left(\frac{cd}{a} - a^2 - c^2\right) = a^2 + \frac{(d-ac)^2}{a^2} > 0,$$

which means that (6.35) is implied by (6.36). Moreover, if $b^2 + d^2 = 0$, then $b = d = 0$, which is contradictory to (6.36). Therefore, each root of (6.26) has a negative real part if and only if both (6.34) and (6.36), or equivalently (6.27) and (6.28), hold. □

Lemma 6.8 *Assume* $p = \mathrm{Re}(-\mu) < 0$, $q = \mathrm{Im}(-\mu)$, $\gamma_0 \geq 0$ *and* $\alpha \geq 0$. *Then, (6.27) and (6.28) hold if and only if*

$$\gamma_0 > \frac{\beta p}{p^2 + q^2}, \tag{6.38}$$

$$\gamma_1 > \frac{\alpha}{p} + \frac{\gamma_0|q|(\alpha|q| + \sqrt{\alpha^2 q^2 + 4\beta^\star p})}{2\beta^\star p}, \tag{6.39}$$

where $\beta^\star = (\beta - p\gamma_0)p - \gamma_0 q^2 = \beta p - \gamma_0(p^2 + q^2)$.

Proof. If $q = 0$ or $\gamma_0 = 0$, the lemma is obvious. We assume $q \neq 0$ and $\gamma_0 \neq 0$ in the following. Since $p < 0$, inequality (6.27) can be rewritten as $\gamma_1 > \alpha/p$. Moreover, if (6.27) holds, then (6.28) can be rewritten as

$$(\beta - p\gamma_0)(\alpha - p\gamma_1)^2 + \gamma_0\gamma_1 q^2(\alpha - p\gamma_1) - \gamma_0^2 q^2 > 0,$$

namely,

$$\beta^\star p\gamma_1^2 - (2\beta^\star + \gamma_0 q^2)\alpha\gamma_1 + (\beta - p\gamma_0)\alpha^2 - \gamma_0^2 q^2 > 0. \tag{6.40}$$

(*Necessity*) Let

$$h(\gamma) = \frac{\gamma_0^2 q^2}{(\alpha - p\gamma)^2} - \frac{\gamma_0 \gamma q^2}{\alpha - p\gamma} + p\gamma_0.$$

With simple calculations, we obtain

$$\frac{\mathrm{dh}}{\mathrm{d}\gamma} = \frac{-\gamma_0 q^2(\alpha - p\gamma)[\alpha(\alpha - p\gamma) - 2p\gamma_0]}{(\alpha - p\gamma)^4} < 0$$

for all $\gamma \in (\frac{\alpha}{p}, +\infty)$, which implies $h(\gamma)$ is a decreasing function. By (6.28) we have

$$\beta > h(\gamma_1) > \lim_{\gamma \to +\infty} h(\gamma) = p\gamma_0 + \frac{\gamma_0 q^2}{p}.$$

Thus, (6.38) holds and $\beta^\star < 0$. Denote by $g(\gamma_1)$ the left side of (6.40), which is a quadratic polynomial with respect to γ_1 with coefficient $\beta^\star p > 0$ and the discriminant

$$\begin{aligned}\Delta &= [(2\beta^\star + \gamma_0 q^2)\alpha]^2 - 4\beta^\star p[(\beta - p\gamma_0)\alpha^2 - \gamma_0^2 q^2] \\ &= \gamma_0^2 q^2 (q^2\alpha^2 + 4\beta^\star p) > 0.\end{aligned}$$

Solving (6.40) yields $\gamma_1 > \rho_1$ or $\gamma_1 < \rho_2$, where

$$\rho_{1,2} = \frac{\alpha}{p} + \frac{\gamma_0 |q|(\alpha|q| \pm \sqrt{q^2\alpha^2 + 4\beta^\star p})}{2p\beta^\star} \tag{6.41}$$

are the two roots of polynomial $g(\gamma_1)$. From (6.41), we can see that $\rho_2 < \alpha/p < \rho_1$. Thus, from $\gamma_1 > \alpha/p$, we obtain $\gamma_1 > \rho_1$, that is (6.39).

(*Sufficiency*) It is easy to see that (6.39) implies $\gamma_1 > \alpha/p$, that is, (6.27) holds. From (6.38), we have $\beta^\star < 0$, which implies $\beta^\star p > 0$. Thus, (6.40) is obtained from $\gamma_1 > \rho_1$. Hence (6.28) holds. □

Remark. For the case of $\alpha \geq 0$, $\beta = 0$ and $\gamma_1 > 0$, the necessary and sufficient condition of Lemma 6.8 reduces to the form:

$$\gamma_1 > \frac{\alpha}{p} + \frac{\alpha q^2 + |q|\sqrt{q^2\alpha^2 - 4p(p^2 + q^2)\gamma_0}}{-2p(p^2 + q^2)}. \tag{6.42}$$

In particular, if $q = 0$, then (6.42) holds for any $\gamma_1 > 0$, which implies Lemma 4.2 of Ren and Atkins (2007). Let $\alpha = \beta = 0$, $\gamma_1 > 0$ and $\gamma_0 > 0$. Then, (6.42) further reduces to the form:

$$\gamma_1 > \frac{\sqrt{\gamma_0}|\mathrm{Im}(-\mu)|}{|\mu|\sqrt{-\mathrm{Re}(-\mu)}}, \tag{6.43}$$

which includes the sufficient condition obtained by Ren and Atkins (2007) as a special case. Let $\alpha = \beta = 0$, $p < 0$ and $\gamma_1 = 1$. Then, by Lemma 6.7, polynomial (6.26) is stable if and only if

$$0 < \gamma_0 < \frac{-(p^2+q^2)p}{q^2}, \tag{6.44}$$

which is just the Lemma 2 of Lin *et al.* (2007).

By Theorem 6.5 and Lemma 6.7, we obtain the following theorem:

Theorem 6.9 *Let* $p_i = \mathrm{Re}(-\mu_i)$ *and* $q_i = \mathrm{Im}(-\mu_i)$. *Then the consensus is achieved asymptotically in system (6.13) with protocol (6.14) if and only if*

$$\max_{2\le i\le n}(p_i\gamma_1) < \alpha, \tag{6.45}$$

$$\max_{2\le i\le n}\left(\frac{\gamma_0^2 q_i^2}{(\alpha - p_i\gamma_1)^2} - \frac{\gamma_0\gamma_1 q_i^2}{\alpha - p_i\gamma_1} + p_i\gamma_0\right) < \beta. \tag{6.46}$$

Remark 1. Inequalities (6.45) and (6.46) describe all admissible parameters of system (6.13) and control gains of protocol (6.14) for consensus. If $\alpha > 0$ and $\beta > 0$, conditions (6.45) and (6.46) imply that all the eigenvalues of Γ have negative real parts. In this case, the multi-agent system is asymptotically stable, in other words, it achieves the so-called trivial consensus.

Remark 2. When $\alpha \le 0$ or $\beta \le 0$, then the non-trivial consensus is achieved if and only if conditions (6.45) and (6.46) hold. One may wonder why it is not assumed that the digraph G contains a globally reachable node (or equivalently say, G has a spanning tree) as stated in Ren and Beard (2005). As a matter of fact, this condition is implied by (6.45) and (6.46) as $\alpha \le 0$ or $\beta \le 0$. This is shown in the necessity part of the proof of next theorem.

Theorem 6.10 *Suppose that* $\alpha \le 0$ *or* $\beta \le 0$. *Then, the system (6.13) is consentable under protocol (6.14) if and only if the topology graph has a globally reachable node.*

Proof. (*Necessity*) Suppose the system (6.13) is consentable under protocol (6.14), but G has no globally reachable node. Then, by Theorem 1.9, there is a $\mu_i = 0$ $(i \ge 2)$, i.e., $p_i = q_i = 0$. Thus (6.45) and (6.46) imply $\alpha > 0$ and $\beta > 0$, which is a contradiction.

(*Sufficiency*) Suppose that G contains a globally reachable node. Then $\max_{2\le i\le n} p_i < 0$. Thus, (6.45) is equivalent to

$$\gamma_1 > \frac{\alpha}{\max_{2\le i\le n} p_i} \tag{6.47}$$

as $\gamma_1 > 0$. Therefore, for any given α and γ_0, there always exist sufficiently large numbers γ_1 and β such that (6.47) and (6.46) hold, that is, the consensus is achieved. □

When the condition that graph G contains a globally reachable node is assumed, the next theorem provides a necessary and sufficient condition for the design of control gains of the consensus protocol.

Theorem 6.11 *Let $p_i = \mathrm{Re}(-\mu_i)$ and $q_i = \mathrm{Im}(-\mu_i)$. Assume $\gamma_0 \geq 0$, $\alpha \geq 0$ and graph G contains a globally reachable node. Then the consensus is achieved in system (6.13) with protocol (6.14) if and only if*

$$\gamma_0 > \max_{2\leq i\leq n} \frac{\beta p_i}{p_i^2 + q_i^2}, \tag{6.48}$$

$$\gamma_1 > \max_{2\leq i\leq n} \left(\frac{\alpha}{p_i} + \frac{\gamma_0 \alpha q_i^2 + \gamma_0 |q_i| \sqrt{q_i^2 \alpha^2 + 4\beta_i^\star p_i}}{2\beta_i^\star p_i} \right), \tag{6.49}$$

where $\beta_i^\star = \beta p_i - \gamma_0(p_i^2 + q_i^2)$.

Proof. Since G contains a globally reachable node, by Theorem 1.9, we have $p_i < 0$ for all $i = 2, 3, \ldots, n$. Hence, by Theorem 6.5, Lemma 6.7 and Lemma 6.8, the theorem is proved. □

Corollary 6.12 *Assume $\gamma_0 > 0$, $\alpha \geq 0$, $\beta = 0$ and graph G contains a globally reachable node. Then the consensus is achieved in system (6.13) with protocol (6.14) if and only if*

$$\gamma_1 > \max_{2\leq i\leq n} \left(\frac{\alpha}{p_i} + \frac{\alpha q_i^2 + |q_i| \sqrt{q_i^2 \alpha^2 - 4 p_i \gamma_0 (p_i^2 + q_i^2)}}{-2 p_i (p_i^2 + q_i^2)} \right). \tag{6.50}$$

In particular, for $\alpha = \beta = 0$, the consensus is achieved by the protocol (6.14) if and only if

$$\gamma_1 > \max_{2\leq i\leq n} \frac{\sqrt{\gamma_0}|\mathrm{Im}(-\mu_i)|}{|\mu_i|\sqrt{-\mathrm{Re}(-\mu_i)}}. \tag{6.51}$$

Remark 1. Corollary 6.12 includes Theorem 4.2 and 4.3 of Ren and Beard (2005) as special cases. For the case of $\alpha = \beta = 0$, Corollary 6.12 is equivalent to Theorem 1 of Lin *et al.* (2007).

Remark 2. From (6.19), we know that α and β determine the consensus dynamics poles. If $\beta < 0$, then one consensus pole is positive. In this case, the positive exponential consensus is also achieved. As γ_0, α and β are given, γ_1 determines the error dynamics poles and the convergence speed. For the problem how to choose γ_1 in $(\hat{\gamma}_1, +\infty)$ such that the maximum convergence speed of the error dynamics is achieved, readers are referred to Zhu, Tian and Kuang (2009).

6.2.3 Periodic Consensus Solutions

By Theorem 6.5, the dynamics of the consensus solution is determined by each agent itself. Suppose that $4\beta - \alpha^2 > 0$. Then, from (6.19) it follows that two eigenvalues of the consensus

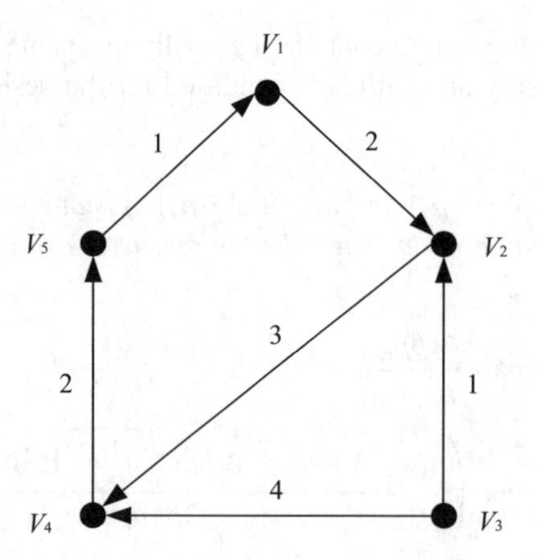

Figure 6.1 The directed graph of the multi-agent system.

dynamics are conjugate complex numbers with real part as $\mathrm{Re}\lambda_{11} = \mathrm{Re}\lambda_{12} = -\alpha/2$. Obviously, if $\alpha < 0$, then the positive exponential consensus is achieved. If $\alpha = 0$, then the periodic consensus is achieved. In particular, we let $\gamma_0 = 0$, then (6.47) and (6.46) are reduced to $\gamma_1 > 0$ and $\beta > 0$. For this special case, we have following theorem.

Theorem 6.13 *Suppose $\alpha = \gamma_0 = 0$. Then, periodic consensus is achieved in system (6.13) with protocol (6.14) if and only if $\gamma_1 > 0$, $\beta > 0$ and G has a spanning tree. Moreover, the vibration frequency of the periodic consensus dynamics is $\omega = \sqrt{\beta}$ and γ_1 determines the error dynamics poles.*

Proof. Substituting $\alpha = \gamma_0 = 0$ into (6.19), we obtain the consensus poles $\lambda_{11,12} = \pm\sqrt{\beta}\mathrm{j}$. By Theorem 6.5 and the proof of Theorem 6.10, we know that the periodic consensus is achieved if and only if $\gamma_1 > 0$, $\beta > 0$ and G contains a globally reachable node. Thus the vibration frequency of the periodic consensus dynamics is $\omega = \sqrt{\beta}$. From (6.18), we know that λ_{i1} and λ_{i2} are the roots of $\lambda^2 + \gamma_1\mu_i\lambda + \beta$ $(i \in \overline{2,n})$. Thus, as the vibration frequency is given, the positions of the other eigenvalues except λ_{11} and λ_{12} are determined by γ_1. Therefore, the error dynamics poles are determined by γ_1. □

6.2.4 Simulation Study

Consider the multi-agent systems with second-order agents given by (6.13) with $\alpha = 0$ and digraph shown in Figure 6.1. Obviously, the Laplacian matrix of the digraph is

$$L = \begin{bmatrix} 2 & -2 & 0 & 0 & 0 \\ 0 & 3 & 0 & -3 & 0 \\ 0 & -1 & 5 & -4 & 0 \\ 0 & 0 & 0 & 2 & -2 \\ -1 & 0 & 0 & 0 & 1 \end{bmatrix}.$$

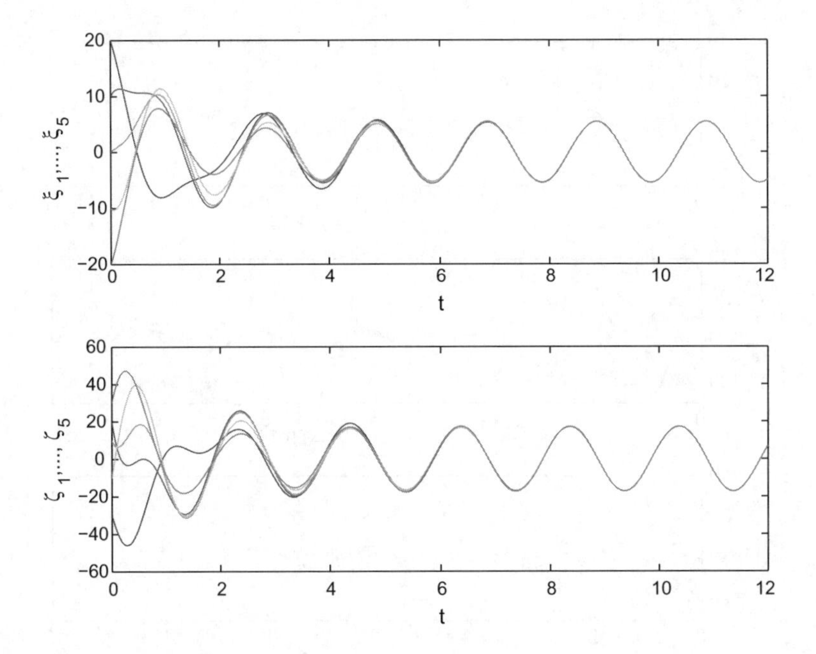

Figure 6.2 The periodic consensus with desired period 2. (Reprinted from *Linear Algebra and its Applications*, **431**, Zhu J., Tian Y.-P. and Kuang J., "On the general consensus protocol of multi-agent systems with double-integrator dynamics," 701–715, 2009, with permission from Elsevier.)

Let $\gamma_0 = 0$, $\gamma_1 = 1$ in the consensus protocol (6.14). Then, periodic consensus should be achieved according to Theorem 6.13. Suppose the desired period of consensus dynamics is $T = 2$. Then, $\beta = (2\pi/T)^2$ by Theorem 6.13. Set the initial value vector as $[\xi^{\mathrm{T}}(0)\ \ \zeta^{\mathrm{T}}(0)]^{\mathrm{T}}$ = $[20\ \ 10\ \ 0\ \ -10\ \ -20\ \ -30\ \ 20\ \ 10\ \ -10\ \ 30]^{\mathrm{T}}$. Figure 6.2 shows the achieved periodic consensus of ξ_i and ζ_i.

If we regard the graph shown in Figure 6.1 as an undirected graph. Then the Laplacian matrix of G is

$$L = \begin{bmatrix} 3 & -2 & 0 & 0 & -1 \\ -2 & 6 & -1 & -3 & 0 \\ 0 & -1 & 5 & -4 & 0 \\ 0 & -3 & -4 & 9 & -2 \\ -1 & 0 & 0 & -2 & 3 \end{bmatrix}.$$

With simple calculations, we obtain $\mu_2 = 2.7639$ $\mu_5 = 12.5826$. Hence, according to Zhu, Tian and Kuang (2009), the maximum convergence speed is achieved as $\gamma_1 = 0.7985$. The simulation can be seen in Figure 6.3 for different values of γ_1 with the same initial value vector as $[\xi^{\mathrm{T}}(0)\ \ \zeta^{\mathrm{T}}(0)]^{\mathrm{T}} = [40\ \ 10\ \ 0\ \ -10\ \ -20\ \ -30\ \ 20\ \ 20\ \ -10\ \ 40]^{\mathrm{T}}$.

Let the directed graph be still given by Figure 6.1. Assume the desired poles of the consensus dynamics are 0.1 and -2. Then $\alpha = 1.9 > 0$, $\beta = -0.2$. With simple calculation, the right side

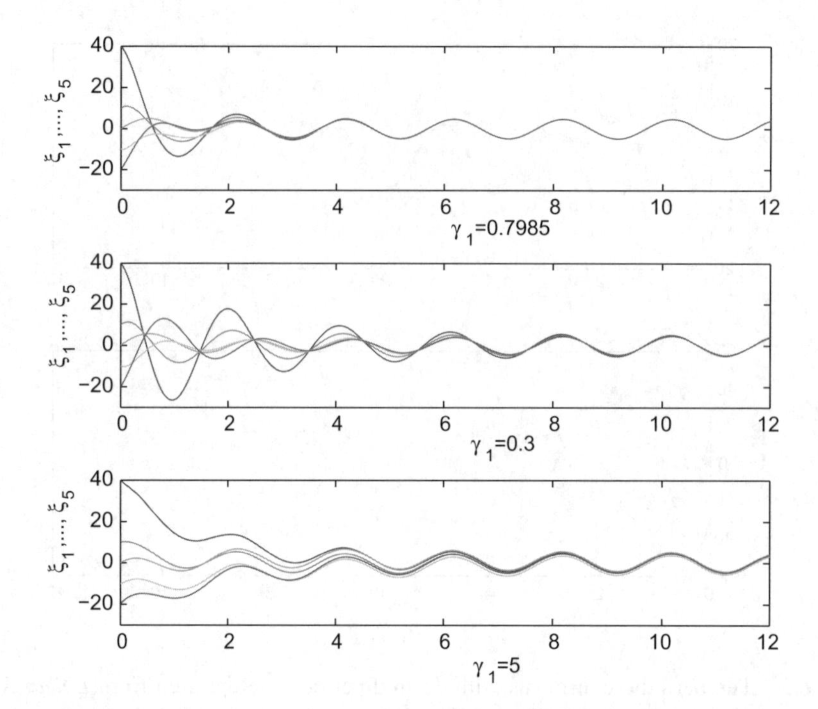

Figure 6.3 The maximum convergence speed is achieved as $\gamma_1 = 0.7985$. (Reprinted from *Linear Algebra and its Applications*, **431**, Zhu J., Tian Y.-P. and Kuang J., "On the general consensus protocol of multi-agent systems with double-integrator dynamics," 701–715, 2009, with permission from Elsevier.)

of (6.48) is equal to 0.0571. Now, we let $\gamma_0 = 1 > 0.0571$. Then the right side of (6.49) equals -0.2107. Hence, by Theorem 6.11, the positive exponential consensus is achieved for any $\gamma_1 > -0.2107$. Figure 6.4 shows the simulation for $\gamma_1 = -0.1$ and the initial value vector as $[0\ -10\ 12\ 4\ 25\ -4\ 6\ 0\ 16\ 3]^{\mathrm{T}}$.

Let $\alpha = \beta = 0$ and $\gamma_1 = 1$. Then, the right side of (6.51) equals 0.4629. Hence, by Corollary 6.12, the consensus is achieved if and only if $\gamma_1 > 0.4629$. Using (11) in Theorem 4.2 of Ren and Atkins (2007), one obtains $\gamma_1 > 1$, which is only a sufficient condition. Figure 6.5 and Figure 6.6 show the phenomena from disagreement to the agreement.

6.3 High-Order Agent System

6.3.1 System Model

Consider a multi-agent system with n identical agents. The ith agent is a linear time-invariant (LTI) dynamic system given by

$$\begin{aligned} \dot{x}_i &= Ax_i(t) + Bu_i(t) \\ y_i(t) &= Cx_i(t), \quad i \in \overline{1, n}, \end{aligned} \tag{6.52}$$

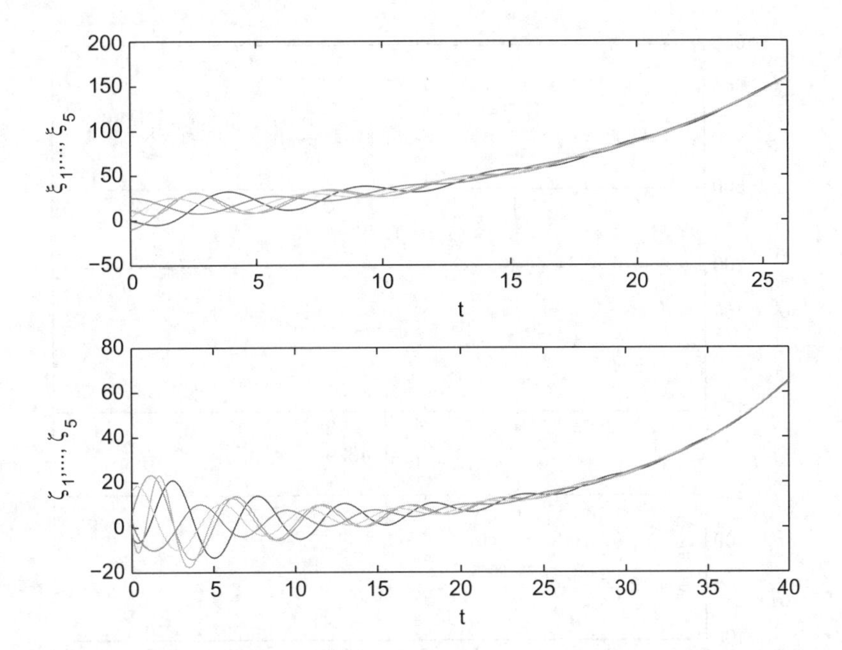

Figure 6.4 The positive exponential consensus. (Reprinted from *Linear Algebra and its Applications*, **431**, Zhu J., Tian Y.-P. and Kuang J., "On the general consensus protocol of multi-agent systems with double-integrator dynamics," 701–715, 2009, with permission from Elsevier.)

where $x_i \in \mathbb{R}^n$, $u_i(t) \in \mathbb{R}^r$ and $y_i(t) \in \mathbb{R}^m$ are the state, input and output of the ith agent, respectively; $A \in \mathbb{R}^{n\times n}$, $B \in \mathbb{R}^{n\times r}$ and $C \in \mathbb{R}^{m\times n}$ are constant matrices.

The output-based consensus protocol has the following form:

$$u_i = K \sum_{j\in N_i} a_{ij}(y_j(t) - y_i(t)), \quad i \in \overline{1, n}, \tag{6.53}$$

where $K \in \mathbb{R}^{r\times m}$ is a gain matrix for output feedback.

If $C = I$, (6.53) becomes the following state-based consensus protocol:

$$u_i = K_s \sum_{j\in N_i} a_{ij}(x_j(t) - x_i(t)), \quad i \in \overline{1, n}, \tag{6.54}$$

where $K_s \in \mathbb{R}^{r\times n}$ is a gain matrix for state feedback.

Definition 6.14 *It is said that the system (6.52) with the consensus protocol (6.53) (or (6.53)) achieves a consensus asymptotically if*

$$\lim_{t\to\infty} \|x_j(t) - x_i(t)\| = 0, \quad \forall i, j \in \overline{1, n}, \ i \neq j \tag{6.55}$$

holds for any initial value $x_i(0) \in \mathbb{R}^n$, $i \in \overline{1, n}$. The system (6.52) is said to be consentable under the consensus protocol (6.53) (or (6.54)) if there is a gain matrix $K_o \in \mathbb{R}^{r\times m}$ (or $K_s \in$

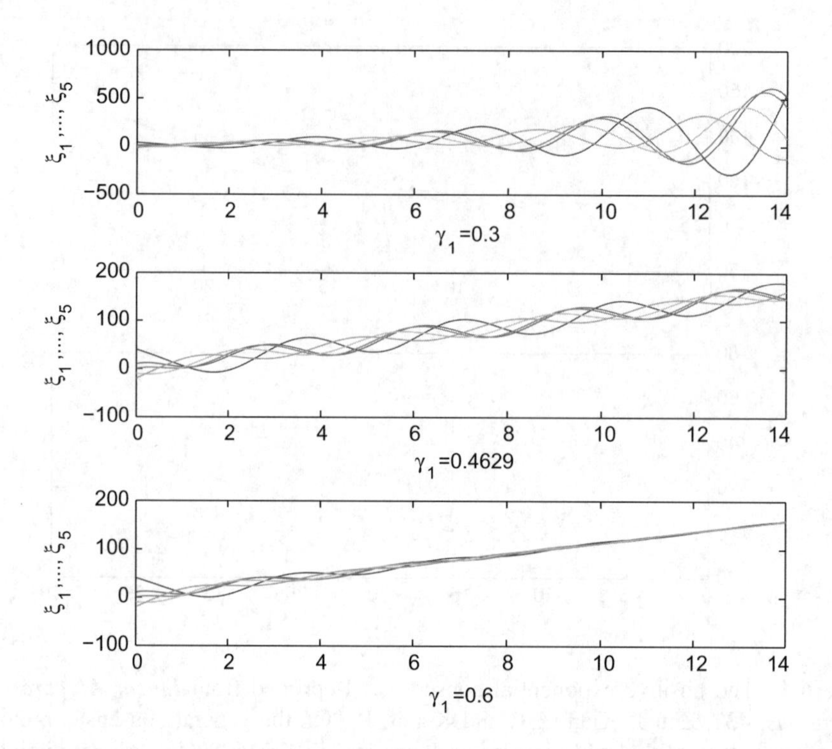

Figure 6.5 From the disagreement to the agreement of ξ_i. (Reprinted from *Linear Algebra and its Applications*, **431**, Zhu J., Tian Y.-P. and Kuang J., "On the general consensus protocol of multi-agent systems with double-integrator dynamics," 701–715, 2009, with permission from Elsevier.)

$\mathbb{R}^{r\times m}$*) such that the closed-loop system of (6.52) with (6.53) (or (6.54)) achieves a consensus asymptotically.*

6.3.2 Consensus Condition

Denote

$$x(t) = \left[x_1^{\mathrm{T}}(t), x_2^{\mathrm{T}}(t), \dots, x_n^{\mathrm{T}}(t)\right]^{\mathrm{T}}, \tag{6.56}$$

$$\tilde{x}(t) = \left[x_2^{\mathrm{T}}(t), \dots, x_n^{\mathrm{T}}(t)\right]^{\mathrm{T}}, \tag{6.57}$$

$$e(t) = \tilde{x}(t) - \mathbf{1}_{n-1} \otimes x_1(t), \tag{6.58}$$

where $\otimes$ denotes the Kronecker product, which satisfies the following properties: (1) $(A \otimes B)(C \otimes D) = AC \otimes BD$; (2) $(A \otimes B)^{\mathrm{T}} = A^{\mathrm{T}} \otimes B^{\mathrm{T}}$ (Horn and Johnson 1985).

By definition (6.58) it is easy to see that (6.55) is equivalent to

$$\lim_{t\to\infty} \|e(t)\| = 0.$$

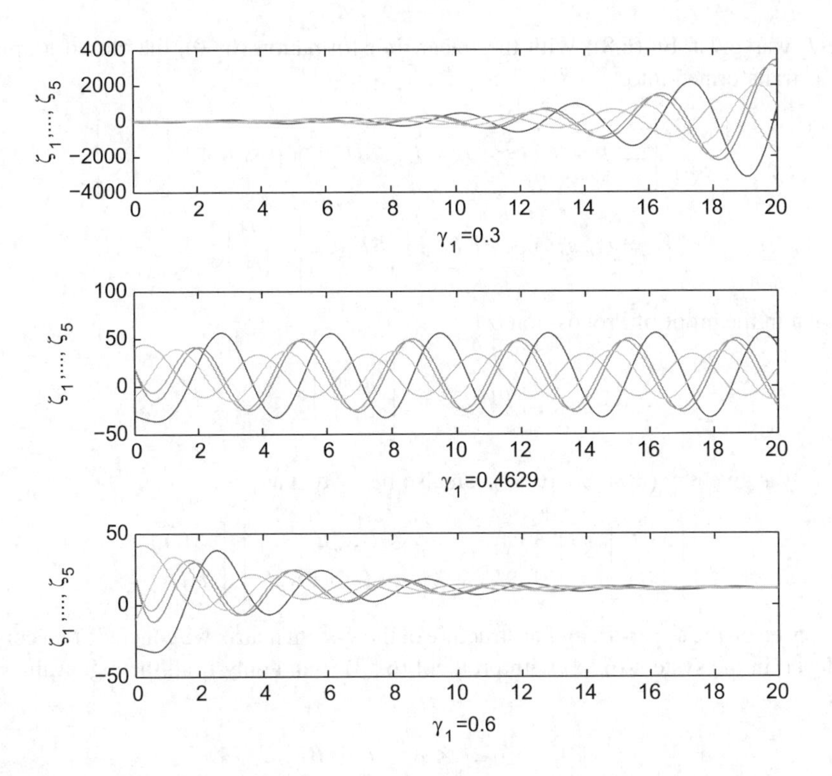

Figure 6.6 From the disagreement to the agreement of ζ_i. (Reprinted from *Linear Algebra and its Applications*, **431**, Zhu J., Tian Y.-P. and Kuang J., "On the general consensus protocol of multi-agent systems with double-integrator dynamics," 701–715, 2009, with permission from Elsevier.)

In other words, the consensus is achieved asymptotically if and only if the error system with state $e(t)$ is asymptotically stabilized. Now, let us derive the state space equation of the error system.

First, we can write the closed-loop system (6.52) with consensus protocol (6.53) as follows

$$\dot{x}(t) = (I_n \otimes A - L \otimes BK_oC)x(t), \tag{6.59}$$

where L is the Laplacian matrix of the topology graph of the system. From (6.56), (6.57) and (6.58) we have

$$\begin{aligned}\begin{bmatrix} x_1(t) \\ e(t) \end{bmatrix} &= \begin{bmatrix} I_n & 0 \\ -\mathbf{1}_{n-1} \otimes I_n & I_{n-1} \otimes I_n \end{bmatrix} x(t) \\ &= (P \otimes I_n)x(t),\end{aligned} \tag{6.60}$$

where P was given by (6.8). With the linear transformation (6.60), the closed-loop system (6.59) is transformed into

$$\begin{aligned}\begin{bmatrix}\dot{x}_1(t)\\ \dot{e}(t)\end{bmatrix} &= P\otimes I_n(I_n\otimes A - L\otimes BK_oC)(P\otimes I_n)^{-1}\begin{bmatrix}x_1(t)\\ e(t)\end{bmatrix}\\ &= (I_n\otimes A - PLP_{-1}\otimes BK_oC)\begin{bmatrix}x_1(t)\\ e(t)\end{bmatrix}.\end{aligned} \tag{6.61}$$

As shown in the proof of Proposition 6.1,

$$PLP^{-1} = \begin{bmatrix}0 & \phi\\ 0 & \tilde{L}\end{bmatrix}$$

where $\tilde{L}$ was given by (6.7). So, (6.61) can also be written as

$$\begin{bmatrix}\dot{x}_1(t)\\ \dot{e}(t)\end{bmatrix} = \begin{bmatrix}A & -\phi\otimes BK_oC\\ 0 & I_{n-1}\otimes A - \tilde{L}\otimes BK_oC\end{bmatrix}\begin{bmatrix}x_1(t)\\ e(t)\end{bmatrix}. \tag{6.62}$$

Hence, based on the upper-triangular structure of the system matrix we conclude that consensus is achieved in the system (6.52) with protocol (6.53) if and only if all the eigenvalues of the matrix

$$\tilde{\Gamma} := I_{n-1}\otimes A - \tilde{L}\otimes BK_oC \tag{6.63}$$

lie inside the LHP; and moreover, the dynamics of the consensus solution are determined by the equation of the open-loop system of (6.52).

In Proposition 6.1 we have shown that the eigenvalues of $\tilde{L}$ are the remaining $n-1$ eigenvalues of L excluding zero, denoted by $\mu_2, \ldots, \mu_n$. Let us write $\tilde{L}$ in upper triangular Jordan blocks, $T^{-1}JT$, with $\mu_2, \ldots, \mu_n$ as diagonal elements of J. Then, we get the characteristic equation of matrix Γ defined by (6.63) as

$$\det(sI - I_{n-1}\otimes A - \tilde{L}\otimes BK_oC) = \det(sI - I_{n-1}\otimes A - J\otimes BK_oC). \tag{6.64}$$

Considering the upper triangular structure of J we have arrived at the following theorem.

Theorem 6.15 *Consensus is achieved in system (6.52) with protocol (6.53) if and only if all the eigenvalues of* $A - \mu_i BK_oC, i\in\overline{2,n}$ *lie inside the LHP. And moreover, the dynamics of the consensus solution are determined by the equation of the open-loop system of (6.52), i.e.,*

$$\dot{x}(t) = Ax(t),$$

where $x(t)\in\mathbb{R}^n$.

To avoid the so-called trial consensus solution we make the following assumption for system (6.52).

Assumption 6.16 *In system (6.52) A is not a Hurwitz matrix.*

6.3.3 Consentability

Theorem 6.15 also implies that system (6.52) is consentable under protocol (6.53) if and only if $n-1$ systems $(A, \mu_i B, C)$ (or $(A, B, \mu_i C)$), $i \in \overline{2, n}$, can be simultaneously stabilized by an output feedback controller.

Since we have assumed that A is not a Hurwitz matrix, the consentability implies that each system $(A, \mu_i B, C)$ or $(A, B, \mu_i C)$ is stabilizable by output feedback, which further implies that $(A, \mu_i B)$ is stabilizable and (C, A) is detectable (or, (A, B) is stabilizable and $(A, \mu_i C)$ is detectable). Obviously, in any case $\mu_i \neq 0$ is a necessary condition. By Theorem 1.9 we know that $\mu_i \neq 0$, $\forall i \in \overline{2, n}$, only if the topology graph of the system contains at least one globally reachable node. So, the following proposition is obviously true.

Proposition 6.17 *If system (6.52) is consentable under protocol (6.53), then the topology graph G contains a globally reachable node.*

Noticing that for any non-zero $\mu_i \in \mathbb{C}$, the stabilizability of $(A, \mu_i B)$ is equivalent to the stabilizability of (A, B), and the detectability of $(A, \mu_i C)$ is equivalent to the detectability of (A, C). So, considering Proposition 6.17 the following proposition is also obviously true.

Proposition 6.18 *If system (6.52) is consentable under protocol (6.53), then (A, B) is stabilizable and (A, C) is detectable.*

Propositions 6.17 and 6.18 provide necessary conditions of the consentability of system (6.52) under protocol (6.53). Now, let us further try to find some necessary and sufficient conditions.

The open-loop transfer function matrix of each agent is

$$G_i(s) = \frac{Y_i(s)}{U_i(s)} = \mu_i C(sI - A)^{-1} B \triangleq \mu_i G(s), \quad i \in \overline{2, n}, \tag{6.65}$$

where $\mu_i \in \mathbb{C}$, $i \in \overline{2, n}$, are the $n-1$ eigenvalues of the Laplacian matrix L except the first zero eigenvalue corresponding to the eigenvector $\mathbf{1}_n$. Then, an equivalent formulation of the consentability problem is given as follows.

Simultaneous stabilization problem:

Does there exist a $K \in \mathbb{R}^{r \times m}$ such that all the systems given by (6.65) can be simultaneously stabilized by a common output feedback controller

$$U_i(s) = -KY_i(s)? \tag{6.66}$$

Now, we first study this problem for the case when the topology graph is undirected. In this case, by Theorem 1.9, all the eigenvalues of Laplacian matrix L are non-negative real numbers, which in an ascending order are written as $0 = \mu_1 \leq \mu_2 \leq \cdots \leq \mu_n$.

Suppose $G(s) = C(sI - A)^{-1} B$ is output stabilizable and $U(s) = -K_0 Y(s)$ is such an output feedback stabilizer. By the continuity of the system poles on entries of K_0, there exists an interval (κ_1, κ_2) such that $G(s)$ is stabilized by $U(s) = -\gamma K_0 Y(s)$ for all $\gamma \in (\kappa_1, \kappa_2)$, i.e., the equation

$$I + \gamma G(s) K_0 = 0$$

has no RHP zero. Let $\kappa_1^\star$ and $\kappa_2^\star$ be the extreme values of κ_1 and κ_2 respectively, i.e., for all $\gamma \in (\kappa_1^\star, \kappa_2^\star)$, γK_0 stabilizes $G(s)$, but both $\kappa_1^\star K_0$ and $\kappa_2^\star K_0$ marginally stabilize $G(s)$. Denote

$$r^{\text{op}} = \sup_{K_0 \text{ stabilizes } G(s)} \frac{\kappa_2^\star}{\kappa_1^\star}. \tag{6.67}$$

Then, we have the following theorem.

Theorem 6.19 *Suppose that*

$$\mu_n / \mu_2 < r^{\text{op}}. \tag{6.68}$$

Then system (6.52) is consentable under protocol (6.53) if and only if the undirected topology graph G is connected.

Proof. Necessity is already ensured by Proposition 6.17. We need to just prove the sufficiency. Since G is connected, $\mu_i > 0, \forall i \in \overline{2, n}$. By (6.68) we can find an $\epsilon > 0$ such that $\frac{\mu_n}{\mu_2} < \frac{\kappa_2^\star}{\kappa_1^\star} - \epsilon < \frac{\kappa_2^\star}{\kappa_1^\star}$. Take $\gamma^\star = \kappa_2^\star - \epsilon \kappa_1^\star$. Then, $\kappa_1^\star < \gamma^\star \mu_i < \kappa_2^\star$, $\forall i \in \overline{2, n}$. Hence, the equation

$$I + \gamma^\star \mu_i G(s) K_0 = 0$$

has no RHP zeros for all $i \in \overline{2, n}$. □

When $\kappa_2^\star$ can be arbitrarily large (high-gain stabilizable) or $\kappa_1^\star$ can be arbitrarily small (low-gain stabilizable), $r^{\text{op}} \to \infty$ and hence, (6.68) always holds. So, the following corollary of Theorem 6.19 is obvious.

Corollary 6.20 *Suppose that system (A, B, C) is high-gain output stabilizable or low-gain output stabilizable. Then, system (6.52) is consentable under protocol (6.53) if and only if its undirected topology graph is connected.*

The following lemma given in Dragan and Hahalany (1999) shows that a square minimum phase system is high-gain stabilizable.

Lemma 6.21 *Suppose system (A, B, C) satisfies*

(1) CB is square and invertible;

(2) the roots of the equation

$$\det \begin{pmatrix} \lambda I - A & -B \\ C & 0 \end{pmatrix} = 0 \tag{6.69}$$

$\text{Re}[\lambda] \leq -2\alpha_2 < 0.$

Let H be a matrix whose eigenvalues satisfies $\text{Re}[\lambda] \leq -2\alpha_1 < 0$. *Then, there exists $\varepsilon_0 > 0$ such that for arbitrary $\varepsilon \in (0, \varepsilon_0)$ and some $c > 0$ independent of ε, the state and the output of the closed-loop system (A, B, C) with output feedback $u = \frac{1}{\varepsilon}(CB)^{-1}y$ satisfy*

$$\|x(t, \varepsilon)\| \leq c\mathrm{e}^{-\alpha_2 t}\|x(0)\|;$$
$$\|y(t, \varepsilon)\| \leq c\left[\mathrm{e}^{-\frac{\alpha_1 t}{\varepsilon}} + \varepsilon\mathrm{e}^{-\alpha_2 t}\right]\|x(0)\|.$$

By Definition 2.11, condition (1) in Lemma 6.21 implies that the system (A, B, C) ia square and has relative degree as one, and condition (2) implies that the system (A, B, C) is minimum phase. From the above lemma and Corollary 6.20 we directly obtain the following theorem on consentability.

Theorem 6.22 *Suppose square system (A, B, C) is minimum phase and CB is invertible. Then, system (6.52) is consentable under protocol (6.53) if and only if the topology graph G is connected.*

Next, we will extend the conclusion given by Corollary 6.20 to the case when the interconnection topology graph is directed. In this case, by Theorem 1.9, all the eigenvalues of Laplacian matrix L have non-negative real parts, i.e., $\mu_i \in \mathbb{C}$, and $0 = \mu_1 \leq \text{Re}(\mu_2) \leq \cdots \leq \text{Re}(\mu_n)$.

Suppose $\mu_i = \alpha_i + \mathrm{j}\beta_i$, $i \in \overline{2, n}$, where $\alpha, \beta \in \mathbb{R}$. Denote

$$\mu_0 = \frac{1}{2}\left(\min_{i \in \overline{2,n}}\{|\mu_i|\} + \max_{i \in \overline{2,n}}\{|\mu_i|\}\right), \tag{6.70}$$

$$\delta_i = \frac{\mu_i - \mu_0}{\mu_0}, \quad i \in \overline{2, n}, \tag{6.71}$$

$$\delta_m = \frac{\max_{i \in \overline{2,n}}\left\{\sqrt{(\alpha_i - \mu_0)^2 + \beta_i^2}\right\}}{\mu_0}. \tag{6.72}$$

Then, we have

$$\mu_i = \mu_0(1 + \delta_i), \tag{6.73}$$
$$|\delta_i| \leq \delta_m, \quad i \in \overline{2, n}. \tag{6.74}$$

Then, the simultaneous stabilization problem is solvable if the following robust stabilization problem is solvable.

Robust stabilization problem:

For any Δ satisfying $\|\Delta\|_\infty \leq \delta_m$, does there exist a $K \in \mathbb{R}^{r \times m}$ such that the uncertain system $\sigma(I + \Delta)G(s)$ can be robustly stabilized by the output feedback controller $U(s) = -KY(s)$?

This robust stabilization problem is sketched in Figure 6.7, where

$$P(s) = \begin{bmatrix} 0 & -G(s) \\ 0 & -G(s) \end{bmatrix}.$$

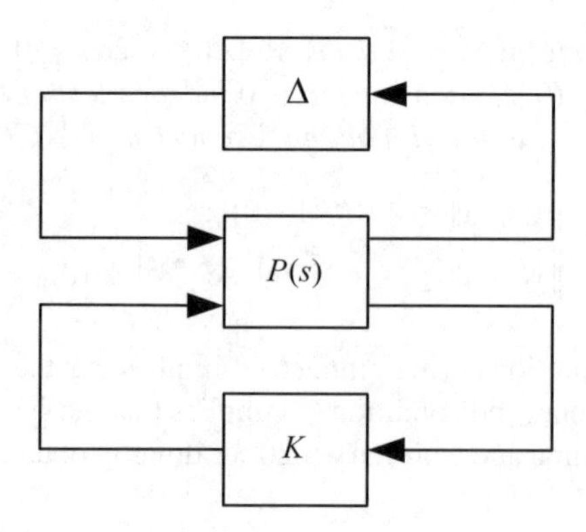

Figure 6.7 Robust stabilization problem.

By the robust control theory (Feng, Tian and Xin 1996; Zhou, Doyle and Glover 1996), the problem is solvable if there exists a $K \in \mathbb{R}^{r\times m}$ such that

$$\|\sigma G(s)K(I + \sigma G(s)K)^{-1}\|_{\infty} < \delta_m^{-1}. \tag{6.75}$$

Suppose that the topology digraph G contains a globally reachable node. Then, $|\mu_i| \neq 0$ for all $i \in \overline{2, n}$, which implies that $\delta_m < \infty$. If $G(s)$ is low-gain output stabilizable, then there exists an $K_0 \in \mathbb{R}^{r\times m}$ such that $G(s)$ can be stabilized by $U(s) = -\kappa K_0 Y(s)$, where $\kappa \in \mathbb{R}^+$ can be arbitrarily close to zero. So, in this case we can always find a $K = \kappa K_0$ such that (6.75) holds, because

$$\|\sigma G(s)\kappa K_0(I + \sigma G(s)\kappa K_0)^{-1}\|_{\infty} \to 0$$

when $\kappa \to 0$. Therefore, we have proved the following theorem.

Theorem 6.23 *Suppose that system (A, B, C) is low-gain output stabilizable. Then, system (6.52) is consentable under protocol (6.53) if and only if its topology digraph contains a globally reachable node.*

The system (A, B) is low-gain stabilizable by state feedback if the system is stabilizable and all the eigenvalues of A are in the closed LHP (Lin 1998). Based on this fact we have the following theorem on consentability.

Theorem 6.24 *Suppose all the eigenvalues of A are in the closed LHP. Then, system (6.52) is consentable under protocol (6.54) if and only if its topology digraph G contains a globally reachable node and (A, B) is stabilizable.*

Now, we study high-gain output stabilizable systems for the case when the interconnection topology is directed.

If (A, B) is stabilizable, then by linear system theory (Cheng and Ma 2006), the Riccati equation

$$A^{\mathrm{T}}P + PA - PBB^{\mathrm{T}}P + I_n = 0 \tag{6.76}$$

has a unique semi-positively definite solution P. If the following rank condition holds

$$\text{rank}(C) = \text{rank}\begin{pmatrix} C \\ B^{\text{T}}P \end{pmatrix}, \tag{6.77}$$

then the matrix equation $XC = B^{\text{T}}P$ has at least a solution denoted by K_0, i.e.,

$$K_0 C = B^{\text{T}}P. \tag{6.78}$$

Then, the system (A, B, C) can be stabilized by output feedback $u = -\gamma K_0 y$ for arbitrary $\gamma \in [1, \infty)$, i.e., the system is high-gain stabilizable (Ma 2009). This can be proved by taking a Lyapunov function $V(x) = x^{\text{T}}Px$ for that closed-loop system. The derivative of the Lyapunov function is

$$\begin{aligned} \dot{V}(x) &= x^{\text{T}}(A^{\text{T}}P + PA - \gamma PBB^{\text{T}}P)x \\ &= x^{\text{T}}(-I - (\gamma - 1)PBB^{\text{T}}P)x \\ &< 0. \end{aligned}$$

It can be further shown that under the condition (6.77), the system $(A, \mu B, C)$ or $(A, B, \mu C)$ with $\mu \in \mathbb{C}$ is also high-gain stabilizable. In this case, since the system contains a complex parameter, the state of the system is a vector in $\mathbb{C}^{n\times 1}$. So, we can take a Lyapunov function as $V = x^* Px$, where x^* is the conjugate transpose of x. Then, under the output feedback $u = -\gamma K_0 y$, where K_0 satisfies (6.78), the derivative of the Lyapunov function along the system solution is

$$\begin{aligned} \dot{V}(x) &= x^*(A^{\text{T}}P + PA - 2\gamma \text{Re}(\mu) PBB^{\text{T}}P)x \\ &= x^*(-I - (2\gamma \text{Re}(\mu) - 1)PBB^{\text{T}}P)x. \end{aligned}$$

Obviously, $\dot{V}(x) < 0$ as long as $\gamma > \frac{1}{2\text{Re}(\mu)}$.

Therefore, we have the following theorem on consentability, which was first obtained in Ma and Zhang (2010).

Theorem 6.25 *Suppose Riccati equation (6.76) has a positively semi-definite solution P such that (6.77) holds. Then, system (6.52) is consentable under protocol (6.53) if and only if the topology digraph G contains a globally reachable node and (A, B) is stabilizable.*

By doing the following exercise the reader will understand that under the condition (6.77) the system (A, B, C) is minimum phase.

Exercise 6.26 *Suppose (A, B) is stabilizable and P is the solution of the Reccati equation (6.76). Show (A, B, C) is minimum phase if the rank condition (6.77) holds.*

Note that when state feedback consensus protocol is used, i.e., $C = I$, then the rank condition (6.77) automatically holds. We have the following corollary of Theorem 6.25.

Theorem 6.27 *System (6.52) is consentable under protocol (6.54) if and only if the topology digraph G contains a globally reachable node and (A, B) is stabilizable.*

6.4 Notes and References

Consensus problems have a long history in statistics (DeGroot 1974), distributed computation (Bertsekas, and Tsitsiklis 1989; Cybenko 1989; Lynch 1997) and distributed estimation (Borkar and Varaiya 1982). Consensus is also one of fundamental principles in the design of sensor networks (Kar and Moura 2008, 2009) and multi-robot coordination control systems (Cortés *et al.* 2004; Fax and Murray 2004; Olfati-Saber, Fax and Murray 2007). Vicsek *et al.* (1995) proposed a discrete-time model of multi-autonomous agents, and demonstrated that without any central control, all the agents move in the same direction when the density is large and the noise is small. Jadbabaie, Lin and Morse (2003) studied the linearized Vicsek's model and proved that all the agents converge to a common steady state provided that the interconnection graph is jointly connected. Based on algebraic graph theory, Olfati-Saber and Murray (2003, 2004) analyzed the consensus problem for networks of first-order integrator agents with fixed and switching topologies, and showed that the consensus can be achieved under the assumption of strong connection of digraphs. Moreau (2005), Ren and Beard (2005) generalized the results of Jadbabaie, Lin and Morse (2003), Olfati-Saber and Murray (2004), and in particular, Ren and Beard (2005), presented a more relaxed condition of the solvability of the consensus problem, i.e., the existence of a spanning tree in the topology digraph. Similar results were obtained by Lin, Francis and Maggiore (2005). Some special second-order consensus protocols were proposed and some sufficient consensus conditions were obtained by (Lin *et al.* 2007; Ren and Atkins 2007). In particular, Ren and Atkins (2007) proposed two kinds of second-order consensus protocols, under which the state of each agent converges to a constant or a linear function with respect to the time.

Section 6.3 tries to unify most results on consensus seeking ability (or consentabilty in this book) in the framework of high-gain and low-gain stabilization.

The consensus seeking ability problem of multi-agent systems (MASs) with continuous-time high-order LTI dynamic agents was investigated by Wang, Cheng and Hu (2008), Xiao and Wang (2007) and Seo *et al.* (2009), which showed that for the system with semi-stable agents the problem is solvable if the interconnection topology has a globally reachable node. This result basically relies on the low-gain stabilizability of semi-stable systems (Lin 1998, 2009). Zhang and Tian (2010, 2012) extended such results to stochastic MASs.

The word "consentability" appeared for the first time in the work Zhang and Tian (2009). It introduced the concept of consentability under two kinds of second-order consensus protocols. For discrete MASs with double-integrator dynamics, Zhang and Tian (2009) obtained necessary and sufficient conditions of the consentability for fixed and stochastic switching topologies. The results of Zhang and Tian (2009) much benefit from the technique shown by Proposition 6.1 and Theorem 6.2, which successfully converts the consensus problem of the original MAS into an ordinary stability problem of a reduced-order system, and hence the consentability problem to a stabilizability problem. Using this technique and the Routh-Hurwitz stability criterion, Zhu, Tian and Kuang (2009) obtained a necessary and sufficient condition for the consensus of general second-order continuous-time MASs, which forms the main body of Section 6.2. Ma and Zhang (2010) used the same technique to handle the consentability problem of MASs with high-order LTI dynamic agents. (By the way, the manuscript of Zhang and Tian (2009) was sent to one of the authors of Ma and Zhang (2010) before its publication for academic exchange.) The main contribution of Ma and Zhang (2010) includes uncovering the link of the high-gain stabilization technique to the consentability problem and gives a constructive condition of the high-gain stabilization of continuous-time LTI systems based on the Riccati equation. Compared with the high-gain stabilization condition of Dragan and Hahalany (1999), this condition is less conservative because it does not require that the open system be square and has the relative degree as one; moreover, it is applicable to systems with a complex-number parameter. Ma and Zhang (2010) also pointed out that such a high-gain technique is not applicable to discrete-time systems. Generally, discrete-time LTI systems is neither low-gain stabilizable nor high-gain stabilizable even under state feedback. You and Xie (2010) noticed the work of Ma and Zhang (2010) and gave a necessary and sufficient condition of the consentability of discrete-time MASs with high-order SISO agents with the help of a crucial tool developed in Fu and Xie (2005). For discrete-time MASs with MIMO agents Zhang and Tian (2012) gave a sufficient condition of consentability and extended the result to stochastic MASs.

MASs with identical agents and uniform communication delays (see, e.g., Cao, Morse and Anderson 2006; Olfati-Saber and Murray 2004) are also homogeneous systems. But, just as in many other control problems, the compromise between delays and gains rather than the consentability problem occupies the central place in the analysis and design of such systems. So, they are not included in this chapter, but left to the next chapter as special cases of heterogeneous systems with diverse delays.

References

Bertsekas DP and Tsitsiklis JN (1989). *Parallel and Distributed Computation: Numerical Methods*. Prentice-Hall, Upper Saddle River, NJ. 1989

Borkar V and Varaiya P (1982). Asymptotic agreement in distributed estimation. *IEEE Transactions on Automatic Control*, AC-27(3), 650–655.

Cheng Z and Ma S (2006). *Linear System Theory*. Science Press, Beijing.

Cao M, Morse AS and Anderson BDO (2006). Reaching an agreement using delayed information. *IEEE Conference on Decision and Control*, San Diego, CA, USA, 3375–3380.

Cortés J, Martinez S, Karatas T, and Bullo F (2004). Coverage control for mobile sensing networks. *IEEE Transactions on Robotics and Automation*, 20(2), 243–255.

Cybenko G (1989). Load balancing for distributed memory multiprocessors. *Journal of Parallel and Distributed Computing*, 7(2), 279–301.

DeGroot MH (1974). Reaching a consensus. *Journal of American Statistical Association*, 69(345), 118–121.

Dragan V and Halanay A (1999). *Stabilization of Linear Systems*. Birkhäuser, Boston.

Fax JA and Murray RM (2004). Information flow and cooperative control of vehicle formations. *IEEE Transactions on Automatic Control*, 49, 1465–1476.

Feng C-B, Tian Y-P and Xin X (1996). *Robust Control System Design* (in Chinese). Southeast University Press, Nanjing.

Fu M and Xie L (2005). The sector bound approach to quantized feedback control. *IEEE Transactions on Automatic Control*, 50(11), 1698–1711.

Horn RA and Johnson CR (1985). *Matrix analysis*, 1st edn. Cambridge University Press, New York.

Jadbabaie A, Lin J and Morse AS (2003). Coordination of groups of mobile autonomous agents using nearest neighbor rules. *IEEE Transactions on Automatic Control*, 48, 988–1001.

Kar S and Moura J (2008). Sensor networks with random links: Topology design for distributed consensus. *IEEE Transactions on Signal Processing*, 56(7), 3315–3326.

Kar S and Moura J (2009). Distributed consensus algorithms in sensor networks with imperfect communication: Link failures and channel noise. *IEEE Transactions on Signal Processing*, 57(1), 355–369.

Lin P, Jia Y, Du J and Yuan S (2007). Distributed consensus control for second-order agents with fixed topology and time-delay. *Proceeding of the 26th Chinese Control Conference*, Zhangjiajie, Hunan, China, 577–581.

Lin Z, Francis B and Maggiore M (2005). Necessary and sufficient graphical conditions for formation control of unicycles. *IEEE Transactions on Automatic Control*, 50, 121–127.

Lin Z (1998). *Low Gain Feedback*. Lecture Notes in Control and Information Sciences, 240, Springer, London.

Lin Z (2009). Low gain and low-and-high gain feedback: a review and some recent results. *Proceedings of 2009 Chinese Control and Decision Conference*, lii–lxi, Guilin, China.

Lynch NA (1997). *Distributed Algorithms*. Morgan Kaufmann, San Francisco.

Ma C (2009). System analysis and control synthesis of linear multi-agent systems. *Ph.D. dissertation*, Academy of Mathematics and System Sciences, Chinese Academy of Sciences, Beijing.

Ma C and Zhang J (2010). Necessary and sufficient condition for consensusability of linear multi-agent systems. *IEEE Transactions on Automatic Control*, 55, 1263–1268.

Moreau L (2005). Stability of multiagent systems with time-dependent communication links. *IEEE Transactions on Automatic Control*, 50, 169–182.

Olfati-Saber R and Murray RM (2003). Consensus protocols for networks of dynamic agents. *Proceedings of American Control Conference*, 951–956.

Olfati-Saber R and Murray RM (2004). Consensus problems in networks of agents with switching topology and time-delays. *IEEE Transactions on Automatic Control*, 49, 1520–1533.

Olfati-Saber R and Murray RM (2006). Flocking for multi-agent dynamic systems: algorithms and theory. *IEEE Transactions on Automatic Control*, 51, 401–420.

Olfati-Saber R, Fax JA and Murray RM (2007). Consensus and cooperation in networked multi-agent systems. *Proceedings of the IEEE*, 95(1), 215–223.

Ren W and Beard RW (2005). Consensus seeking in multiagent systems under dynamically changing interaction topologies. *IEEE Transactions on Automatic Control*, 50, 655–661.

Ren W and Atkins E (2007). Distributed multi-vehicle coordinated control via local information exchange. *International Journal of Robust and Nonlinear Control*, 17, 1002–1033.

Seo JH, Shim H and Back J (2009). Consensus of high-order linear systems using dynamic output feedback compensator: low gain approach. *Automatica*, 45, 2659–2664.

Vicsek T, Czirok A, Ben Jacob E, *et al.* (1995). Novel type of phase transitions in a system of self-driven particles. *Physical Review Letters*, 75, 1226–1229.

Wang J, Cheng D and Hu X (2008). Consensus of multi-agent linear dynamic systems. *Asian Journal of Control*, 10, 144–155.

Xiao F and Wang L (2007). Consensus problems for high-dimensional multi-agent systems. *IET Control Theory and Applications*, 1, 830–837.

You K and Xie L (2010). Consensusability of discrete-time multi-agent systems via relative output feedback. *Proceedings of the 11th International Conference on Control, Automation, Robotics and Vision*, Singapore, 1239–1244.

Zhang Y and Tian Y-P (2009). Consentability and protocol design of multi-agent systems with stochastic switching topology. *Automatica*, 45, 1195–1201.

Zhang Y and Tian Y-P (2010). Consensus of data-sampled multi-agent systems with random communication delay and packet loss. *IEEE Transactions on Automatic Control*, 55(4), 939–943.

Zhang Y and Tian Y-P (2012). Maximum allowable loss probability for consensus of multi-agent systems over random weighted lossy networks. *IEEE Transactions on Automatic Control*, in press.

Zhou K, Doyle JC and Glover K (1996). *Robust and Optimal Control*. Prentice-Hall, Upper Saddle River, NJ.

Zhu J, Tian Y-P and Kuang J (2009). On the general consensus protocol of multi-agent systems with double-integrator dynamics. *Linear Algebra and its Applications*, 431, 701–715.

7

Consensus in Heterogeneous Multi-Agent Systems

Gentlemen can live harmoniously together even though they have heterogeneous characters, but non-gentlemen of the same character live in discord.

Confucius (551–479 BC), Analects of Confucius

Multi-agent systems (MASs) may contain heterogeneous network channels and/or heterogeneous agent dynamics. Even for systems with identical plants in agents, diverse input delays generate heterogeneous agent dynamics, and diverse communication delays result in heterogeneous network channels. Consensus problems in heterogeneous MASs are studied in this chapter. High-order consensus is defined for a class of high-order heterogeneous MASs. A necessary and sufficient condition is given for the existence of high-order consensus solutions to the considered class of systems. The condition shows that for systems with diverse communication delays, high-order consensus does not require that the self-delay of each agent to be equal to the corresponding communication delay. The frequency-domain scalability analysis method developed in Chapter 3 is applied to first-order and second-order MASs with diverse input delays and communication delays.

7.1 Integrator Agent System with Diverse Input and Communication Delays

In this section, we first consider the consensus problem for the integrator agent system with diverse input delays based on undirected graphs. Due to the heterogeneousness caused by the diverse input delays, the consensus problem of such a system can not be converted into an ordinary stability problem by using the technique developed in Chapter 6. However, we will show that it can still be considered as a semi-stability problem if the interconnection topology graph is connected. Therefore, using the frequency-domain analysis theory developed in Chapter 3, we develop various scalable consensus conditions, which uses only local information of each agent. Finally, by considering the consensus problem for digraph-based systems with both

Frequency-Domain Analysis and Design of Distributed Control Systems, First Edition. Yu-Ping Tian.

diverse communication delays and diverse input delays, we will show that consensus condition is dependent on input delays but independent of communication delays when the digraph contains a globally reachable node.

7.1.1 Consensus in Discrete-Time Systems

Consider a discrete-time multi-agent system with topology graph $G = (V, E, A)$ and integrator agents given by

$$x_i(k+1) = x_i(k) + u_i(k), \quad i \in \overline{1, n}, \tag{7.1}$$

where $x_i(k) \in \mathbb{R}$ and $u_i(k) \in \mathbb{R}$ denote the state and the control input of agent i at time instant k, respectively. The topology graph $G = (V, E, A)$ of the system can be directed or undirected, depending on context. The consensus protocol for the system is given by

$$u_i(k) = \sum_{j \in N_i} a_{ij}(x_j(k) - x_i(k)), \tag{7.2}$$

where N_i denotes the neighbors of agent i, and $a_{ij} > 0$ is the adjacency element of A in the graph $G = (V, E, A)$.

When each agent is subject to an input delay D_i, system (7.1) becomes

$$x_i(k+1) = x_i(k) + u_i(k - D_i), \quad i \in \overline{1, n}. \tag{7.3}$$

Under diverse communication delays, the consensus protocol becomes

$$u_i(k) = \sum_{j \in N_i} a_{ij}(x_j(k - \tau_{ij}) - x_i(k)), \tag{7.4}$$

where τ_{ij} represents the communication delay from agent j to agent i.

Under protocol (7.4), multi-agent system (7.3) is said to achieve a consensus asymptotically if

$$\lim_{k \to \infty} x_i(k) = c, \quad \forall i \in \overline{1, n},$$

where $c \in \mathbb{R}$ is a constant.

The closed-loop system of (7.3) and (7.4) is

$$x_i(k+1) = x_i(k) + \sum_{j \in N_i} a_{ij}(x_j(k - \tau_{ij} - D_i) - x_i(k - D_i)), i \in \overline{1, n}. \tag{7.5}$$

Let $x(k) = [x_1(k), \cdots, x_n(k)]^{\mathrm{T}}$, and

$$\begin{aligned} d_{m1} &= \tau_{ij} + D_i, \quad & m_1 &= 1, \cdots, n^2, \\ d_{m2} &= D_i, & m_2 &= n^2 + 1, \cdots, n(n+1). \end{aligned}$$

Then, equation (7.5) can be rewritten as a time-delayed system in a vector form

$$x(k+1) = x(k) + \sum_{i=1}^{n_d} A_i x(k - d_i), \tag{7.6}$$

where $A_i \in \mathbb{R}^{n\times n}$, and $n_d = n(n+1)$. Obviously, $\sum_{i=1}^{n_d} A_i = L$, which is the Laplacian matrix of the topology graph.

The characteristic equation of system (7.6) is given by

$$\det\left((z-1)I - \sum_{i=1}^{n_d} A_i z^{-d_i}\right) = 0. \tag{7.7}$$

The equilibrium set of system (7.6) is defined by

$$X_e = \{x \in \mathbb{R}^n : Lx = 0\}.$$

When L is singular, X_e is a continuum of equilibrium points. Assume that the interconnection topology of the system is described by a connected undirected graph or a digraph containing a globally reachable node. Then, by Theorem 1.9, the Laplacian matrix L has a simple eigenvalue 0, i.e., $\det(L) = 0$ and $\text{rank}(L) = n-1$. By the definition of L we also have $L\mathbf{1}_n = 0$. So, all the elements in X_e can be represented as $c\mathbf{1}_n$ where c is any constant. Therefore, system (7.6) achieves a consensus asymptotically, if the solution of the system starting from any given initial states $x(-k) \in \mathbb{R}^n$, $k = 0, 1, \cdots, n_d - 1$, asymptotically converges to an element in X_e. According to this analysis the following lemma can be easily proved.

Lemma 7.1 *If the solutions of equation (7.7) have modulus less than unity except for a root at $z = 1$, then system (7.6) with a connected undirected graph or a digraph containing a globally reachable node achieves a consensus asymptotically.*

This lemma implies that under the assumption that the graph is connected or the digraph contains a globally reachable node the first-order agent system with diverse input delays and communication delays achieves a consensus asymptotically if the closed-loop system is steady semi-stable with $z = 1$ as a simple pole.

7.1.2 Consensus under Diverse Input Delays

In this subsection we consider the consensus problem for multi-agent systems with input delays only. In this case, the closed-loop form (7.5) of the system reduces to

$$x_i(k+1) = x_i(k) + \sum_{j\in N_i} a_{ij}(x_j(k-D_i) - x_i(k-D_i)), \ i \in \overline{1,n}. \tag{7.8}$$

The following theorem gives a scalable consensus condition for multi-agent systems with diverse input delays

Theorem 7.2 *Assume that system (7.8) of n agents is based on an undirected and connected graph $G = (V, E, A)$ with symmetric weights. The system achieves a consensus asymptotically if*

$$\sum_{j\in N_i} a_{ij} < \sin\frac{\pi}{2(2D_i+1)}, \quad \forall\, i \in \overline{1,n}. \tag{7.9}$$

Proof. Taking the z-transformation of system (7.8) and writing it in vector form, we get

$$(z-1)X(z) = -\text{diag}\left\{z^{-D_i}, i \in \overline{1,n}\right\} LX(z). \tag{7.10}$$

Note that L in (7.10) is a positively semi-definite matrix since an undirected graph is considered. The characteristic equation is

$$\det\left((z-1)I + \text{diag}\left\{z^{-D_i}, i \in \overline{1,n}\right\} L\right) = 0. \tag{7.11}$$

Define $p(z) = \det((z-1)I + \text{diag}\{z^{-D_i}, i \in \overline{1,n}\}L)$. Then, we will prove that all the zeros of $p(z)$ have modulus less than unity except for a zero at $z = 1$.

Let $z = 1$, then $p(1) = \det(L)$. Since $G = (V, E, A)$ is connected, by Theorem 1.9, zero is a simple eigenvalue of L, i.e., $\det(L) = 0$ and $\text{rank}(L) = n - 1$. Thus, $p(z)$ indeed has a zero at $z = 1$.

To prove that the system is steady semi-stable with $z = 1$ as a simple pole it suffices to prove that the zeros of $\det\left(I + \text{diag}\left\{\frac{z^{-D_i}}{z-1}, i \in \overline{1,n}\right\} L\right)$ have modulus less than unity. By the general Nyquist stability criterion for discrete-time systems (Theorem 2.22), this is the case if the eigenloci of

$$F(\text{j}\omega) = \text{diag}\left\{\frac{\text{e}^{-\text{j}\omega D_i}}{\text{e}^{\text{j}\omega} - 1}, i \in \overline{1,n}\right\} L$$

do not enclose the point $(-1, \text{j}0)$ for all $\omega \in [-\pi, \pi]$. To show this we rewrite $F(\text{j}\omega)$ as

$$F(\text{j}\omega) = \text{diag}\left\{k_i \frac{\text{e}^{-\text{j}\omega D_i}}{\text{e}^{\text{j}\omega} - 1}\right\} \text{diag}\left\{k_i^{-1}\right\} L,$$

where $k_i = 2\sin\frac{\pi}{2(2D_i+1)}$. Since

$$\begin{aligned}
&0 \cup \sigma\left(\text{diag}\left\{k_i \frac{\text{e}^{-\text{j}\omega D_i}}{\text{e}^{\text{j}\omega} - 1}\right\} \text{diag}\{k_i^{-1}\}L\right) \\
&= 0 \cup \sigma\left(\text{diag}\left\{k_i \frac{\text{e}^{-\text{j}\omega D_i}}{\text{e}^{\text{j}\omega} - 1}\right\} \text{diag}\left\{k_i^{-\frac{1}{2}}\right\} L\text{diag}\left\{k_i^{-\frac{1}{2}}\right\}\right),
\end{aligned}$$

by Lemma 2.34 we have

$$\begin{aligned}
&\lambda\left(\text{diag}\left\{k_i \frac{\text{e}^{-\text{j}\omega D_i}}{\text{e}^{\text{j}\omega} - 1}\right\} \text{diag}\left\{k_i^{-1}\right\} L\right) \\
&\in \rho\left(\text{diag}\left\{k_i^{-\frac{1}{2}}\right\} L\text{diag}\left\{k_i^{-\frac{1}{2}}\right\}\right) \text{Co}\left(0 \cup \left\{\text{diag}\{k_i \frac{\text{e}^{-\text{j}\omega D_i}}{\text{e}^{\text{j}\omega} - 1}\}\right\}\right).
\end{aligned}$$

Since the spectral radius of any matrix is bounded by its maximum absolute row sum according to Corollary 2.31 of Gershgorin's disc lemma, it follows from the condition (7.9) that

$$\rho\left(\text{diag}\left\{k_i^{-\frac{1}{2}}\right\} L\text{diag}\left\{k_i^{-\frac{1}{2}}\right\}\right) < 1.$$

Now, from Theorem 3.16 we conclude that eigenloci of $F(\mathrm{j}\omega)$ do not enclose the point $(-1, \mathrm{j}0)$ for all $\omega \in [-\pi, \pi]$, which implies that the zeros of $p(z)$ have modulus less than unity except for a zero at $z = 1$. Theorem 7.2 is thus proved by Lemma 7.1. □

Remark. By using Corollary 2.32 instead of Corollary 2.31 of Gershgorin's disc lemma in the proof, we know that system (7.8) achieves a consensus asymptotically if there exists $H = \mathrm{diag}\{h_k > 0, k \in \overline{1,n}\}$ such that

$$\frac{h_i}{h_j}\sum_{j \in N_i} a_{ij} < \sin\frac{\pi}{2(2D_i+1)}, \quad \forall\, i \in \overline{1,n}. \tag{7.12}$$

Of course, this condition can be less conservative than (7.9) and reduces to it if $H = I$.

Theorem 7.2 gives a scalable delay-dependent consensus condition. This condition suggests that under large input delays, small interconnection weights and small numbers of neighbors increase the possibility of achieving consensus if the interconnection topology graph is connected.

Now, we apply the result of Theorem 7.2 to study the effect of diverse input delays on some important systems that have been extensively investigated in literature.

First, let us consider Vicsek's model, which describes a group of agents moving in the plane with the same line velocity (Vicsek *et al.* 1995). When the headings of the agents are close to each other, the local updating rule for the headings may be approximated by the following linearized equation (Jadbabaie, Lin and Morse 2003)

$$x_i(k+1) = x_i(k) + \frac{1}{1+n_i}\sum_{j \in N_i}(x_j(k) - x_i(k)), \quad i \in \overline{1,n} \tag{7.13}$$

with $x_i \in \mathbb{R}$ and where n_i denotes the number of the neighbors of agent i. By applying Theorem 7.2 to this model, it is easy to get its consensus condition as follows:

$$\frac{n_i}{1+n_i} < \sin\frac{\pi}{2(2D_i+1)}, \quad \forall\, i \in \overline{1,n}. \tag{7.14}$$

Obviously, when $n > 1$, condition (7.14) holds only if $D_i = 0$. This implies that Vicsek's model is very sensitive to input delays. To overcome this problem we proposed a modified linearized Vicsek's model with input delays as

$$x_i(k+1) = x_i(k) + \frac{\varepsilon_i}{1+n_i}\sum_{j \in N_i}(x_j(k-D_i) - x_i(k-D_i)), \quad i \in \overline{1,n}, \tag{7.15}$$

where $\varepsilon_i > 0$ is an adjustable interconnection gain. From Theorem 7.2, we get the following corollary.

Corollary 7.3 *Suppose that the interconnection topology graph of system (7.15) is connected. Then, the system achieves a consensus asymptotically if*

$$\frac{n_i\varepsilon_i}{1+n_i} < \sin\frac{\pi}{2(2D_i+1)}, \quad \forall\, i \in \overline{1,n}. \tag{7.16}$$

Remark 1. When $D_i = 0$, from (7.16) it follows that

$$n_i(\varepsilon_i - 1) < 1,$$

which always holds if $\varepsilon_i \leq 1$. This implies that the linearized Vicsek's model in its original form ($\varepsilon_i = 1$) can achieve a consensus asymptotically if and only if the interconnection topology graph of the system is connected. When $D_i \geq 1$, inequality (7.16) holds only if $\varepsilon_i < 1$. So, the introduction of small ε_i is necessary for enhancing the robustness of the linearized Vicsek's model against input delays.

Remark 2. Corollary 7.3 clearly shows the relationship of input delays, interconnection weights and number of neighbors: for large input delays one should use small interconnection weights or have small numbers of neighbors when the graph is kept connected.

Similarly, we can also apply Theorem 7.2 to Moreau's model (Moreau 2005) with input delays

$$x_i(k+1) = x_i(k) + \frac{1}{1 + \sum_{j \in N_i} w_{ij}} \sum_{j \in N_i} w_{ij}(x_j(k - D_i) - x_i(k - D_i)), \quad i \in \overline{1, n}, \quad (7.17)$$

where w_{ij} denotes the positive weight corresponding to the edge e_{ij} in the weighted graph G. The following result is a direct corollary of Theorem 7.2.

Corollary 7.4 *Suppose that the interconnection topology graph of system (7.17) is connected and has symmetric weights. Then, the system achieves a consensus asymptotically if*

$$\frac{\sum_{j \in N_i} w_{ij}}{1 + \sum_{j \in N_i} w_{ij}} < \sin \frac{\pi}{2(2D_i + 1)}, \quad \forall\, i \in \overline{1, n}. \quad (7.18)$$

Remark. Obviously, (7.18) always holds if $D_i = 0$. This implies that a symmetric Moreau's model can achieve a consensus asymptotically if and only if the interconnection topology graph of the system is connected. For $D_i \geq 1$, from (7.18) we get a sufficient consensus condition of Moreau's model as

$$\sum_{j \in N_i} w_{ij} < 1. \quad (7.19)$$

7.1.3 Consensus under Diverse Communication Delays and Input Delays

In this subsection, we consider multi-agent systems with both communication delays and input delays. Actually, if the communication delays τ_{ij} are symmetric, i.e., they satisfy the requirement (2.26), then all the results given in Section 7.1.2 can be extended to systems with both communication delays and input delays without any difficulty.

In general, however, the diversity of communication delays destroys the symmetry of the system even if the graph is undirected with symmetric weights. This implies that the tools used for Theorem 7.2, which are mainly referred to Lemma 2.34 and Theorem 3.16, are no longer applicable. So, in the following analysis we will use Greshgorin's disk theorem (Lemma 2.30) to estimate matrix eigenvalues, which does not require the symmetry. Note that the interconnection topology studied in this subsection can be a digraph with asymmetric weights.

Let us first introduce the following lemma as a preliminary result.

Lemma 7.5 *The following inequality*

$$\frac{\sin\left(\frac{2D+1}{2}\omega\right)}{\sin\left(\frac{\omega}{2}\right)} \le 2D + 1 \tag{7.20}$$

holds for all non-negative integers D and all $\omega \in [-\pi, \pi]$.

Proof. First of all, we claim that

$$\sin\left(\frac{\pi}{2(2D+1)}\right) \ge \frac{1}{2D+1} \tag{7.21}$$

holds for any non-negative integer D. Indeed, by denoting $x = \frac{1}{2D+1}$, we have $x \in (0, 1]$ for any non-negative integer D. Thus, inequality (7.21) is equivalent to the well-known inequality $\sin(\frac{\pi}{2}x) \ge x$, where $x \in (0, 1]$.

Now, we note that $\lim_{\omega\to 0} \frac{\sin\left(\frac{2D+1}{2}\omega\right)}{\sin(\frac{\omega}{2})} = 2D + 1$. We just need to prove (7.20) for all $\omega \in (0, \pi]$ because the left-hand side of (7.20) is an even function for $\omega \in [-\pi, \pi]$.

When $\omega \in (0, \frac{\pi}{2D+1}]$, let $h(\omega) = \sin\left(\frac{2D+1}{2}\omega\right) - (2D + 1)\sin\left(\frac{\omega}{2}\right)$. Calculating the derivative of $h(\omega)$ with respect to ω yields $\dot{h}(\omega) = \frac{2D+1}{2}\left(\cos\left(\frac{2D+1}{2}\omega\right) - \cos\left(\frac{\omega}{2}\right)\right)$. Obviously, we have $\dot{h}(\omega) \le 0$, i.e., $h(\omega)$ is not increasing for all $\omega \in \left(0, \frac{\pi}{2D+1}\right]$. Since $h(0) = 0$, we have $h(\omega) \le 0$, i.e., $\sin\left(\frac{2D+1}{2}\omega\right) \le (2D + 1)\sin\left(\frac{\omega}{2}\right)$ for all $\omega \in \left(0, \frac{\pi}{2D+1}\right]$. Since $\sin(\frac{\omega}{2}) > 0$, we get $\frac{\sin\left(\frac{2D+1}{2}\omega\right)}{\sin(\frac{\omega}{2})} \le 2D + 1$.

When $\omega \in \left(\frac{\pi}{2D+1}, \pi\right]$, we have $\sin\left(\frac{\omega}{2}\right) > \sin\left(\frac{\pi}{2(2D+1)}\right) > 0$ for all non-negative integers D. So, from (7.21), we get

$$\frac{\sin\left(\frac{2D+1}{2}\omega\right)}{\sin\left(\frac{\omega}{2}\right)} \le \frac{1}{\sin\left(\frac{\omega}{2}\right)} < \frac{1}{\sin\left(\frac{\pi}{2(2D+1)}\right)} \le 2D + 1$$

for all $\omega \in \left(\frac{\pi}{2D+1}, \pi\right]$ and all non-negative integers D. The lemma is proved. □

Theorem 7.6 *Consider multi-agent system (7.3) with protocol (7.4). Assume that the interconnection topology digraph $G = (V, E, A)$ of the system has a globally reachable node. Then the system achieves a consensus asymptotically if*

$$\sum_{j\in N_i} a_{ij} < \frac{1}{2D_i + 1}, \forall\, i \in \overline{1, n}. \tag{7.22}$$

Proof. The closed-loop system of (7.3) with (7.4) is given by (7.5). Taking the z-transformation of the system (7.5), we get

$$zX_i(z) = X_i(z) + \sum_{j \in N_i} a_{ij}(X_j(z)z^{-\tau_{ij}-D_i} - X_i(z)z^{-D_i}), i \in \overline{1,n}, \tag{7.23}$$

where $X_i(z)$ is the z-transformation of $x_i(k)$. Define an $n \times n$ matrix $\tilde{L}(z) = \{\tilde{l}_{ij}(z)\}$ as follows:

$$\tilde{l}_{ij}(z) = \begin{cases} -a_{ij}z^{-\tau_{ij}-D_i}, & j \in N_i; \\ \left(\sum_{k \in N_i} a_{ik}\right) z^{-D_i}, & j = i; \\ 0, & \text{otherwise.} \end{cases}$$

Obviously, $\tilde{L}(1) = L$, which is the Laplacian matrix. Then, (7.23) can be written as $zX(z) = X(z) - \tilde{L}(z)X(z)$, where $X(z) = [X_1(z), \cdots, X_n(z)]^{\mathrm{T}}$. Define

$$p(z) = \det((z-1)I + \tilde{L}(z)).$$

Then, we will prove that all the zeros of $p(z)$ have modulus less than unity except for a zero at $z = 1$ in the following.

Let $z = 1$, $p(1) = \det(\tilde{L}(1)) = \det(L)$. Since $G = (V, E, A)$ has a globally reachable node, by Theorem 1.9, zero is a simple eigenvalue of L, i.e., $\det(L) = 0$ and $\mathrm{rank}(L) = n - 1$. Thus, $p(z)$ indeed has a simple zero at $z = 1$.

Now, we prove that the zeros of $f(z) = \det(I + \frac{1}{z-1}\tilde{L}(z))$ have modulus less than unity. Based on the general Nyquist stability criterion (Corollary 2.20), the zeros of $f(z)$ have modulus less than unity, if the eigenloci of $\frac{1}{\mathrm{e}^{\mathrm{j}\omega}-1}\tilde{L}(\mathrm{e}^{\mathrm{j}\omega})$, i.e., $\lambda(\frac{1}{\mathrm{e}^{\mathrm{j}\omega}-1}\tilde{L}(\mathrm{e}^{\mathrm{j}\omega}))$, do not enclose the point $(-1, \mathrm{j}0)$ for $\omega \in [-\pi, \pi]$. By Greshgorin's disk lemma (Lemma 2.30), we have $\lambda(\frac{1}{\mathrm{e}^{\mathrm{j}\omega}-1}\tilde{L}(\mathrm{e}^{\mathrm{j}\omega})) \in \bigcup_{i \in \overline{1,n}} \mathcal{D}_i$ for all $\omega \in [-\pi, \pi]$, where

$$\mathcal{D}_i = \left\{ \zeta : \zeta \in \mathbb{C}, \left| \zeta - \left(\sum_{j \in N_i} a_{ij} \right) \frac{\mathrm{e}^{-\mathrm{j}\omega D_i}}{\mathrm{e}^{\mathrm{j}\omega} - 1} \right| \leq \sum_{j \in N_i} \left| a_{ij} \frac{\mathrm{e}^{-\mathrm{j}\omega(\tau_{ij}+D_i)}}{\mathrm{e}^{\mathrm{j}\omega} - 1} \right| \right\}.$$

Further, we can show that

$$\mathcal{D}_i = \left\{ \zeta : \zeta \in \mathbb{C}, \left| \zeta - K_i \frac{\mathrm{e}^{-\mathrm{j}\omega D_i}}{\mathrm{e}^{\mathrm{j}\omega} - 1} \right| \leq \left| K_i \frac{\mathrm{e}^{-\mathrm{j}\omega D_i}}{\mathrm{e}^{\mathrm{j}\omega} - 1} \right| \right\},$$

where $K_i = \sum_{j \in N_i} a_{ij}$. Now, define

$$\begin{aligned} G_i(\omega) &= K_i \frac{\mathrm{e}^{-\mathrm{j}\omega D_i}}{\mathrm{e}^{\mathrm{j}\omega} - 1} \\ &= -K_i \frac{\sin\left(\frac{2D_i+1}{2}\omega\right)}{2\sin\left(\frac{\omega}{2}\right)} - \mathrm{j}K_i \frac{\cos\left(\frac{2D_i+1}{2}\omega\right)}{2\sin\left(\frac{\omega}{2}\right)}. \end{aligned} \tag{7.24}$$

The Nyquist plot of $G_i(\omega)$ for $\omega \in [-\pi, \pi]$ is illustrated by Figure 7.1. Note that $G_i(\omega)$ is just the center of the disc $\mathcal{D}_i$. So, $\lambda\left(\frac{1}{\mathrm{e}^{\mathrm{j}\omega}-1}\tilde{L}(\mathrm{e}^{\mathrm{j}\omega})\right)$ does not enclose the point $(-1, \mathrm{j}0)$ for all $\omega \in [-\pi, \pi]$ as long as the point $(-a, \mathrm{j}0)$ with $a \geq 1$ is not in the disc $\mathcal{D}_i$ for all $\omega \in [-\pi, \pi]$, i.e., $|-a + \mathrm{j}0 - K_i \frac{\mathrm{e}^{-\mathrm{j}\omega D_i}}{\mathrm{e}^{\mathrm{j}\omega}-1}| > |K_i \frac{\mathrm{e}^{-\mathrm{j}\omega D_i}}{\mathrm{e}^{\mathrm{j}\omega}-1}|$ holds for all $\omega \in [-\pi, \pi]$ when $a \geq 1$.

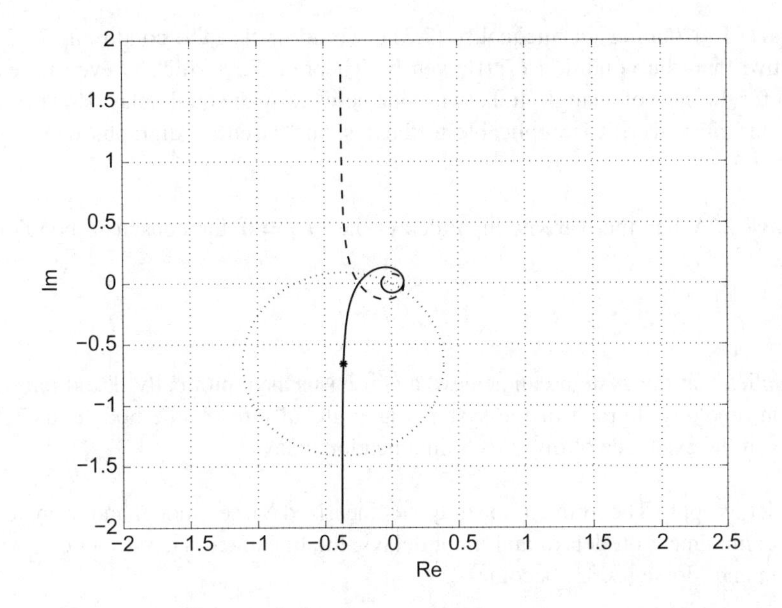

Figure 7.1 Nyquist plot of $G_i(\omega)$. (Reproduced with permission from Tian Y.-P. and Liu C.-L., "Consensus of multi-agent systems with diverse input and communication delays," *IEEE Transactions on Automatic Control*, **53**, 9, 2122–2128, 2008. © 2008 IEEE.)

From (7.24), we have

$$\left|-a + \mathrm{j}0 - K_i \frac{\mathrm{e}^{-\mathrm{j}\omega D_i}}{\mathrm{e}^{\mathrm{j}\omega} - 1}\right|^2 - \left|K_i \frac{\mathrm{e}^{-\mathrm{j}\omega D_i}}{\mathrm{e}^{\mathrm{j}\omega} - 1}\right|^2 = a\left(a - K_i \frac{\sin\left(\frac{2D_i+1}{2}\omega\right)}{\sin\left(\frac{\omega}{2}\right)}\right).$$

Because $\frac{\sin\left(\frac{2D_i+1}{2}\omega\right)}{\sin\left(\frac{\omega}{2}\right)} \leq 2D_i + 1$ holds for $\omega \in [-\pi, \pi]$ by Lemma 7.5, it follows from (7.22) that

$$K_i \frac{\sin\left(\frac{2D_i+1}{2}\omega\right)}{\sin\left(\frac{\omega}{2}\right)} \leq K_i(2D_i + 1) < 1 \leq a.$$

Thus,

$$\left|a + \mathrm{j}0 - K_i \frac{\mathrm{e}^{-\mathrm{j}\omega D_i}}{\mathrm{e}^{\mathrm{j}\omega} - 1}\right|^2 - \left|K_i \frac{\mathrm{e}^{-\mathrm{j}\omega D_i}}{\mathrm{e}^{\mathrm{j}\omega} - 1}\right|^2 > 0,$$

i.e.,

$$\left|a + \mathrm{j}0 - K_i \frac{\mathrm{e}^{-\mathrm{j}\omega D_i}}{\mathrm{e}^{\mathrm{j}\omega} - 1}\right| > \left|K_i \frac{\mathrm{e}^{-\mathrm{j}\omega D_i}}{\mathrm{e}^{\mathrm{j}\omega} - 1}\right|$$

holds for all $\omega \in [-\pi, \pi]$ when $a \geq 1$.

Now, we have proved that the zeros of $p(z)$ have modulus less than unity except for a zero at $z = 1$. Therefore, Theorem 7.6 is proved by Lemma 7.1. □

Remark 1. Noticing the inequality (7.21), we know that the condition (7.22) is more conservative than the condition (7.9) given by Theorem 7.2, which is even necessary and sufficient for the case of a single-link, two-node network with equal delay. However, it is still scalable, and moreover, it is applicable to the systems based on digraphs with asymmetric weights.

Remark 2. When there are no input delays, i.e., $D_i = 0$, the consensus condition (7.22) reduces to

$$\sum_{j \in N_i} a_{ij} < 1, \tag{7.25}$$

which implies that the system can achieve a consensus asymptotically if and only the interconnection topology digraph of the system has a globally reachable node and (7.25) holds regardless of the existence of diverse communication delays.

Now, let us apply Theorem 7.6 to study the linearized Vicsek model and Moreau's model.

With communication delays and input delays, the linearized Vicsek model given in Jadbabaie, Lin and Morse (2003) becomes

$$x_i(k+1) = x_i(k) + \frac{1}{1+n_i}\left(\sum_{j \in N_i} (x_j(k - \tau_{ij} - D_i) - x_i(k - D_i)\right), \tag{7.26}$$

where n_i denotes the number of the neighbors of agent i.

From Theorem 7.6 we get the following corollary for Vicsek's model (7.26).

Corollary 7.7 *Assume that the interconnection topology of system (7.26) has a globally reachable node. System (7.26) achieves a consensus asymptotically if*

$$\frac{n_i}{1+n_i} < \frac{1}{2D_i + 1}, \quad \forall\, i \in \overline{1, n}. \tag{7.27}$$

Remark. On the one hand, when $D_i = 0$, inequality (7.27) holds automatically. This implies that the convergence of the consensus protocol given by Vicsek's model is independent of communication delays provided the graph has a globally reachable node. This coincides with the result given in Wang and Slotine (2006). Cao *et al.* (2006) extended this result to the case when the graph is jointly rooted. On the other hand, inequality (7.27) holds only for $D_i = 0$. This may suggest that Vicsek's model is very sensitive to input delays. To enhance its robustness against input delays, one should introduce small weights as shown in model (7.15).

Similarly, Moreau's model given in (Moreau 2005) with communication delays and input delays can be written as

$$\begin{aligned} x_i(k+1) &= x_i(k) \\ &+ \frac{1}{1+\sum_{j \in N_i} w_{ij}} \sum_{j \in N_i} w_{ij}(x_j(k - \tau_{ij} - D_i) - x_i(k - D_i)), \quad i \in \overline{1, n}. \end{aligned} \tag{7.28}$$

The following result is a direct corollary of Theorem 7.6.

Corollary 7.8 *If the interconnection topology digraph of system (7.28) has a globally reachable node, then the system achieves a consensus asymptotically if*

$$\frac{\sum_{j\in N_i} w_{ij}}{1+\sum_{j\in N_i} w_{ij}} < \frac{1}{2D_i+1}, \quad \forall \quad i \in \overline{1,n}. \tag{7.29}$$

Remark. Obviously, (7.29) can be rewritten as $\sum_{j\in N_i} w_{ij} < \frac{1}{2D_i}$. So, it always holds if $D_i = 0$. But for $D_i > 0$ it holds only for some appropriately designed weights w_{ij}.

7.1.4 Continuous-Time System

Consider a continuous-time multi-agent system with diverse input delays and communication delays

$$\dot{x}_i(t) = \sum_{j\in N_i} a_{ij}(x_j(t-\tau_{ij}-T_i) - x_i(t-T_i)), \quad i \in \overline{1,n}. \tag{7.30}$$

It is easy to show that under the assumption that the topology graph is connected or the topology digraph contains a globally reachable node, the system achieves a consensus asymptotically if it is steady semi-stable with $s = 0$ as a simple pole.

Using the theory developed in Chapter 3 (Theorem 3.10), in a similar way to that shown in the proof of Theorem 7.2 one can get the consensus condition for continuous-time systems with input delays.

Theorem 7.9 *Suppose that the topology graph of system (7.30) is connected with symmetric weights. Then, the system achieves a consensus asymptotically if*

$$T_i \sum_{j\in N_i} a_{ij} < \frac{\pi}{4}, \quad \forall\, i \in \overline{1,n}. \tag{7.31}$$

Remark. It is easy to see that in the case when all the delays T_i are the same for all $i \in \overline{1,n}$, the condition (7.31) reduces to the result given by Olfati-Saber and Murray (2004).

The continuous-time system with diverse input and communication delays can be written as

$$\dot{x}_i(t) = \sum_{j\in N_i} a_{ij}(x_j(t-\tau_{ij}-T_i) - x_i(t-T_i)), \quad i \in \overline{1,n}. \tag{7.32}$$

Through a similar procedure to that used in the proof of Theorem 7.6, one can get a consensus condition for continuous-time systems with diverse communication and input delays.

Theorem 7.10 *If the interconnection topology digraph of system (7.32) has a globally reachable node, then the system achieves a consensus asymptotically if*

$$T_i \sum_{j\in N_i} a_{ij} < \frac{1}{2}, \quad \forall\, i \in \overline{1,n}. \tag{7.33}$$

Remark. When there is no input delays, i.e., $T_i = 0$, the consensus condition (7.33) always holds, which implies that the system can achieve a consensus asymptotically if and only the interconnection topology digraph of the system has a globally reachable node regardless of the existence of diverse communication delays. This conclusion coincides with existing results in references such as Blondel *et al.* (2005) and Wang and Slotine (2006).

7.1.5 Simulation Study

Example 7.11 *Symmetric system.*

Consider a system of eighty agents described by the modified linearized Vicsek model (7.15). The interconnection topology for the agents is a closed ring (Figure 7.2). Note that under non-zero input delays, the linearized Vicsek's model with unity weights ($\varepsilon_i = 1$) has no consensus. In simulations, we choose the coupling weights as

$$\begin{aligned}
\varepsilon_1 &= \cdots = \varepsilon_{10} = 0.05,\\
\varepsilon_{11} &= \cdots = \varepsilon_{20} = 0.1,\\
\varepsilon_{21} &= \cdots = \varepsilon_{30} = 0.15,\\
\varepsilon_{31} &= \cdots = \varepsilon_{40} = 0.2,\\
\varepsilon_{41} &= \cdots = \varepsilon_{50} = 0.25,\\
\varepsilon_{51} &= \cdots = \varepsilon_{60} = 0.3,\\
\varepsilon_{61} &= \cdots = \varepsilon_{70} = 0.4,\\
\varepsilon_{71} &= \cdots = \varepsilon_{80} = 0.45.
\end{aligned}$$

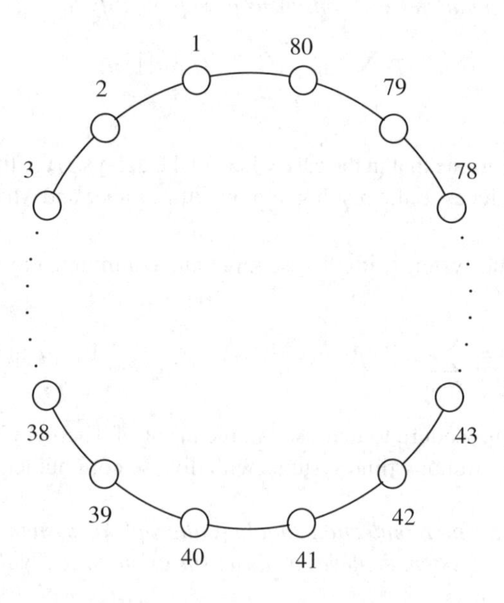

Figure 7.2 Undirected graph: a closed ring. (Reproduced with permission from Tian Y.-P. and Liu C.-L., "Consensus of multi-agent systems with diverse input and communication delays," *IEEE Transactions on Automatic Control*, **53**, 9, 2122–2128, 2008. © 2008 IEEE.)

By Corollary 7.3, the admissible values of the input delays are

$$
\begin{aligned}
&D_i \leq 23, \quad i = 1, \cdots, 10;\\
&D_i \leq 11, \quad i = 11, \cdots, 20;\\
&D_i \leq 7, \quad i = 21, \cdots, 30;\\
&D_i \leq 5, \quad i = 31, \cdots, 40;\\
&D_i \leq 4, \quad i = 41, \cdots, 50;\\
&D_i \leq 3, \quad i = 51, \cdots, 60;\\
&D_i \leq 2, \quad i = 61, \cdots, 70;\\
&D_i \leq 2, \quad i = 71, \cdots, 80.
\end{aligned}
$$

Simulations validate that under the admissible input delays the system achieves consensus asymptotically. We depict the boundary values of the input delay versus the coupling weights in Figure 7.3.

Note that the bound of input delay determined by Corollary 7.3 may not apply to asymmetric systems. For example, with each agent's input delay at boundary, we introduce an identical communication delay, $T \geq 1$, which destroys the conjugate symmetry of the system matrix in the frequency domain, between each pair of neighboring agents. Then, the simulation shows that the system has no asymptotic consensus.

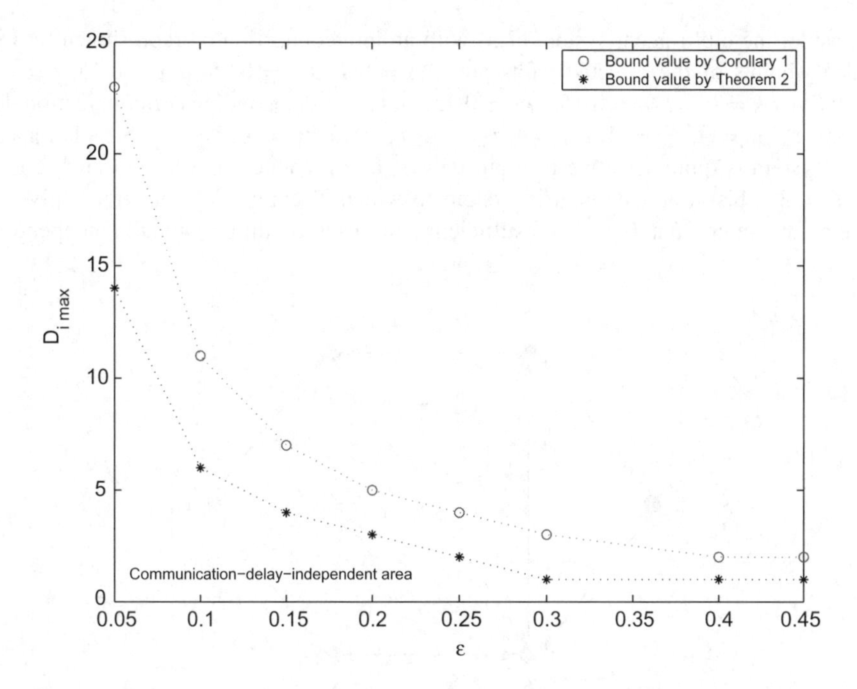

Figure 7.3 Bounds of input delays. (Reproduced with permission from Tian Y.-P. and Liu C.-L., "Consensus of multi-agent systems with diverse input and communication delays," *IEEE Transactions on Automatic Control*, **53**, 9, 2122–2128, 2008. © 2008 IEEE.)

Now, we can use Theorem 7.6 to get a consensus condition for the system with both input delays and communication delays. By Theorem 7.6, the admissible values of the input delays are

$$
\begin{aligned}
&D_i \leq 14, \quad i = 1, \cdots, 10;\\
&D_i \leq 6, \quad i = 11, \cdots, 20;\\
&D_i \leq 4, \quad i = 21, \cdots, 30;\\
&D_i \leq 3, \quad i = 31, \cdots, 40;\\
&D_i \leq 2, \quad i = 41, \cdots, 50;\\
&D_i \leq 1, \quad i = 51, \cdots, 60;\\
&D_i \leq 1, \quad i = 61, \cdots, 70;\\
&D_i \leq 1, \quad i = 71, \cdots, 80.
\end{aligned}
$$

The boundary values of input delays estimated by Theorem 7.6 are also shown in Figure 7.3. The boundary determined by Theorem 7.6 is lower than the boundary given by Corollary 7.3, which implies that Theorem 7.6 is more conservative than Corollary 7.3 for symmetric systems. However, with input delays in the area determined by Theorem 7.6, the asymptotic consensus of the system is robust to arbitrary communication delays.

Example 7.12 *Asymmetric system.*

Consider the multi-agent system (7.5) with an interconnection digraph shown by Figure 7.4. The weights of the directed paths are: $a_{12} = 0.1$, $a_{16} = 0.05$, $a_{23} = 0.15$, $a_{36} = 0.1$, $a_{43} = 0.05$, $a_{45} = 0.1$, $a_{56} = 0.15$, $a_{62} = 0.15$, and the corresponding communication delays are: $\tau_{12} = 5$, $\tau_{16} = 3$, $\tau_{23} = 4$, $\tau_{36} = 4$, $\tau_{43} = 4$, $\tau_{45} = 6$, $\tau_{56} = 6$, $\tau_{62} = 5$. Simulation shows that the system is quite sensitive to input delays, and it cannot converge to any consensus when $T_i > 3$. This is an asymmetric system to which Theorem 7.2 does not apply. Using Theorem 7.6, we get that $T_i \leq 2$ is a sufficient consensus condition which is independent of

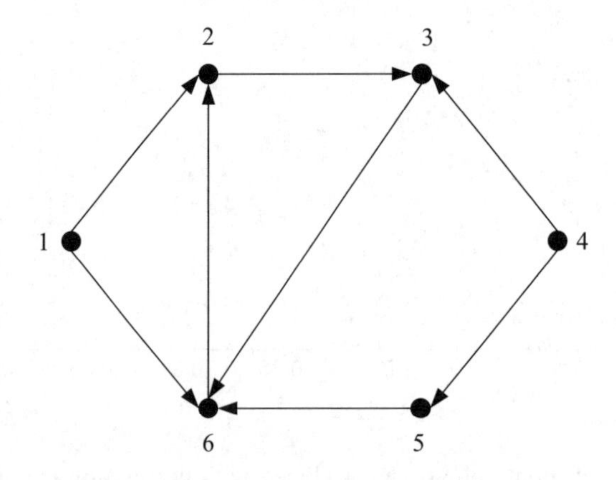

Figure 7.4 Digraph of a group of six agents.

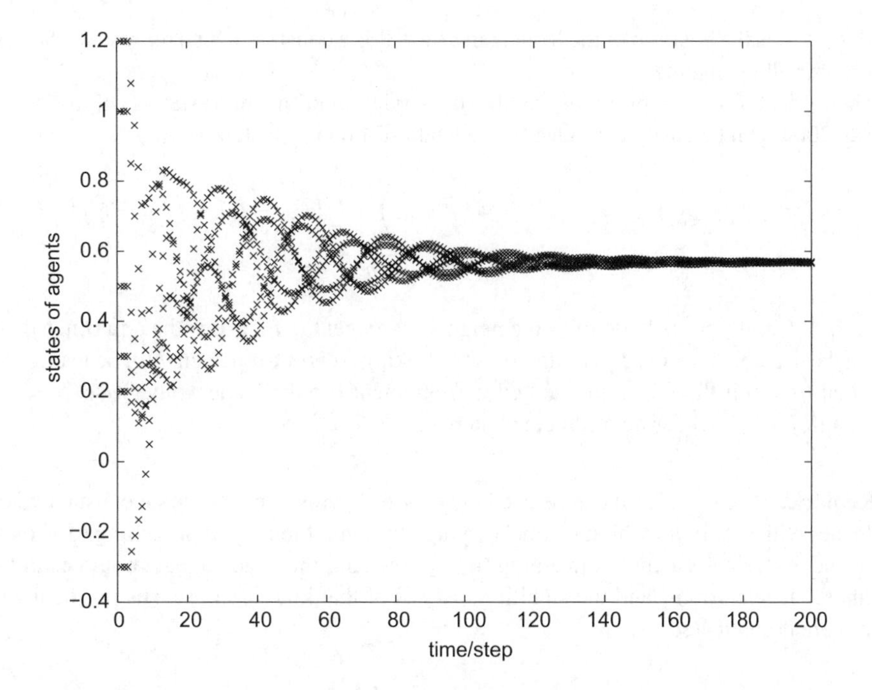

Figure 7.5 Consensus with communication and input delays. (Reproduced with permission from Tian Y.-P. and Liu C.-L., "Consensus of multi-agent systems with diverse input and communication delays," *IEEE Transactions on Automatic Control*, **53**, 9, 2122–2128, 2008. © 2008 IEEE.)

communication delays. We choose $T_i = 2, i = 1, 2, 3, 4, 5, 6$, in the simulation. The multi-agent system converges to a consensus as shown by Figure 7.5.

7.2 Double Integrator System with Diverse Input Delays and Interconnection Uncertainties

7.2.1 Leader-Following Consensus Algorithm

Consider a multi-agent system with interconnection topology graph $G = (V, E, A)$. There are n agents with diverse input delays

$$\begin{aligned} \dot{\xi}_i &= \zeta_i, \\ \dot{\zeta}_i &= u_i(t - T_i), \quad i \in \overline{1, n}, \end{aligned} \tag{7.34}$$

where $\xi_i \in \mathbb{R}$, $\zeta_i \in \mathbb{R}$, $u_i \in \mathbb{R}$, and $T_i > 0$ are the position, velocity, acceleration and input delay, respectively, of agent i.

For system (7.34), the leader-following coordination control strategy is adopted in this section. Let the dynamics of the leader be determined by

$$\dot{\xi}_0 = \zeta_0, \tag{7.35}$$

where $\xi_0 \in \mathbb{R}$ is the position of the leader, and $\zeta_0 \in \mathbb{R}$ is a constant which represents the desired velocity for all the agents.

Then, the consensus protocol for the first-order multi-agent system (Olfati-Saber and Murray 2004) can be easily extended to the leader-following system as follows

$$u_i = k_i\Big(\sum_{j\in N_i} a_{ij}\big((\zeta_j - \zeta_i) + \gamma(\xi_j - \xi_i)\big) + b_i\big((\zeta_0 - \zeta_i) + \gamma(\xi_0 - \xi_i)\big)\Big),\quad i \in \overline{1,n}, \tag{7.36}$$

where $k_i > 0$ and $\gamma > 0$, N_i denotes the neighbors of agent i, $a_{ij} > 0$ is the adjacency element of A in the digraph $G = (V, E, A)$, and b_i is the linking weight from agent i to the leader (7.35). Note that $b_i > 0$ if there is a directed edge from agent i to the leader; otherwise, $b_i = 0$. Let $B = \mathrm{diag}\{b_i, i \in \overline{1,n}\}$ for notation convenience.

Remark. Protocol (7.36) can be used only for following a leader with a constant velocity, or a leader with a velocity which is time-varying but asymptotically approaching to a constant. If the leader's velocity is not converging to any constant, then each agent should estimate its neighbors' accelerations, and the stability analysis of that kind of consensus protocol will be much more complicated.

With consensus protocol (7.36), the closed-loop form of system (7.34) is given by

$$\begin{aligned}
\dot{\xi}_i &= \zeta_i,\\
\dot{\zeta}_i &= k_i\Big(\sum_{j\in N_i} a_{ij}\big((\zeta_j(t-T_i) - \zeta_i(t-T_i)) + \gamma(\xi_j(t-T_i) - \xi_i(t-T_i))\big)\\
&\quad + b_i\big((\zeta_0 - \zeta_i(t-T_i)) + \gamma(\xi_0(t-T_i) - \xi_i(t-T_i))\big)\Big),\quad i \in \overline{1,n}.
\end{aligned} \tag{7.37}$$

The following lemma gives some structural property of the leader-following system.

Lemma 7.13 *Assume that the interconnection topology graph of n agents together with the leader in system (7.37) has the leader as a globally reachable node. Then, the matrix $L + B$ has no zero eigenvalues, where L is the Laplacian matrix of the interconnection topology of n agents without the leader.*

Proof. Consider the interconnection topology graph with $n + 1$ nodes corresponding to the n agents of system (7.37) and the leader. Obviously the Laplacian matrix of this topology is given by

$$\hat{L} = \begin{bmatrix} 0 & 0 \\ -\hat{b} & L + B \end{bmatrix},$$

where $\hat{b} = [b_1, \cdots, b_n]^{\mathrm{T}}$. Since the leader is a globally reachable node in the graph, we have $\mathrm{rank}(\hat{L}) = n$, by Theorem 1.9. Taking elementary column transforms for $\hat{L}$ by adding all the other columns to the first column as follows

$$\begin{bmatrix} 0 & 0 \\ -\hat{b} & L+B \end{bmatrix} \rightarrow \begin{bmatrix} 0 & 0 \\ 0 & L+B \end{bmatrix},$$

we get $\text{rank}(L+B) = n$. The lemma is proved. □

7.2.2 Consensus Condition under Symmetric Coupling Weights

Let

$$\begin{aligned} \overline{\xi}_i &= \xi_i - \xi_0, \\ \overline{\zeta}_i &= \zeta_i - \zeta_0, \quad i \in \overline{1,n}. \end{aligned}$$

Then, from (7.37) it follows that

$$\begin{aligned} \dot{\overline{\xi}}_i &= \overline{\zeta}_i, \\ \dot{\overline{\zeta}}_i &= k_i \Bigg(\sum_{j \in N_i} a_{ij} \Big((\overline{\zeta}_j(t-T_i) - \overline{\zeta}_i(t-T_i)) + \gamma(\overline{\xi}_j(t-T_i) - \overline{\xi}_i(t-T_i)) \Big) \\ &\quad - b_i \Big(\overline{\zeta}_i(t-T_i) + \gamma \overline{\xi}_i(t-T_i) \Big) \Bigg), \quad i \in \overline{1,n}. \end{aligned} \tag{7.38}$$

Taking the Laplace transform of (7.38), one gets

$$\begin{aligned} s\overline{\xi}_i(s) &= \overline{\zeta}_i(s), \\ s\overline{\zeta}_i(s) &= k_i \Bigg(\sum_{j \in N_i} a_{ij} \Big((\overline{\zeta}_j(s)\mathrm{e}^{-sT_i} - \overline{\zeta}_i(s)\mathrm{e}^{-sT_i}) + \gamma(\overline{\xi}_j(s)\mathrm{e}^{-sT_i} - \overline{\xi}_i(s)\mathrm{e}^{-sT_i}) \Big) \\ &\quad - b_i \Big(\overline{\zeta}_i(s)\mathrm{e}^{-sT_i} + \gamma \overline{\xi}_i(s)\mathrm{e}^{-sT_i} \Big) \Bigg). \quad i \in \overline{1,n} \end{aligned} \tag{7.39}$$

Denote

$$H_i(s) = \frac{k_i \left(\sum_{j \in N_i} a_{ij} \right) (s+\gamma)\mathrm{e}^{-sT_i}}{s^2 + k_i \left(\sum_{j \in N_i} a_{ij} \right) (s+\gamma)\mathrm{e}^{-sT_i}}. \tag{7.40}$$

Then, using the framework of Lee and Spong (2006) one can get a sufficient condition of consensus as

$$|H_i(\mathrm{j}\omega)| < 1, \forall \omega > 0, \ \forall i \in \overline{1,n}. \tag{7.41}$$

However, it can be shown that this condition is so conservative that it gives an empty set of available control parameters for the second-order multi-agent system with input delays. Let us rewrite (7.40) as

$$H_i(s) = \frac{\kappa_i W_i(s)}{1 + \kappa_i W_i(s)},$$

where $\kappa_i = k_i \left(\sum_{j \in N_i} a_{ij}\right)$, and

$$W_i(s) = \frac{(s+\gamma)\mathrm{e}^{-sT_i}}{s^2}, i \in \overline{1,n}. \tag{7.42}$$

Then, the condition (7.41) is equivalent to

$$\mathrm{Re}[\kappa_i W_i(\mathrm{j}\omega)] > -1/2, \forall \omega > 0. \tag{7.43}$$

Actually, such a condition never holds for any $\kappa_i > 0$ when $\gamma > 0$, $T_i > 0$.

Let us derive a less conservative consensus condition for system (7.37) with symmetric coupling weights.

For all $i \in \overline{1,n}$, we denote

$$D_i = T_i \gamma,$$

$$\omega_0(i) = \frac{\sqrt{\frac{3D_i - D_i^2 + \sqrt{(3D_i - D_i^2)^2 + 8D_i(1-D_i)}}{2}}}{T_i}.$$

As shown in Chapter 3 (Proposition 3.26), $\omega_0(i)$ is the critical point of the frequency response of $W_i(s)$ from clockwise part to anti-clockwise part.

Let $\hat{i} \in \overline{1,n}$ be the agent which has the maximal input delay constant $T_{\hat{i}}$, i.e.,

$$\hat{i} = \arg \max_{i \in \overline{1,n}} T_i. \tag{7.44}$$

Now, we are in a position to present some sufficient consensus conditions for the second-order multi-agent system with input delays.

Theorem 7.14 *Assume that system (7.37) is composed of n agents and a leader with a static interconnection topology that has the leader as a globally reachable node, and the topology graph has symmetric weights, i.e., $a_{ij} = a_{ji}$. For each agent the following preconditions are assumed:*

$$D_i < 0.4495, \quad \forall i \in \overline{1,n}, \tag{7.45}$$

$$\omega_0(i) \le T_{\hat{i}}^{-1} \arctan\left(\frac{\omega_0(i)}{\gamma}\right), \quad \forall i \in \overline{1,n}. \tag{7.46}$$

Then, all the agents in the system asymptotically converge to the leader's state, if

$$k_i (G_i^M)^{-1} (2 \sum_{j \in N_i} a_{ij} + b_i) < 1, \quad i \in \overline{1,n}, \tag{7.47}$$

where G_i^M is the gain margin of the transfer function $W_i(s)$ defined in (7.42).

Proof. Writing (7.39) in the vector form, we can get the characteristic equation of system (7.38) as

$$\det\left(s^2 I + \mathrm{diag}\{k_i, i \in \overline{1,n}\} \mathrm{diag}\{(s+\gamma)\mathrm{e}^{-sT_i}, i \in \overline{1,n}\}(L+B)\right) = 0, \tag{7.48}$$

where L is the Laplacian matrix corresponding to the interconnection topology for all the agents without the leader.

Define $F(s) = \det(s^2 I + \text{diag}\{k_i, i \in \overline{1,n}\}\text{diag}\{(s+\gamma)\text{e}^{-sT_i}, i \in \overline{1,n}\}(L+B))$. To prove Theorem 7.14 it suffices to prove that all the zeros of $F(s)$ are in the open left half of the complex plane.

Let $s = 0$. Then $F(0) = \det(\gamma\text{diag}\{k_i\}(L+B))$. Because the interconnection topology composed of the n agents together with the leader has the leader as a globally reachable node, $F(0) \neq 0$ by Lemma 7.13.

Now, define $p(s) = \det(I + \text{diag}\{k_i\}\text{diag}\left\{\frac{(s+\gamma)\text{e}^{-sT_i}}{s^2}\right\}(L+B))$. We will prove that all the zeros of $p(s)$ are inside the LHP. Based on the general Nyquist stability criterion (Corollary 2.20), all the zeros of $p(s)$ lie inside the LHP, if the eigenloci

$$\lambda\left(\text{diag}\{k_i\}\text{diag}\left\{\frac{(\text{j}\omega+\gamma)\text{e}^{-\text{j}\omega T_i}}{(\text{j}\omega)^2}\right\}(L+B))\right)$$

do not enclose the point $(-1, \text{j}0)$ for all $\omega \in \mathbb{R}$.

For the symmetric weights $(a_{ij} = a_{ji})$, $L + B = (L+B)^{\text{T}}$. Hence, based on Lemma 2.34, we have

$$\begin{aligned}
&\lambda\left(\text{diag}\{k_i\}\text{diag}\left\{\frac{(\text{j}\omega+\gamma)\text{e}^{-\text{j}\omega T_i}}{(\text{j}\omega)^2}\right\}(L+B)\right) \\
&= \lambda\left((L+B)\text{diag}\{k_i\}\text{diag}\left\{\frac{(\text{j}\omega+\gamma)\text{e}^{-\text{j}\omega T_i}}{(\text{j}\omega)^2}\right\}\right) \\
&= \lambda\left(\text{diag}\left\{\sqrt{k_i(G_i^M)^{-1}}\right\}(L+B)\text{diag}\left\{\sqrt{k_i(G_i^M)^{-1}}\right\}\text{diag}\left\{G_i^M\frac{(\text{j}\omega+\gamma)\text{e}^{-\text{j}\omega T_i}}{(\text{j}\omega)^2}\right\}\right) \\
&\in \rho\left(\text{diag}\left\{\sqrt{k_i(G_i^M)^{-1}}\right\}(L+B)\text{diag}\left\{\sqrt{k_i(G_i^M)^{-1}}\right\}\right)\text{Co}\left(0 \cup \left\{G_i^M\frac{(\text{j}\omega+\gamma)\text{e}^{-\text{j}\omega T_i}}{(\text{j}\omega)^2}\right\}\right).
\end{aligned}$$

Since the spectral radius of any matrix is bounded by its largest absolute row sum, it follows from the condition (7.47) that

$$\begin{aligned}
&\rho\left(\text{diag}\left\{\sqrt{k_i\left(G_i^M\right)^{-1}}\right\}(L+B)\text{diag}\left\{\sqrt{k_i\left(G_i^M\right)^{-1}}\right\}\right) \\
&= \rho\left(\text{diag}\left\{k_i\left(G_i^M\right)^{-1}\right\}(L+B)\right) \\
&\leq \max_{i\in\overline{1,n}} k_i\left(G_i^M\right)^{-1}\left(\left|\sum_{j\in N_i} a_{ij} + b_i\right| + \sum_{j\in N_i}|a_{ij}|\right) \\
&= \max_{i\in\overline{1,n}} k_i\left(G_i^M\right)^{-1}\left(2\sum_{j\in N_i} a_{ij} + b_i\right) \\
&< 1.
\end{aligned}$$

Therefore, from Theorem 3.32 it follows that

$$(-1, 0) \notin \rho\left(\mathrm{diag}\left\{\sqrt{k_i(G_i^M)^{-1}}\right\}(L+B)\mathrm{diag}\left\{\sqrt{k_i(G_i^M)^{-1}}\right\}\right)\mathrm{Co}\left(0 \cup \left\{G_i^M \frac{(\mathrm{j}\omega+\gamma)\mathrm{e}^{-\mathrm{j}\omega T_i}}{(\mathrm{j}\omega)^2}\right\}\right),$$

i.e., the eigenloci of $\mathrm{diag}\{k_i\}\mathrm{diag}\left\{\frac{(\mathrm{j}\omega+\gamma)\mathrm{e}^{-\mathrm{j}\omega T_i}}{(\mathrm{j}\omega)^2}\right\}(L+B))$ do not enclose the point $(-1, \mathrm{j}0)$ for all $\omega \in \mathbb{R}$, which implies that the zeros of $F(s)$ are all inside the LHP. Theorem 7.14 is thus proved. □

Remark. The result of Theorem 7.14 can be extended to systems with both communication delays and input delays without any difficulty if the diverse communication delays τ_{ij} are symmetric, i.e., they satisfy the requirement (2.26).

7.2.3 Robust Consensus under Asymmetric Perturbations

The consensus condition given by Theorem 7.14 depends on the strict symmetry of the Laplacian matrix L. In practice, however, perturbations of coupling weights may occur and destroy the symmetry. In the following, we study the robustness of the consensus protocol against asymmetric perturbations.

Suppose that the symmetric coupling weights of system (7.37) are subject to some asymmetric perturbations, denoted by $\delta_{ij} \in \mathbb{R}$ for each one. Then the system becomes

$$\begin{aligned}\dot{\xi}_i &= \zeta_i,\\ \dot{\zeta}_i &= k_i\Bigg(\sum_{j\in N_i}(a_{ij}+\delta_{ij})\big((\zeta_j(t-T_i)-\zeta_i(t-T_i))+\gamma(\xi_j(t-T_i)-\xi_i(t-T_i))\big)\\ &\quad + b_i\big((\zeta_0-\zeta_i(t-T_i))+\gamma(\xi_0(t-T_i)-\xi_i(t-T_i))\big)\Bigg), \quad i \in \overline{1,n}, \end{aligned} \tag{7.49}$$

where $a_{ij} = a_{ji}$, and $a_{ij} + \delta_{ij} > 0$ hold for $j \in N_i$.

A robust consensus condition of the perturbed system is given by the following theorem.

Theorem 7.15 *Assume that the nominal part of system (7.49), i.e., the system without asymmetric weight perturbations δ_{ij}, converges to the leader's states asymptotically. Let*

$$M(s) = \left(s^2 I + s^2 K D(s)(L+B)\right)^{-1} s^2 K D(s), \tag{7.50}$$

where $K = \mathrm{diag}\{k_i, i \in \overline{1,n}\}$ and

$$D(s) = \mathrm{diag}\left\{\frac{(s+\gamma)\mathrm{e}^{-sT_i}}{s^2}, \quad i \in \overline{1,n}\right\}.$$

Then, the agents in the perturbed system (7.49) converge to the leader's states asymptotically, if

$$\overline{\sigma}(\Delta)\overline{\sigma}(M(\mathrm{j}\omega)) < 1, \quad \forall\omega \in \mathbb{R}, \tag{7.51}$$

where $\overline{\sigma}(\cdot)$ denotes the largest singular value of matrix, and $\Delta = \{\Delta_{ij}\}$ is the asymmetric perturbation matrix, which is defined as follows

$$\Delta_{ij} = \begin{cases} \sum_{j \in N_i} \delta_{ij}, & j = i, \\ -\delta_{ij}, & j \in N_i, \\ 0, & \text{otherwise.} \end{cases}$$

Proof. Under the same variable transformation as used in the previous subsection

$$\overline{\xi}_i = \xi_i - \xi_0, \quad \overline{\zeta}_i = \zeta_i - \zeta_0, \quad i \in \overline{1, n},$$

it is easy to get the characteristic equation of system (7.49) as

$$\det\left(s^2 I + \mathrm{diag}\{k_i, i \in \overline{1,n}\}\mathrm{diag}\{(s+\gamma)\mathrm{e}^{-sT_i}, \quad i \in \overline{1,n}\}(L + B + \Delta)\right) = 0. \qquad (7.52)$$

Since the system (7.49) without asymmetric weight perturbations δ_{ij} converges to the leader's states asymptotically, the roots of the characteristic equation (7.48) all lie inside the LHP, i.e., the zeros of $\det(s^2 I + s^2 KD(s)(L + B))$ lie inside the LHP, and $\det(L + B) \neq 0$.

In the following, we will prove that the roots of equation (7.52) are all inside the LHP.

First we show that equation (7.52) has no roots at $s = 0$. Indeed, by setting $\omega = 0$ we get from (7.51) that $\overline{\sigma}(\Delta)\overline{\sigma}((\gamma K(L + B))^{-1}\gamma K) < 1$. This implies that

$$\overline{\sigma}[(\gamma K(L + B))^{-1}\gamma K\Delta] < 1.$$

So, it follows that

$$\det(I + (\gamma K(L + B))^{-1}\gamma K\Delta) \neq 0,$$

or equivalently,

$$\det(\gamma K(L + B + \Delta)) \neq 0.$$

This proves that equation (7.52) has no roots at $s = 0$. Therefore, the characteristic equation (7.52) can be equivalently rewritten as

$$\det\left(I + KD(s)(L + B + \Delta)\right) = 0. \qquad (7.53)$$

The feedback diagram corresponding to the characteristic equation (7.53) is demonstrated in Figure 7.6. Using the linear fractional transformation, the diagram in Figure 7.6 can be

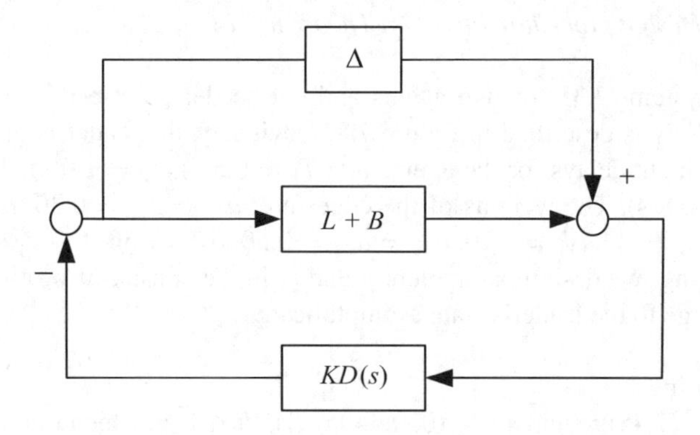

Figure 7.6 System with asymmetric perturbation.

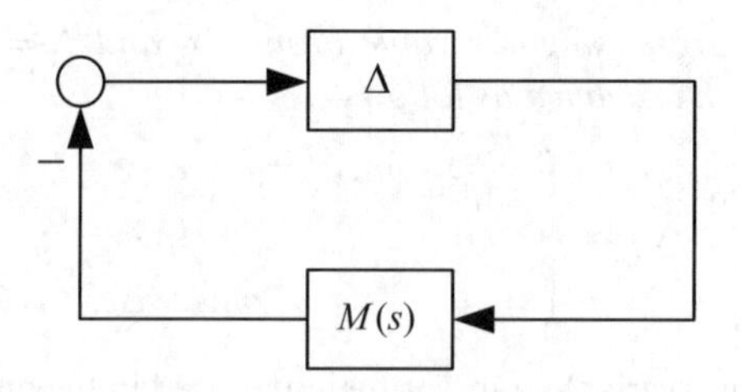

Figure 7.7 Transformed system.

equivalently transformed into the form shown by Figure 7.7, where $M(s)$ is given by equation (7.50).

The characteristic equation of the closed-loop system in Figure 7.7 is

$$\det(I + \Delta M(s)) = 0. \tag{7.54}$$

Obviously, $D(s)$ has no poles in the open RHP. Thus, $\Delta M(s)$ has no poles in the open RHP. According to the general Nyquist stability criterion (Corollary 2.20), the roots of the characteristic equation (7.54) all lie inside the LHP, as long as the eigenloci of $\Delta M(s)$, i.e., $\lambda(\Delta M(\mathrm{j}\omega))$, do not enclose the point $(-1, \mathrm{j}0)$ for $\omega \in \mathbb{R}$.

From condition (7.51) it follows that

$$\begin{aligned}\rho(\Delta M(\mathrm{j}\omega)) &\leq \overline{\sigma}(\Delta M(\mathrm{j}\omega)) \\ &\leq \overline{\sigma}(\Delta)\overline{\sigma}(M(\mathrm{j}\omega)) \\ &< 1, \ \forall \omega \in \mathbb{R}.\end{aligned} \tag{7.55}$$

Hence, $\lambda(\Delta M(\mathrm{j}\omega))$ does not enclose the point $(-1, \mathrm{j}0)$ for all $\omega \in \mathbb{R}$, i.e., the roots of the characteristic equation (7.54) all lie inside the LHP. Therefore, the closed-loop system in Figure 7.7 is asymptotically stable, and the agents in (7.49) converge to the leader's states asymptotically. Theorem 7.15 is proved. □

7.2.4 Simulation Study

Example 7.16 *Design procedure based on Theorem 7.14*

Consider a system (7.37) of five agents and one leader described by (7.35). The interconnection topology is described in Figure 7.8. Obviously, the leader is globally reachable. Assume that the input delays for the agents are: $T_1 = 0.5$(s), $T_2 = 1.0$(s), $T_3 = 0.7$(s), $T_4 = 0.6$(s) and $T_5 = 0.8$(s). The weights of the edges are: $a_{12} = a_{21} = 0.30$, $a_{25} = a_{52} = 0.70$, $a_{13} = a_{31} = 0.10$, $a_{34} = a_{43} = 1.10$, $a_{42} = a_{24} = 0.50$, $b_5 = 1.50$.

In the following, we design parameters γ and k_i in the consensus protocol (7.36) so that the agents converge to the leader's state asymptotically.

Step 1: Choosing γ.

The condition (7.45) requires $\gamma \in (0, 0.4495/T_i)$, $\forall i \in \overline{1, n}$, which implies that

$$\gamma \in (0, 0.4495/T_2) = (0, 0.4495). \tag{7.56}$$

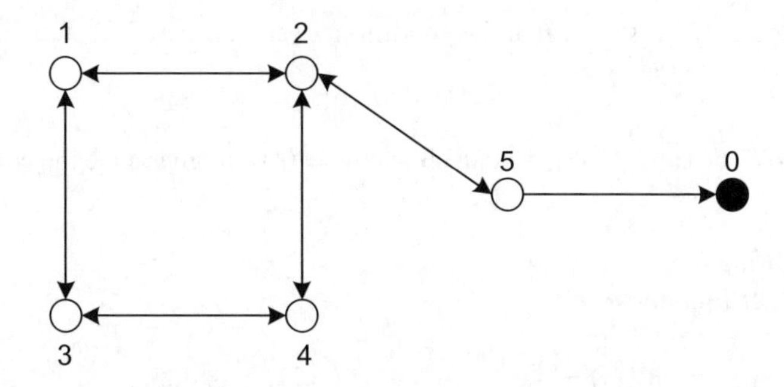

Figure 7.8 Network of five agents and a leader. (Reprinted from *Automatica*, **45**, 5, Tian Y.-P. and Liu C.-L., "Robust consensus of multi-agent systems with diverse input delays and asymmetric interconnection perturbations," 1347–1353, 2009, with permission from Elsevier.)

Since T_2 is the maximal input delay, we have $\hat{i} = 2$. Denote $E_i = T_2^{-1} \arctan(\frac{\omega_0(i)}{\gamma}) - \omega_0(i), i = 1, 2, 3, 4, 5,$ and the condition (7.46) can be represented as $E_i > 0, i = 1, 2, 3, 4, 5$. With the given input delays, E_i is a function of γ. The curves of E_i on γ are shown in Figure 7.9.

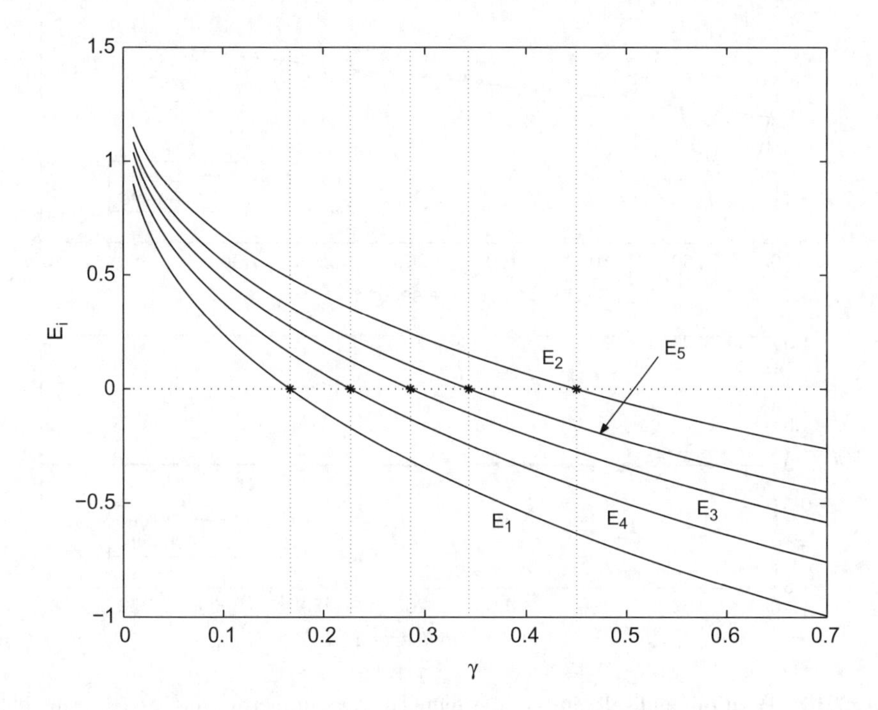

Figure 7.9 Choosing parameter γ. (Reprinted from *Automatica*, **45**, 5, Tian Y.-P. and Liu C.-L., "Robust consensus of multi-agent systems with diverse input delays and asymmetric interconnection perturbations," 1347–1353, 2009, with permission from Elsevier.)

From Figure 7.9 it is clear that the condition $E_i > 0, i = 1, 2, 3, 4, 5$, holds if

$$\gamma \in (0, 0.16]. \tag{7.57}$$

According to (7.56) and (7.57), we can choose $\gamma = 0.10$ to guarantee the conditions (7.45) and (7.46).

Step 2: Choosing k_i.
For the transfer functions

$$W_i(s) = \frac{(s+\gamma)e^{-T_i s}}{s^2}, \quad i = 1, 2, 3, 4, 5,$$

using the MATLAB® simulator we obtain the inverses of their gain margins as $(G_1^M)^{-1} \simeq 0.33, (G_2^M)^{-1} \simeq 0.67, (G_3^M)^{-1} \simeq 0.46, (G_4^M)^{-1} \simeq 0.39, (G_5^M)^{-1} \simeq 0.53$. From the condition (7.47), the constraints on k_i can be calculated as $k_1 \in (0, 3.788), k_2 \in (0, 0.498), k_3 \in (0, 0.906), k_4 \in (0, 0.801), k_5 \in (0, 0.651)$, We choose $k_1 = 3.4, k_2 = k_3 = k_4 = k_5 = 0.4$ for the simulation.

With the parameters chosen above and the initial states generated randomly, the agents in the system (7.37) asymptotically converge to the leader's state as shown in Figure 7.10.

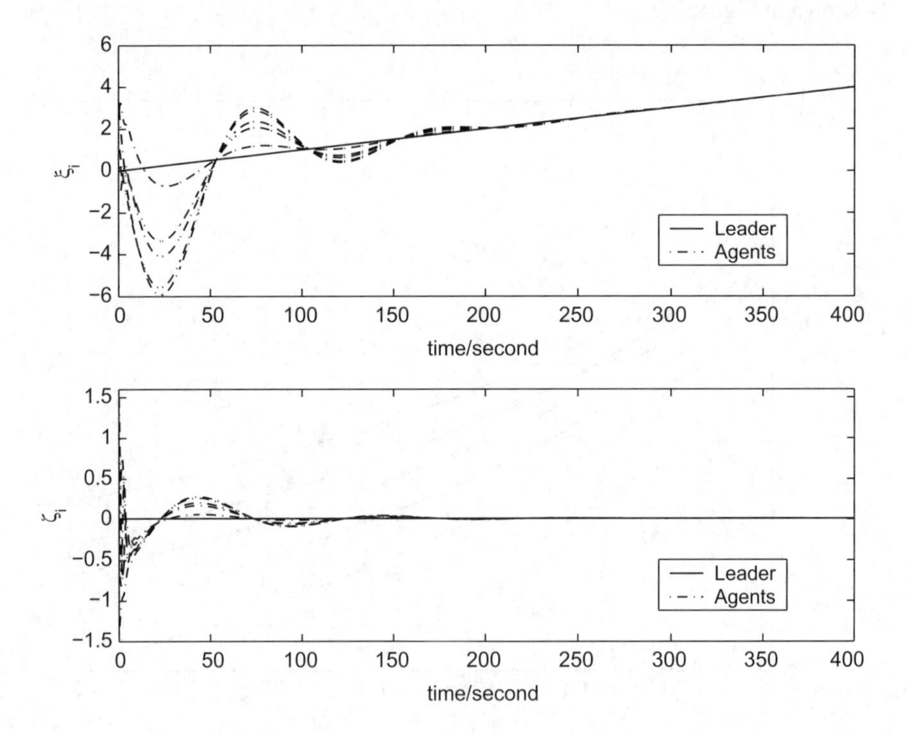

Figure 7.10 Positions and velocities of the agents under symmetric weights. (Reprinted from *Automatica*, **45**, 5, Tian Y.-P. and Liu C.-L., "Robust consensus of multi-agent systems with diverse input delays and asymmetric interconnection perturbations," 1347–1353, 2009, with permission from Elsevier.)

Since there are no other theoretic results to compare with, we test the conservatism of our results by simulation. The procedure is as follows. Setting k_2, k_3, k_4 and k_5 at our theoretic boundary values as 0.498, 0.906, 0.801 and 0.651 respectively, we increase k_1 from our boundary value 3.788 until the system has no consensus. Then we find the computational margin for k_1 is $k_{1m} = 7.519$. Using similar procedures we can obtain the other marginal gains as $k_{2m} = 0.797$, $k_{3m} = 1.088$, $k_{4m} = 0.954$, $k_{5m} = 0.773$. Note that unlike our theoretical results, these computational margins cannot be used in the consensus protocol simultaneously.

Example 7.17 *System with interconnection uncertainties.*

Consider the multi-agent systems (7.49) of five agents and one leader described by (7.35) with the same interconnection topology as Example 7.16 (see Figure 7.8). For simplicity, we choose the same a_{ij}, b_i, input delays T_i and control parameters γ and κ_i, $i \in \overline{1, n}$ as given in Example 7.16. From Theorem 7.14, the system (7.49) without asymmetric weight perturbations converges to the leader's states asymptotically, and the zeros of $\det(s^2 I + s^2 KD(s)(L + B))$ lie inside the LHP. Using the MATLAB® simulator, we obtain that the largest value of $\overline{\sigma}(M(\mathrm{j}\omega))$ on $\omega \in (-\infty, \infty)$ is $\max_{\omega \in (-\infty,\infty)} \overline{\sigma}(M(\mathrm{j}\omega)) \approx 23.2$. From Theorem 7.15, if the largest singular value of the asymmetric disturbance matrix Δ, i.e., $\overline{\sigma}(\Delta)$, satisfies $\overline{\sigma}(\Delta) < 1/23.2$, the closed system in Figure 7.6 with Δ is asymptotically stable. For example, when

$$\Delta = \begin{bmatrix} 0.005 & -0.015 & 0.01 & 0 & 0 \\ 0 & 0.02 & 0 & -0.02 & 0 \\ 0 & 0 & 0.01 & -0.01 & 0 \\ 0 & 0 & -0.02 & 0.02 & 0 \\ 0 & -0.015 & 0 & 0 & 0.015 \end{bmatrix},$$

one can check that $\overline{\sigma}(\Delta) = 0.0375 < 1/23.2$, and $a_{ij} + \delta_{ij} > 0$ for $j \in N_i$. Therefore, with the Laplacian matrix $L + \Delta$ and the initial states generated randomly, the agents in (7.49) converge to the leader's states asymptotically as shown in Figure 7.11.

7.3 High-Order Consensus in High-Order Systems

7.3.1 System Model

Suppose the interconnection topology of the system is described by a digraph $G = (V, E, A)$ with $|V| = n$. We assume that the interconnection topology of the system is a connected undirected graph or a digraph containing a globally reachable node. Then, by Theorem 1.9, the Laplacian matrix L has a simple eigenvalue 0, i.e., $\det(L) = 0$ and $\mathrm{rank}(L) = n - 1$. Moreover, the definition of L implies that $L \cdot \mathbf{1}_n = 0$.

Let the model of the ith agent ($i \in \overline{1, n}$) be given by the following transfer function

$$\begin{aligned} G_i(s) &= \frac{Y_i(s)}{U_i(s)} \\ &= \frac{\mathrm{e}^{-T_i s}}{s^{\nu}(c_{im_i} s^{m_i - \nu} + c_{im_i - 1} s^{m_i - \nu - 1} + \cdots + c_{i\nu})}, \end{aligned} \qquad (7.58)$$

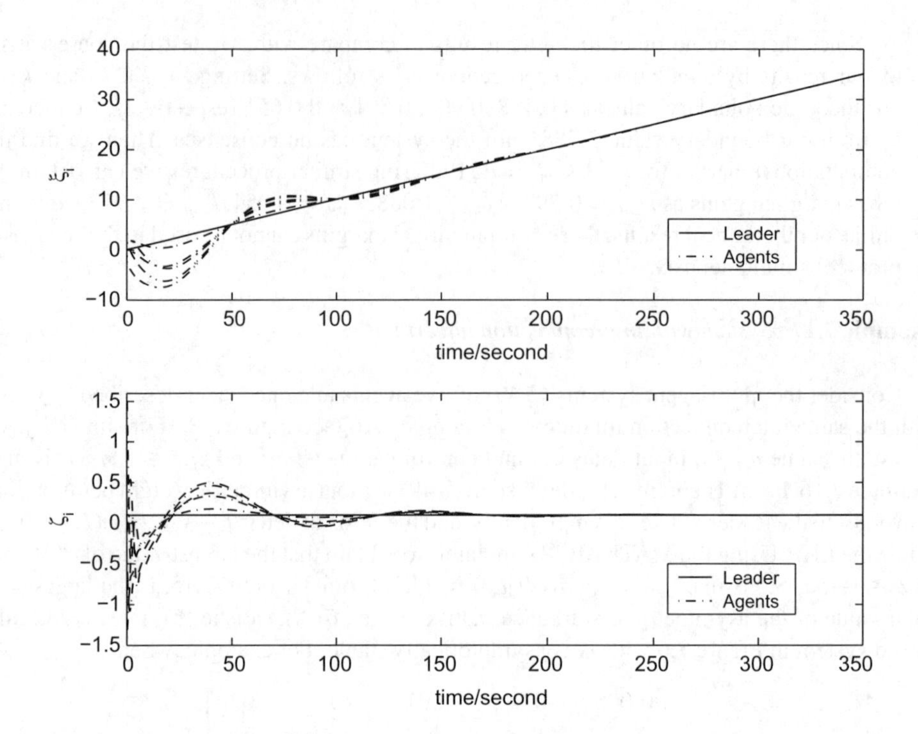

Figure 7.11 Positions and velocities of the agents under asymmetric weights. (Reprinted from *Automatica*, **45**, 5, Tian Y.-P. and Liu C.-L., "Robust consensus of multi-agent systems with diverse input delays and asymmetric interconnection perturbations," 1347–1353, 2009, with permission from Elsevier.)

where $Y_i(s) \in \mathbb{C}$ and $U_i(s) \in \mathbb{C}$ denote the Laplace transformation of the output and input, respectively, of the ith agent; T_i is the input delay; ν and m_i are positive integers satisfying $m_i \geq \nu$; $c_{ik} \in \mathbb{R}$ are system parameters. For any non-negative integer k, denotes by $y_i^{(k)}(t)$ the kth-order derivative of the output $y_i(t)$ of the ith agent.

Definition 7.18 *Multi-agent system (7.58) is said to reach the rth-order consensus asymptotically if*

$$\lim_{t\to\infty} |y_i(t) - y_j(t)| = 0, \quad \forall i, j \in \overline{1, n}, \tag{7.59}$$

$$\lim_{t\to\infty} \left| y_i^{(k)}(t) \right| = 0, \quad \forall i \in \overline{1, n}, \quad \forall k > r \tag{7.60}$$

for system solutions from any admissible initial conditions, and

$$\lim_{t\to\infty} \left| y_i^{(k)}(t) \right| \neq 0, \quad \forall i \in \overline{1, n}, \quad \forall k \in \overline{0, r} \tag{7.61}$$

for system solutions from some admissible initial conditions.

Note that when $r = 0$, the above definition implies that

$$\lim_{t\to\infty} y_i(t) = c, \quad \forall i \in \overline{1,n},$$

where $c \in \mathbb{R}$ is a constant. In this case we say the system achieves a *constant consensus*. So, constant consensus can be regarded as *the zeroth-order consensus* by Definition 7.18. We also note by this definition the constant c can not identically be zero for all initial conditions, i.e., it excludes the case of *trivial consensus* which actually implies that each agent is asymptotically stabilized.

Let τ_{ij} be the communication delay from agent j to agent i. Then, at time instance t the information obtained by agent i from agent j is $y_j^{(k)}(t - \tau_{ij})$ instead of $y_j^{(k)}(t - \tau_{ij})$. Let the consensus protocol be

$$\begin{aligned} u_i(t) = \kappa_i \sum_{k=0}^{r} b_{ik} \left(\sum_{j\in N_i} a_{ij} \left(y_j^{(k)}(t-\tau_{ij}) - y_i^{(k)}(t-\tau'_{ij}) \right) \right) \\ - \kappa_i \sum_{k=r+1}^{m_i} b_{ik} y_i^{(k)}(t) \end{aligned} \tag{7.62}$$

where $\kappa_i > 0$, $b_{ik} \in \mathbb{R}$ are some control parameters, and τ'_{ij} denotes the estimation of communication delay τ_{ij} used by agent i. Sometimes τ'_{ij}, $j \in N_i$ are also referred to as self-delays of agent i. We will show that the high-order consensus can be achieved even though the self-delays are not equal to the communication delays.

7.3.2 Consensus Condition

It is easy to get the closed-loop form of system (7.58) with protocol (7.62) as

$$\begin{aligned} \sum_{k=\nu}^{m_i} c_{ik} y_i^{(k)}(t) = \kappa_i \sum_{k=0}^{r} b_{ik} \sum_{j\in N_i} a_{ij} \left(y_j^{(k)}(t - T_i - \tau_{ij}) - y_i^{(k)}(t - T_i - \tau'_{ij}) \right) \\ - \kappa_i \sum_{k=r+1}^{m_i} b_{ik} y_i^{(k)}(t), \quad i \in \overline{1,n}. \end{aligned} \tag{7.63}$$

Taking the Laplace transform under zero initial condition for the above equation yields

$$\begin{aligned} \sum_{k=\nu}^{m_i} c_{ik} s^k Y_i(s) = \kappa_i \mathrm{e}^{-T_i s} \sum_{k=0}^{r} b_{ik} s^k \sum_{j\in N_i} a_{ij} (Y_j(s)\mathrm{e}^{-\tau_{ij}s} - Y_i(s)\mathrm{e}^{-\tau'_{ij}s}) \\ - \kappa_i \sum_{k=r+1}^{m_i} b_{ik} s^k Y_i(s), \quad i \in \overline{1,n}. \end{aligned} \tag{7.64}$$

Let

$$Y(s) = [Y_1(s), \ldots, Y_n(s)]^{\mathrm{T}}, \tag{7.65}$$

$$c_i(s) = \sum_{k=\nu}^{m_i} c_{ik} s^{k-r-1} + \sum_{k=r+1}^{m_i} \kappa_i b_{ik} s^{k-r-1}, \tag{7.66}$$

$$b_i(s) = \sum_{k=0}^{r} b_{ik} s^k, \tag{7.67}$$

$$l_{ij}(s) = \begin{cases} -a_{ij}\mathrm{e}^{-\tau_{ij}s}, & i \neq j \\ \sum_{j=1}^{n} a_{ij}\mathrm{e}^{-\tau'_{ij}s}, & i = j \end{cases} \tag{7.68}$$

$$\tag{7.69}$$

and

$$H(s) = \mathrm{diag}\left\{\frac{b_i(s)}{c_i(s)}\mathrm{e}^{-T_i s},\ i \in \overline{1, n}\right\}, \tag{7.70}$$

$$K = \mathrm{diag}\{\kappa_i,\ i \in \overline{1, n}\}, \tag{7.71}$$

$$L(s) = \{l_{ij}(s)\}. \tag{7.72}$$

Then, the closed-loop system can be shown by Figure 7.12. The frequency-domain model of the closed-loop system is given by

$$Y(s) = -\frac{1}{s^{r+1}} H(s)KL(s)Y(s), \tag{7.73}$$

and the return difference equation of the system is given by

$$I + \frac{1}{s^{r+1}} H(s)KL(s) = 0.$$

According to (2.42), direct use of the equation

$$\det\left(I + \frac{1}{s^{r+1}} H(s)KL(s)\right) = 0 \tag{7.74}$$

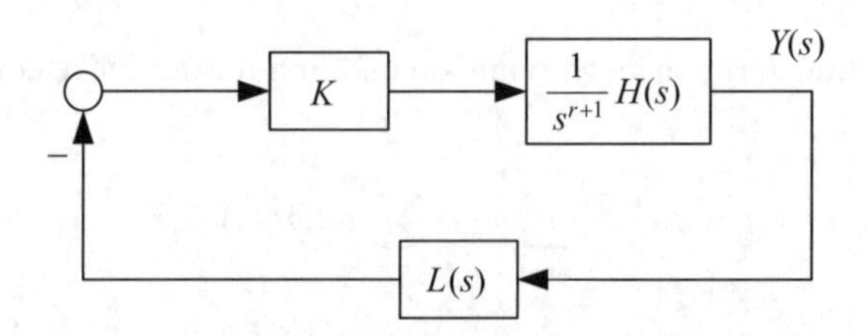

Figure 7.12 Interconnected system.

may ignore cancelation between the poles of the closed-loop system and the poles of the open-loop system at $s = 0$. To avoid these cancelations we consider the following equation

$$\det(s^{r+1}I + H(s)KL(s)) = 0 \tag{7.75}$$

as the characteristic equation of the system. Obviously,

$$\det\left(I + \frac{1}{s^{r+1}}H(s)KL(s)\right) = \det\left(\frac{1}{s^{r+1}}I\right)\det\left(s^{r+1}I + H(s)KL(s)\right). \tag{7.76}$$

Therefore, all the non-zero solutions of (7.75) are contained in the solutions of (7.76), and vice versa.

The following proposition gives an obvious condition for justifying if the open-loop poles at $s = 0$ enter into the set of the closed-loop poles.

Proposition 7.19 *Suppose $H(s)KL(s)$ is analytic at $s = 0$. If* $\text{rank}[H(s)KL(s)] = n - 1$, *for all $s \in \mathbb{C}$ at which $H(s)KL(s)$ is analytic, then*

$$\det(s^{r+1}I + H(s)KL(s)) = s^{r+1}g(s),$$

where $g(s)$ has neither poles nor zeros at $s = 0$.

Proof. The proposition can be easily proved by converting $H(s)KL(s)$ into a diagonal matrix through a similar transformation. □

Proposition 7.19 shows that $s = 0$ is a repeated pole with multiplicity $r + 1$ of the closed-loop system. This also implies that the solution of the closed-loop system can be expressed as

$$y(t) = \sum_{i=0}^{r} C_i t^i + \sum_{k=1}^{M} P_k(t)e^{\lambda_k t}, \tag{7.77}$$

where $C_i \in \mathbb{R}^n$ are constant vectors, $P_i(t) \in \mathbb{R}^n$ are vectors whose elements are polynomials of t, λ_k are zeros of $g(s)$, and M is a finite or infinite positive integer. Furthermore, if $\text{Re}\lambda_i < 0$, we have $y(t) \to \sum_{i=0}^{r} C_i t^i$ as $t \to \infty$. We call $y(t) = \sum_{i=0}^{r} C_i t^i$ the steady state of the system.

Lemma 7.20 *Assume that* $\det(s^{r+1}I + H(s)KL(s)) = s^{r+1}g(s)$, *where $g(s)$ has neither poles nor zeros at $s = 0$. Then, for any $s \neq 0$, $g(s) = 0$ if and only if*

$$\det\left(I + \frac{1}{s^{r+1}}H(s)KL(s)\right) = 0. \tag{7.78}$$

Proof. The lemma is obvious due to the equation (7.76). □

By Definition 2.7, a linear time-invariant system is said to be steady semi-stable if all its characteristic roots are inside the LHP or at the origin of the complex plane. By Lemma 7.20, to verify the steady semi-stability of the system, we just need to check if all the zeros of (7.78) have negative real parts. Note that the generalized Nyquist stability criterion (Theorem 2.19 and Theorem 2.20) can be used for this purpose.

Denote $\hat{L}(s) = H(s)KL(s)$ for convenience, and denote by $\hat{L}^{(k)}(s)$ the kth-order derivative of $\hat{L}(s)$.

Theorem 7.21 *Assume that the multi-agent system (7.58) with consensus protocol (7.62) is steady semi-stable, and the transfer function $\hat{L}(s)$ is analytic in a neighborhood of $s = 0$. Then, the rth-order consensus is reached at the steady state if $\hat{L}^{(k)}(0) \cdot \mathbf{1}_n = 0$, for all $k \in \overline{0, r}$, and* $\operatorname{rank}[\hat{L}(s)] = n - 1$ *for all $s \in \mathbb{C}$ at which $\hat{L}(s)$ is analytic.*

Proof. By Proposition 7.19 we know that the system has zeros at $s = 0$ with multiplicity $r + 1$, and the system solution can be expressed by (7.77). With the assumption of the steady semi-stability, from (7.77) it is easy to see that

$$y^{(j)}(t) \to 0, \quad \forall j \geq r + 1 \tag{7.79}$$

and

$$y(t) \to \sum_{k=0}^{r} C_k t^k \tag{7.80}$$

when $t \to \infty$. So, conditions (7.60) and (7.61) in Definition 7.18 are already satisfied. We just need to check (7.59).

Let s be in the neighborhood D of $s = 0$ in which $\hat{L}(s)$ is analytic. By the Taylor formula for functions of a complex variable we have

$$\hat{L}(s) = L_1(s) + L_2(s)s^{r+1}, \tag{7.81}$$

where

$$L_1(s) = \hat{L}(0) + \hat{L}^{(1)}(0)s + \cdots + \frac{1}{r!}\hat{L}^{(r)}(0)s^r, \tag{7.82}$$

$$L_2(s) = \frac{1}{(r+1)!}\hat{L}^{(r+1)}(0) + \frac{1}{(r+2)!}\hat{L}^{(r+2)}(0)s + \cdots. \tag{7.83}$$

Substituting (7.81) into (7.73) yields

$$s^{r+1}Y(s) = -\left(\hat{L}(0) + \hat{L}^{(1)}(0)s + \cdots + \frac{1}{r!}\hat{L}^{(r)}(0)s^r + \cdots\right)Y(s).$$

Note that the above equation holds in the neighborhood D of $s = 0$. When the system is steady semi-stable, $Y(s)$ is analytic for $\operatorname{Re} s > 0$. Then, by taking the inverse Laplace transformation of the equation, we have

$$\begin{aligned} -y^{(r+1)}(t) = {} & \hat{L}(0)y(t) + \hat{L}^{(1)}(0)y^{(1)}(t) + \cdots + \frac{1}{r!}\hat{L}^{(r)}(0)y^{(r)}(t) \\ & + \frac{1}{(r+1)!}\hat{L}^{(r+1)}(0)y^{(r+1)}(t) + \cdots \end{aligned} \tag{7.84}$$

for $t \to \infty$. Applying (7.79) to the above equation we get

$$\hat{L}(0)y(t) + \hat{L}^{(1)}(0)y^{(1)}(t) + \cdots + \frac{1}{r!}\hat{L}^{(r)}(0)y^{(r)}(t) \to 0 \tag{7.85}$$

when $t \to \infty$.

Differentiating (7.80) r times yields

$$
\begin{aligned}
\dot{y}(t) &\rightarrow \sum_{k=1}^{r} kC_k t^{k-1}, \\
\ddot{y}(t) &\rightarrow \sum_{k=2}^{r} k(k-1)C_k t^{k-2}, \\
&\vdots \\
y^{(r)}(t) &\rightarrow r!C_r.
\end{aligned} \tag{7.86}
$$

Substituting (7.80) and (7.86) into (7.85) yields

$$E_r t^r + E_{r-1}t^{r-1} + \cdots + E_1 t + E_0 \rightarrow 0, \tag{7.87}$$

where

$$
\begin{aligned}
E_r &= \hat{L}(0)C_r, \\
E_{r-1} &= \hat{L}(0)C_{r-1} + r\hat{L}^{(1)}(0)C_r, \\
&\vdots \\
E_{r-j} &= \hat{L}(0)C_{r-j} + C^1_{r-j+1}\hat{L}^{(1)}(0)C_{r-j+1} + C^2_{r-j+2}\hat{L}^{(2)}(0)C_{r-j+2} \\
&\quad + \cdots + C^j_r\hat{L}^{(j)}(0)C_r, \\
&\vdots \\
E_0 &= \hat{L}(0)C_0 + \hat{L}^{(1)}(0)C_1 + \hat{L}^{(2)}(0)C_2 + \cdots + \hat{L}^{(r)}(0)C_r, \\
C^j_r &= \tfrac{r!}{(r-j)!j!}.
\end{aligned} \tag{7.88}
$$

Equation (7.87) holds if and only if $E_{r-j} = 0, \forall j \in \overline{0, r}$.

From $E_r = 0$ we get $C_r \in \text{span}(\mathbf{1}_n)$ because $\text{rank}[\hat{L}(0)] = n - 1$ and $\hat{L}(0)\mathbf{1}_n = 0$. Using the result $C_r \in \text{span}(\mathbf{1}_n)$ and the assumption $\hat{L}^{(1)}(0)\mathbf{1}_n = 0$, from $E_{r-1} = 0$ it follows that $\hat{L}(0)C_{r-1} = 0$ which implies $C_{r-1} \in \text{span}(\mathbf{1}_n)$. Conducting this procedure to the end, i.e., $E_0 = 0$, we get $C_{r-j} \in \text{span}(\mathbf{1}_n), \forall j \in \overline{0, r}$. Thus, we have $y(t) \in \text{span}(\mathbf{1}_n)$ when $t \rightarrow \infty$, which, by (7.80), also implies that $y^{(k)}(t) \in \text{span}(\mathbf{1}_n), \forall k \in \overline{1, r}$. □

7.3.3 Existence of High-Order Consensus Solutions

Theorem 7.21 gives some sufficient consensus conditions for the system. Here we try to find a necessary and sufficient condition of the existence of high-order consensus solutions. First, we show that the condition given by Proposition 7.19 can be further weakened as follows.

Proposition 7.22 *Suppose $\hat{L}(s)$ is analytic in a neighborhood of $s = 0$. Then,*

$$\det(s^{r+1}I + \hat{L}(s)) = s^{r+1}g(s)$$

with $g(s)$ having neither poles nor zeros at $s = 0$, if and only if $\text{rank}[\hat{L}(0)] = n - 1$, *and there exists a non-zero constant vector $\alpha \in \text{span}(\mathbf{1}_n)$ such that $\hat{L}^{(k)}(0) \cdot \alpha = 0, \forall k \in \overline{0, r}$, and $\alpha + \frac{1}{(r+1)!}\hat{L}^{(r+1)}(0) \cdot \alpha \notin \text{span}(\hat{L}(0))$.*

Proof. Let s be in the neighborhood of $s = 0$, denoted by D, in which $\hat{L}(s)$ is analytic. By the Taylor formula for functions of a complex variable, we have

$$\hat{L}(s) = L_1(s) + L_2(s)s^{r+1}, \tag{7.89}$$

where

$$L_1(s) = \hat{L}(0) + \hat{L}^{(1)}(0)s + \cdots + \frac{1}{r!}\hat{L}^{(r)}(0)s^r, \tag{7.90}$$

$$L_2(s) = \frac{1}{(r+1)!}\hat{L}^{(r+1)}(0) + \frac{1}{(r+2)!}\hat{L}^{(r+2)}(0)s + \cdots. \tag{7.91}$$

It is clear that

$$\det\left(s^{r+1}I + L_1(s) + L_2(s)s^{r+1}\right) = s^{r+1}g(s)$$

if and only if there exists a normalized non-singular matrix $T \in \mathbb{R}^{n\times n}$ (i.e., $\det(T) = 1$) such that one column (say without loss of generality, the first one) of $T^{-1}L_1(s)T$ is zero, i.e., $[T^{-1}L_1(s)T](:, 1) = 0$. Denote by α the first column of T. It is easy to get

$$\left[T^{-1}L_1(s)T\right](:, 1) = \hat{L}(0)\alpha + s\hat{L}^{(1)}(0)\alpha + \cdots + \frac{s^r}{r!}\hat{L}^{(r)}(0)\alpha = 0.$$

Obviously, $[T^{-1}L_1(s)T](:, 1) = 0$, $\forall s \in D$ if and only if $\hat{L}^{(k)}(0)\cdot\alpha = 0$, $\forall k \in \overline{0, r}$.

Denoting by p_{ij} the (i, j)th elements of $T^{-1}L_1(s)T$ for $i \in \overline{1, n}$, $j \in \overline{2, n}$, we have

$$T^{-1}L_1(s)T = \begin{bmatrix} 0 & p_{12} & \cdots & p_{1n} \\ 0 & p_{22} & \cdots & p_{2n} \\ \vdots & \vdots & & \vdots \\ 0 & p_{n2} & \cdots & p_{nn} \end{bmatrix}. \tag{7.92}$$

And denoting by g_{ij} the (i, j)th elements of $T^{-1}L_2(s)T$, we have

$$\begin{aligned} &\det(s^{r+1}I + \hat{L}(s)) = \det(s^{r+1}I + T^{-1}\hat{L}(s)T) \\ &= \det\left(\begin{bmatrix} s^{r+1} + s^{r+1}g_{11} & s^{r+1}g_{12} + p_{12} & \cdots & s^{r+1}g_{1n} + p_{1n} \\ s^{r+1}g_{21} & s^{r+1} + s^{r+1}g_{22} + p_{22} & \cdots & s^{r+1}g_{2n} + p_{2n} \\ \vdots & & & \vdots \\ s^{r+1}g_{n1} & s^{r+1}g_{n2} + p_{n2} & \cdots & s^{r+1} + s^{r+1}g_{nn} + p_{nn} \end{bmatrix}\right) \\ &= s^{r+1}\det\left(\begin{bmatrix} 1 + g_{11} & s^{r+1}g_{12} + p_{12} & \cdots & s^{r+1}g_{1n} + p_{1n} \\ g_{21} & s^{r+1} + s^{r+1}g_{22} + p_{22} & \cdots & s^{r+1}g_{2n} + p_{2n} \\ \vdots & & & \vdots \\ g_{n1} & s^{r+1}g_{n2} + p_{n2} & \cdots & s^{r+1} + s^{r+1}g_{nn} + p_{nn} \end{bmatrix}\right) \\ &= s^{r+1}g(s). \end{aligned} \tag{7.93}$$

Since $H(s)KL(s)$ is analytic at $s = 0$, all the elements p_{ij} ($i \in \overline{1, n}$, $j \in \overline{2, n}$) are analytic at $s = 0$. Hence, $g(s)$ is also analytic at $s = 0$.

Now, we prove $g(s)$ has no zeros at $s = 0$, i.e., $g(0) \neq 0$ if and only if $\text{rank}[\hat{L}(0)] = n - 1$ and $\alpha + \frac{1}{(r+1)!}\hat{L}^{(r+1)}(0) \cdot \alpha \notin \text{span}(\hat{L}(0))$. Denote $e_1 = [1, 0, \ldots, 0]^{\mathrm{T}} \in \mathbb{R}^{n \times 1}$. From (7.93) we have

$$g(0) = \det\left[e_1 + \left[T^{-1}\frac{1}{(r+1)!}\hat{L}^{(r+1)}(0)T\right](:, 1), [T^{-1}\hat{L}(0)T](:, \overline{2, n})\right]$$

$$= \det\left[Te_1 + T\left[T^{-1}\frac{1}{(r+1)!}\hat{L}^{(r+1)}(0)T\right](:, 1), T[T^{-1}\hat{L}(0)T](:, \overline{2, n})\right]$$

because $\det(T^{-1}) = 1$. Obviously, $g(0) \neq 0$ if and only if

$$\text{rank}[T[T^{-1}\hat{L}(0)T](:, \overline{2, n})] = n - 1$$

and

$$Te_1 + T\left[T^{-1}\frac{1}{(r+1)!}\hat{L}^{(r+1)}(0)T\right](:, 1) \notin \text{span}(T[T^{-1}\hat{L}(0)T](:, \overline{2, n})).$$

From (7.92) we know that

$$\text{span}(T[T^{-1}\hat{L}(0)T](:, \overline{2, n})) = \text{span}(\hat{L}(0))$$

and hence $\text{rank}[T[T^{-1}\hat{L}(0)T](:, \overline{2, n})] = n - 1$ if and only if $\text{rank}[\hat{L}(0)] = n - 1$. Simple calculation shows that $Te_1 + T[T^{-1}\frac{1}{(r+1)!}\hat{L}^{(r+1)}(0)T](:, 1) = \alpha + \frac{1}{(r+1)!}\hat{L}^{(r+1)}(0) \cdot \alpha$. So $g(0) \neq 0$ if and only if $\text{rank}[\hat{L}(0)] = n - 1$, and $\alpha + \frac{1}{(r+1)!}\hat{L}^{(r+1)}(0) \cdot \alpha \notin \text{span}(\hat{L}(0))$. Finally, we note that the non-zero vector $\alpha \in \text{span}(\mathbf{1}_n)$. This is indeed the case. Since we have proved $\text{rank}[\hat{L}(0)] = n - 1$, the null space of $\hat{L}(0)$ is one-dimensional. From the property of Laplacian matrix, $L(0) \cdot \mathbf{1}_n = 0$, we get $\hat{L}(0) \cdot \mathbf{1}_n = 0$. So, $\alpha \in \text{span}(\mathbf{1}_n)$. The proposition is proved. □

Remark. When $\text{rank}[\hat{L}(0)] = n - 1$, a sufficient condition for $\mathbf{1}_n + \frac{1}{(r+1)!}\hat{L}^{(r+1)}(0) \cdot \mathbf{1}_n \notin \text{span}(\hat{L}(0))$ is $\mathbf{1}_n + \frac{1}{(r+1)!}\hat{L}^{(r+1)}(0) \cdot \mathbf{1}_n = c\mathbf{1}_n$, where c is a non-zero constant. And a more sufficient condition for is $\hat{L}^{(r+1)}(0) \cdot \mathbf{1}_n = 0$.

Exercise 7.23 *Suppose* $\hat{L}(0) \in \mathbb{R}^{n \times n}$, $\text{rank}[\hat{L}(0)] = n - 1$ *and* $\text{rank}[\hat{L}(0)] \cdot \mathbf{1}_n = 0$. *Show* $\alpha := c\mathbf{1}_n$ *with* $c \neq 0$ *is not in* $\text{span}(\hat{L}(0))$.

When $s = 0$ is a repeated pole with multiplicity $r + 1$ of the closed-loop system, the solution of the closed-loop system can be expressed by equation (7.77)

Now, we are ready to present the following theorem.

Theorem 7.24 *Assume that the multi-agent system (7.58) with consensus protocol (7.62) is steady semi-stable and the transfer function* $\hat{L}(s) = H(s)KL(s)$ *is analytic in a neighborhood of* $s = 0$. *Then, the* r*th-order consensus is reached at the steady state if and only if* $\hat{L}^{(k)}(0) \cdot \mathbf{1} = 0$, $\forall k \in \overline{0, r}$, $\text{rank}[\hat{L}(0)] = n - 1$, *and* $\mathbf{1}_n + \frac{1}{(r+1)!}\hat{L}^{(r+1)}(0) \cdot \mathbf{1}_n \notin \text{span}(\hat{L}(0))$.

Proof. (*Sufficiency*) Suppose $\hat{L}^{(k)}(0) \cdot \mathbf{1} = 0$, $\forall k \in \overline{0, r}$, $\text{rank}[\hat{L}(0)] = n - 1$, and $\mathbf{1}_n + \frac{1}{(r+1)!}\hat{L}^{(r+1)}(0) \cdot \mathbf{1}_n \notin \text{span}(\hat{L}(0))$. Then, by Proposition 7.22 we know that the system has zeros at $s = 0$ with multiplicity $r + 1$, and the system solution can be expressed by (7.77). The rest of the sufficiency part of the this proof is as the same as given in the proof of Theorem 7.21.

(*Necessity*) Suppose that the system reaches the consensus defined by (7.59), (7.60) and (7.61). Note that (7.60), (7.61) and the steady semi-stability assumption imply that the system has zeros at $s = 0$ of multiplicity $r + 1$. So, by Proposition 7.22, the necessity is obvious. □

7.3.4 Constant Consensus

For the case of constant consensus, i.e., the case when $r = 0$, without loss of generality, we let $b_{i0} = 1$ for all $i \in \overline{1, n}$. Then, the protocol (7.62) reduces to the following form

$$u_i(t) = \kappa_i \left(\sum_{j \in N_i} a_{ij}(y_j(t - \tau_{ij}) - y_i(t - \tau'_{ij})) \right) - \kappa_i \sum_{k=1}^{m_i} b_{ik} y_i^{(k)}(t), \qquad (7.94)$$

the closed-loop system equation becomes

$$\sum_{k=\nu}^{m_i} c_{ik} s^k Y_i(s) = \kappa_i \mathrm{e}^{-T_i s} \sum_{j \in N_i} a_{ij} \left(Y_j(s)\mathrm{e}^{-\tau_{ij}s} - Y_i(s)\mathrm{e}^{-\tau'_{ij}s} \right) - \kappa_i \sum_{k=1}^{m_i} b_{ik} s^k Y_i(s), \quad i \in \overline{1, n},$$

and $H(s)$ defined in (7.70) takes the form

$$H(s) = \text{diag}\left\{ h_i(s), \quad i \in \overline{1, n} \right\},$$

where

$$h_i(s) = \frac{\mathrm{e}^{-T_i s}}{\sum_{k=\nu}^{m_i} c_{ik} s^{k-1} + \kappa_i \sum_{k=1}^{m_i} b_{ik} s^{k-1}}. \qquad (7.95)$$

Let us apply Theorem 7.24 to the case of constant consensus. For this case, Theorem 7.24 requires: (1) the marginal stability of the closed-loop system, (2) $\text{rank}[\hat{L}(0)] = n - 1$ and $\hat{L}(0) \cdot \mathbf{1} = 0$. Since $\kappa_i \neq 0$, requirement (2) implies that $\text{rank}[L(0)] = n - 1$ and $L(0) \cdot \mathbf{1} = 0$. It is well known that this is equivalent to the connectivity condition for the interconnection graph, i.e., the digraph has a globally reachable node.

To check the marginal stability of the closed-loop system, we may make some loop transformation on the system as shown by Figure 2.8. Denote by $\tilde{L}(s)$ the open-loop transfer function matrix after the loop transformation. Then, based on the extended spectral radius theorem for steady semi-stability (Theorem 2.24), we get the following result immediately.

Theorem 7.25 *Consider the multi-agent system (7.58) with the protocol (7.94). Assume the interconnection digraph has a globally reachable node, then, the system achieves a constant consensus, if*

$$\rho(\tilde{L}(\mathrm{j}\omega)) < 1, \quad \forall \omega \in \mathbb{R}, \quad \omega \neq 0; \qquad (7.96)$$

$$\rho(\tilde{L}(\mathrm{j}\omega)) = 1, \quad \text{and } \det(I + \tilde{L}(\mathrm{j}\omega)) = 0 \quad \text{for } \omega = 0. \qquad (7.97)$$

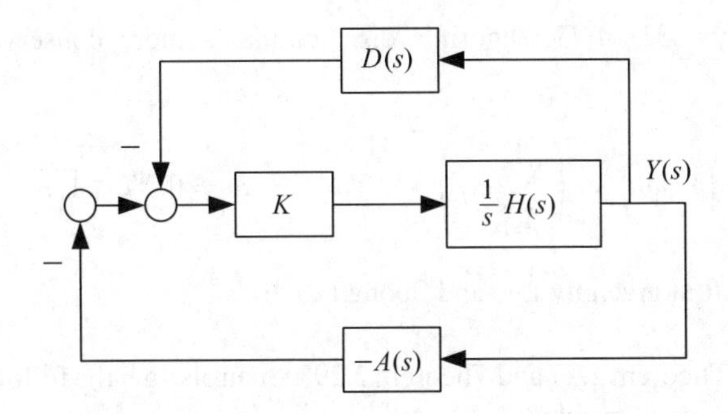

Figure 7.13 Equivalent diagram for constant consensus.

Since $L(s)$ defined in (7.72) can be rewritten as

$$\begin{aligned} L(s) &= D(s) - A(s) \\ &= \mathrm{diag}\{d_i(s),\ i \in \overline{1,n}\} - \{a_{ij}(s)\}, \end{aligned}$$

where

$$d_i(s) = \sum_{j=1}^{n} a_{ij}\mathrm{e}^{-\tau'_{ij}s}.$$

and

$$a_{ij}(s) = \begin{cases} a_{ij}\mathrm{e}^{-\tau_{ij}s}, & i \neq j, \\ 0, & i = j, \end{cases}$$

the system diagram shown in Figure 7.12 can be equivalently transformed as shown in Figure 7.13. In this case we have

$$\tilde{L}(s) = \mathrm{diag}\{g_i(s),\ i \in \overline{1,n}\}(-A(s)), \tag{7.98}$$

where

$$g_i(s) = \frac{\kappa_i h_i(s)}{s + \kappa_i h_i(s) \sum_{j=1}^{n} a_{ij}\mathrm{e}^{-\tau'_{ij}s}}. \tag{7.99}$$

Straightforward calculation shows that

$$\rho(\tilde{L}(\mathrm{j}\omega)) = 1, \text{ and } \det(I + \tilde{L}(\mathrm{j}\omega)) = 0 \text{ for } \omega = 0.$$

So, the requirement (7.97) of Theorem 7.25 is satisfied. Therefore, under the assumption that the interconnection digraph has a globally reachable node, a sufficient condition for achieving a constant consensus is

$$\rho(\mathrm{diag}\{g_i(\mathrm{j}\omega),\ i \in \overline{1,n}\}(-A(\mathrm{j}\omega)) < 1, \ \forall\omega \in \mathbb{R}, \omega \neq 0. \tag{7.100}$$

By using Corollary 2.31 of Gershgorin's disc lemma, a more conservative but scalable condition is

$$|g_i(\mathrm{j}\omega)| < \left(\sum_{j=1}^{n} a_{ij}\right)^{-1}, \ \forall \omega \in \mathbb{R}, \omega \neq 0, \forall i \in \overline{1,n}. \tag{7.101}$$

This condition was first given by Lee and Spong (2006).

Similarly, from Theorem 2.27 and Theorem 2.29 we can also get the following two theorems for the constant consensus problem.

Theorem 7.26 *Consider the multi-agent system (7.58) with the protocol (7.94). Assume the interconnection digraph has a globally reachable node, then, the system achieves a constant consensus, if*

$$\|\tilde{L}(\mathrm{j}\omega)\| < 1, \ \forall \omega \in \mathbb{R}, \omega \neq 0; \tag{7.102}$$

$$\|\tilde{L}(\mathrm{j}\omega)\| = 1, \text{ and } \det(I + \tilde{L}(\mathrm{j}\omega)) = 0, \text{ for } \omega = 0. \tag{7.103}$$

Theorem 7.27 *Consider the multi-agent system (7.58) with the protocol (7.94). Assume the interconnection digraph has a globally reachable node, then, the system achieves a constant consensus, if*

$$1 + \mathrm{Re}(\lambda_i(\tilde{L}(\mathrm{j}\omega))) > 0, \ \forall \omega \in \mathbb{R}, \omega \neq 0, \forall i \in \overline{1,n}; \tag{7.104}$$

$$1 + \mathrm{Re}(\lambda_i(\tilde{L}(\mathrm{j}\omega))) = 0, \text{ for } \omega = 0, i \in \overline{1,n}. \tag{7.105}$$

7.3.5 Consensus in Ideal Networks

Before applying Theorem 7.24 to ideal networks, i.e., networks with zero communication delays and constant channel dynamics, let us review the following fact from graph theory.

Let G be a digraph having at least one globally reachable node. If we choose only one globally reachable node, say v_j, of G, and cut off all the edges from v_j, then obviously v_j is still a globally reachable node of G. But, if we do this operation for a node which is not globally reachable in G, or simultaneously do this operation for two or more globally reachable nodes, then G has no globally reachable node anymore. This is equivalent to saying, if $L \in \mathbb{R}^{n\times n}$ is a Laplacian of a digraph having at least one globally reachable node, then $\mathrm{rank}[\mathrm{diag}\{b_1, \ldots, b_n\}L] = n - 1$ if and only if $b_i \neq 0, \forall i \in \overline{1,n}$, or $b_i \neq 0, \forall i \in \overline{1,n}\backslash j$, $b_j = 0$ and v_j is a globally reachable node of G, where $\overline{1,n}\backslash j$ denotes the set of integers from 1 to n excluding j.

Now, let us consider a multi-agent system based on an ideal network. Suppose the topology digraph G contains at least one globally reachable node. Then, we have $\mathrm{rank}[L] = n - 1$ and $L \cdot \mathbf{1}_n = 0$. In this case we have $\tau_{ij} = \tau'_{ij} = 0$ and $L(s) = L$. From $L \cdot \mathbf{1}_n = 0$ it is easy to get $\hat{L}^{(k)}(0) \cdot \mathbf{1}_n = 0, \forall k \in \overline{0,r}$ and $\mathbf{1}_n + \frac{1}{(r+1)!}\hat{L}^{(r+1)}(0) \cdot \mathbf{1}_n = \mathbf{1}_n \notin \mathrm{span}(\hat{L}(0))$. Finally, by denoting $g_i(s) = \frac{b_i(s)}{c_i(s)}$, we know that $\mathrm{rank}[\hat{L}(0)] = \mathrm{rank}[L] = n - 1$ if and only if $g_i(0) \neq 0$, or $g_i(0) \neq 0, \forall i \in \overline{1,n}\backslash j$, $g_j(0) = 0$ and v_j is a globally reachable node of digraph G. Thus, we get the following result as a corollary of Theorem 7.24.

Theorem 7.28 *In an ideal network with zero communication delays and constant channel dynamics, the rth-order consensus will be achieved for a steady semi-stable system at its steady state if and only if the topology graph contains at least one globally reachable node, and the agents's dynamics and the protocol satisfy* $g_i(0) \neq 0$, *or* $g_i(0) \neq 0, \forall i \in \overline{1,n} \backslash j$, $g_j(0) = 0$ *and* v_j *is a globally reachable node of digraph* G.

7.4 Integrator-Chain Systems with Diverse Communication Delays

7.4.1 Matching Condition for Self-Delay

Consider the multi-agent system, whose agents are chains of integrators, i.e., $m_i = \nu$, and $h_i(s) = 1$. In this case $g_i(s) = b_i(s)$, and thus we have

$$\begin{aligned} \hat{L}(s) &= \text{diag}\{\kappa_i b_i(s)\} L(s), \\ \hat{L}^{(1)}(s) &= \text{diag}\{\kappa_i b_i(s)\} L^{(1)}(s) + \text{diag}\left\{\kappa_i b_i^{(1)}(s)\right\} L(s), \\ &\cdots \\ \hat{L}^{(\nu-1)}(s) &= \text{diag}\{\kappa_i b_i(s)\} L^{(\nu-1)}(s) + \cdots . \end{aligned}$$

Since $\kappa_i > 0$ and $b_{i0} \neq 0$, we have $\text{rank}[\hat{L}(0)] = \text{rank}[L(0)] = n - 1$ when the topology graph contains a globally reachable node. Now, we check the condition $\hat{L}^{(k)}(0) \cdot \mathbf{1} = 0$, $\forall k \in \overline{0, (\nu - 1)}$. For the second-order consensus ($\nu = 2$), this condition implies that

$$\sum_{j=1}^{n} a_{ij}\tau_{ij} = \sum_{j=1}^{n} a_{ij}\tau'_{ij}. \tag{7.106}$$

And generally, the consensus condition for the νth-order consensus can be obtained as

$$\begin{cases} \sum_{j=1}^{n} a_{ij}\tau_{ij} = \sum_{j=1}^{n} a_{ij}\tau'_{ij}, \\ \cdots \\ \sum_{j=1}^{n} a_{ij}\tau_{ij}^{\nu-1} = \sum_{j=1}^{n} a_{ij}(\tau'_{ij})^{\nu-1}. \end{cases} \tag{7.107}$$

Equation (7.106) or (7.107) uncovers a very interesting fact that the second-order or high-order consensus does not necessarily require $\tau'_{ij} = \tau_{ij}$. We call condition (7.107) the *matching condition* for self-delays.

7.4.2 Adaptive Adjustment of Self-Delay

Since (7.106) does not require each self-delay to be equal to the communication delay, for the second-order consensus each agent can set an identical self-delay, i.e., $\tau'_{ij} = \tau'_i$, $\forall j \in N_i$, and adjust τ'_i so that (7.106) is satisfied. In the following, we propose a simple algorithm for the adaptive adjustment of the self-delay.

The performance index can be proposed as

$$E_i = \frac{1}{T}\int_{t0}^{t_0+T} \frac{1}{2}\left[\left(\sum_{j\in N_i} a_{ij}(y_j(t-\tau_{ij}) - y_i(t-\tau_i'))\right)^2 + \left(\sum_{j\in N_i} a_{ij}(\dot{y}_j(t-\tau_{ij}) - \dot{y}_i(t-\tau_i'))\right)^2\right] \mathrm{d}t,\ i \in \overline{1,n}, \tag{7.108}$$

where t_0 is an initial time instant from which the adaptive adjustment begins, $T > 0$ is a large enough constant of time interval.

The algorithm of adaptive adjustment of τ_i' is as follows:

1. Set the tolerance error ϵ, sampling period length Δt, and an initial value $\tau'(0)$. Discretize the performance index (7.108) and write it in the recursive form given as

$$E_i(m) = \frac{m-1}{m}E_i(m-1) + \frac{\Delta t}{2m}\left[\left(\sum_{j\in N_i} a_{ij}(y_j(t_{m-1}-\tau_{ij}) - y_i(t_{m-1}-\tau_i'))\right)^2 + \left(\sum_{j\in N_i} a_{ij}(\dot{y}_j(t_{m-1}-\tau_{ij}) - \dot{y}_i(t_{m-1}-\tau_i'))\right)^2\right]. \tag{7.109}$$

where $t_m = m\Delta t$.

2. Compute the adjustment to τ_i' by

$$\tau_i'(m+1) = \tau_i'(m) - \beta\left.\frac{\partial E_i(m)}{\partial \tau_i'}\right|_{\tau_i'=\tau_i'(m)}, \tag{7.110}$$

where $\beta > 0$ is a proper adaptation parameter, and $\frac{\partial E_i(m)}{\partial \tau_i'}$ can be obtained from (7.109) as follows:

$$\frac{\partial E_i(m)}{\partial \tau_i'} = \frac{m-1}{m}\frac{\partial E_i(m-1)}{\partial \tau_i'} + \frac{1}{m}\left(z_i(t_{m-1})\Delta y_i(t_{m-1}) + \dot{z}_i(t_{m-1})\Delta \dot{y}_i(t_{m-1})\right), \tag{7.111}$$

where

$$z_i(t_{m-1}) = \left(\sum_{j\in N_i} a_{ij}(y_j(t_{m-1}-\tau_{ij}) - y_i(t_{m-1}-\tau_i'))\right)\sum_{j\in N_i} a_{ij}, \tag{7.112}$$

$$\Delta y_i(t_{m-1}) = y_i(t_{m-1}-\tau_i') - y_i(t_{m-2}-\tau_i'). \tag{7.113}$$

Set $m = m + 1$.

3. If $E_i > \epsilon$, go to step 2; otherwise, the algorithm is stopped.

The proposed algorithm is essentially a gradient algorithm which is usually just locally convergent. To get the equilibria of the algorithm, one can set $\frac{\partial E_i(m)}{\partial \tau_i'} \rightarrow 0$. Then, it follows

$$z_i(t_{m-1})\Delta y_i(t_{m-1}) + \dot{z}_i(t_{m-1})\Delta \dot{y}_i(t_{m-1}) \rightarrow 0. \tag{7.114}$$

Since $y_i(t) \rightarrow C_0 + C_1 t$, from (7.114) it follows that

$$\sum_{j\in N_i} a_{ij}\tau_i'(m) \rightarrow \sum_{j\in N_i} a_{ij}\tau_{ij}, \tag{7.115}$$

or

$$C_1 \rightarrow 0. \tag{7.116}$$

Obviously, (7.115) is exactly (7.106) corresponding to the desired equilibrium; (7.116) implies that the velocities of all the agents go to zero, which is the zeroth-order consensus (constant consensus) state for the second-order system.

7.4.3 Simulation Study

In this subsection several simulation experiments are conducted through two numerical examples to verify the obtained theoretical results.

Example 7.29 *Consensus of a system of heterogeneous agents.*

Consider a system of five agents described by

$$\frac{d^4 y_1}{dt^4} + 0.2\frac{d^3 y_1}{dt^3} + 0.1\frac{d^2 y_1}{dt^2} = \frac{du_1}{dt} + u_1,$$
$$\frac{d^2 y_i}{dt^2} = \frac{du_i}{dt} + u_i, \quad i = \overline{2,5}.$$

Obviously, agent 1 is absolutely different from the other four agents which are all double-integrators. The agents are interconnected by a digraph shown in Figure 7.14. The weighted adjacency matrix is given as follows

$$A = \begin{bmatrix} 0 & 1 & 0 & 1 & 0 \\ 0 & 0 & 0 & 1 & 0 \\ 1 & 1 & 0 & 1 & 0 \\ 0 & 0 & 0 & 0 & 1 \\ 1 & 0 & 0 & 0 & 0 \end{bmatrix}.$$

Case 1: no communication delays

Firstly, we consider the ideal network without communication delays. Thus, the consensus protocol is simply given by

$$u_i(t) = \kappa_i \sum_{j\in N_i} a_{ij}(y_j(t) - y_i(t)).$$

Let $\kappa_i = 10, i \in \overline{1,5}$. Then, for a random set of initial states the simulation results are presented in Figures 7.15 to 7.18. From these figures one can see that both positions and velocities of all the agents of the system reach consensus solutions.

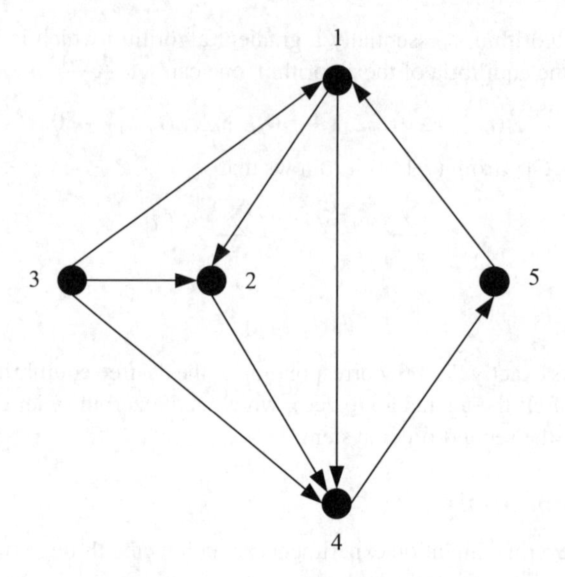

Figure 7.14 Interconnection graph.

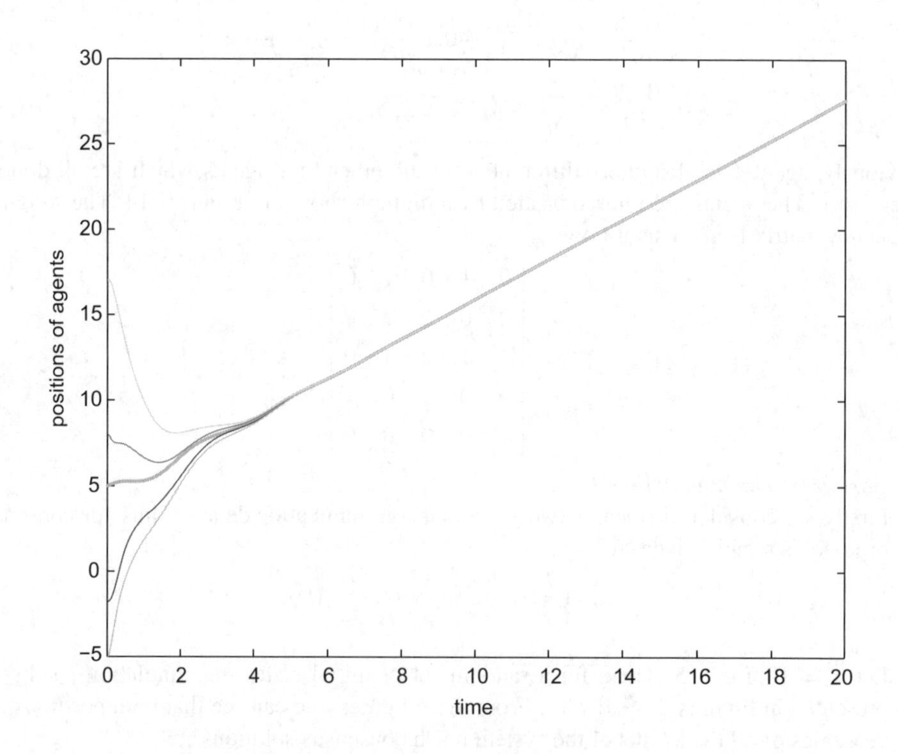

Figure 7.15 Positions of the agents (without delays).

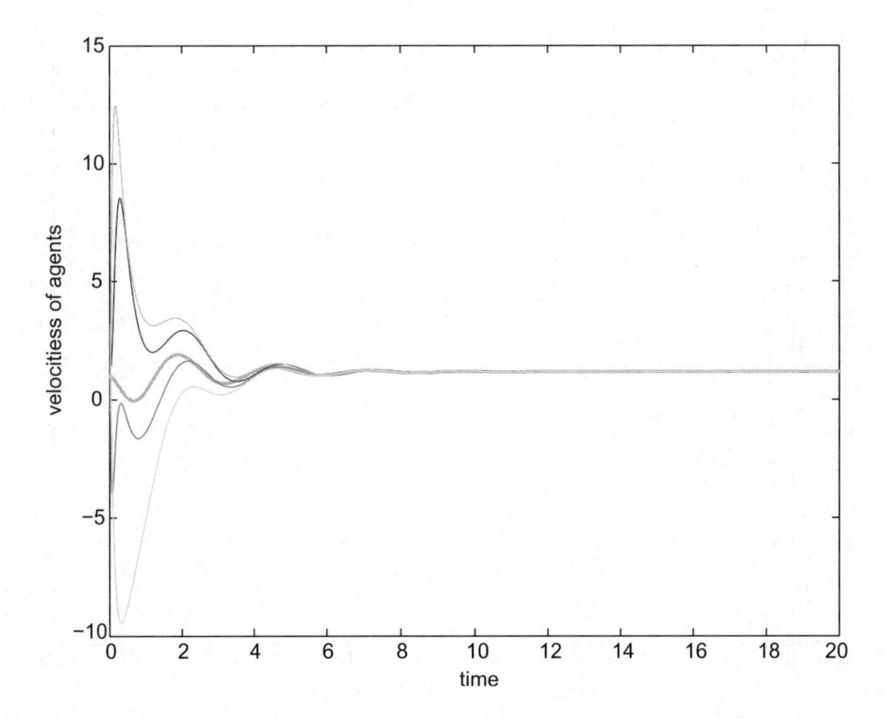

Figure 7.16 Velocities of the agents (without delays).

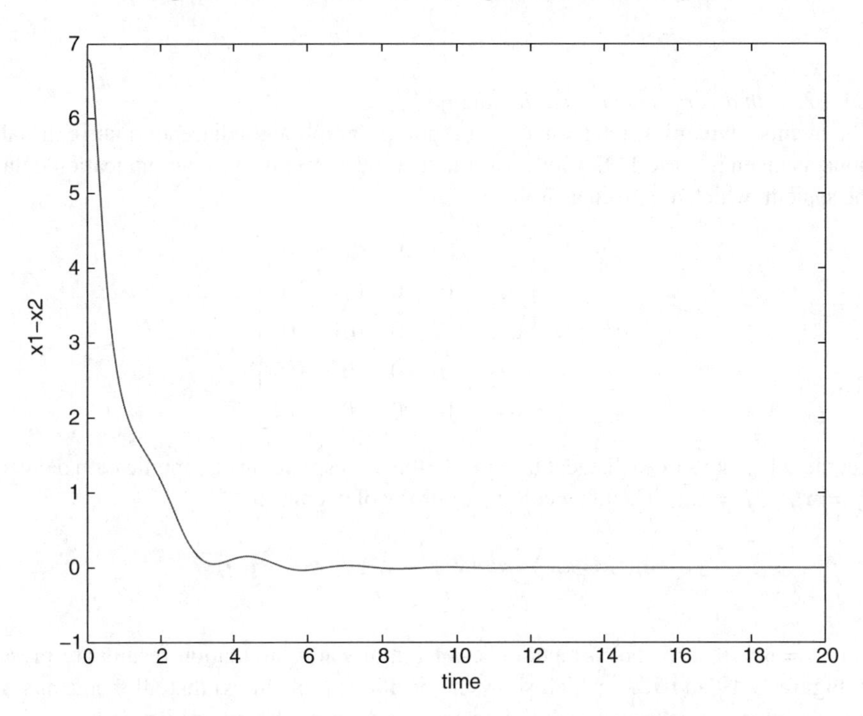

Figure 7.17 Position error between agent 1 and agent 2.

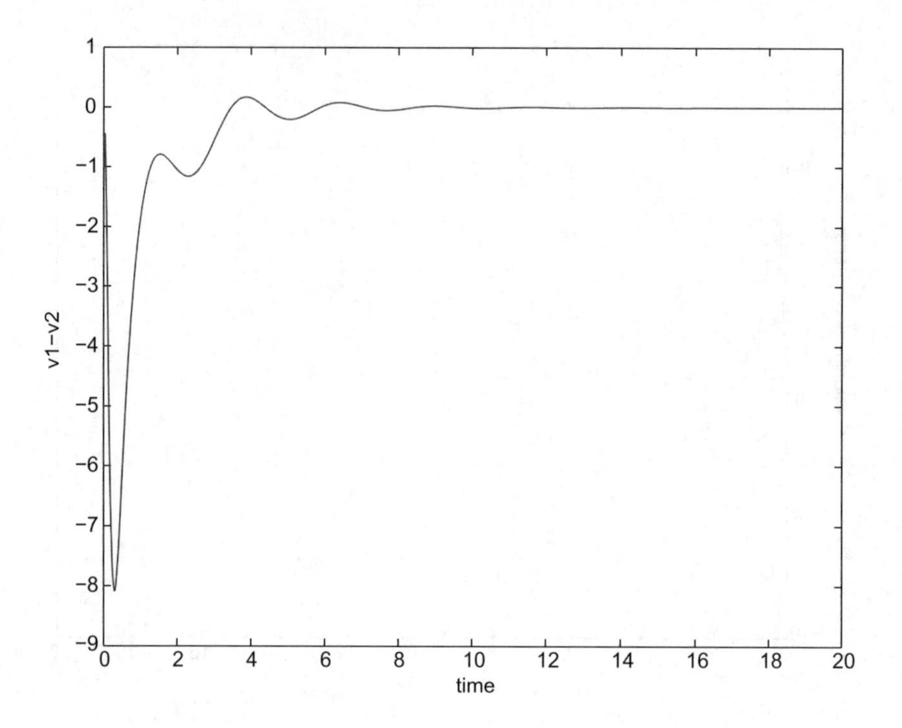

Figure 7.18 Velocity error between agent 1 and agent 2.

Case 2: with diverse communication delays
The agents' dynamics, interconnection topology graph and adjacency matrix are all the same as given in Case 1. Now, we assume that there are diverse communication delays in the system, which are given as follows

$$\{\tau_{ij}\} = \begin{bmatrix} 0 & 0.5 & 0 & 0.3 & 0 \\ 0 & 0 & 0 & 0.7 & 0 \\ 0.2 & 0.2 & 0 & 0.1 & 0 \\ 0 & 0 & 0 & 0 & 0.5 \\ 0.4 & 0 & 0 & 0 & 0 \end{bmatrix}.$$

Let the self-delays of each agent be equal to the corresponding communication delays, i.e., $\tau'_{ij} = \tau_{ij}$, $i, j = \overline{1, 5}$. Then, the consensus protocol becomes

$$u_i(t) = \kappa_i \sum_{j \in N_i} a_{ij}(y_j(t - \tau_{ij}) - y_i(t - \tau_{ij})).$$

Let $\kappa_i = 30$, $i \in \overline{1, 5}$. For a random set of initial states simulation results are presented in Figures 7.19 and 7.20, which show the positions (velocities) that all the agents of the system reach at a consensus solution in spite of diverse communication delays.

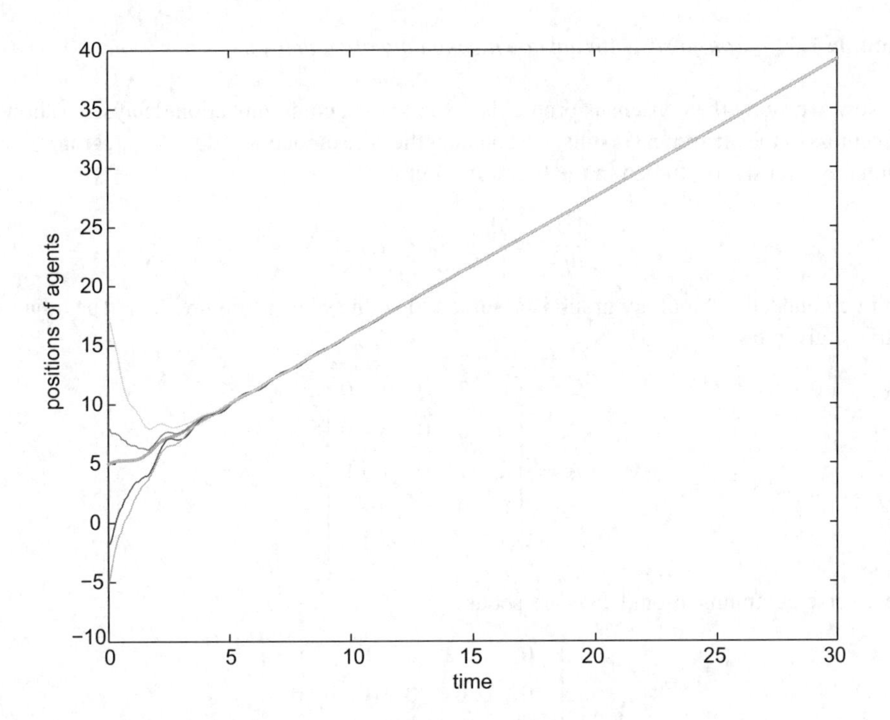

Figure 7.19 Positions of the agents (with diverse delays).

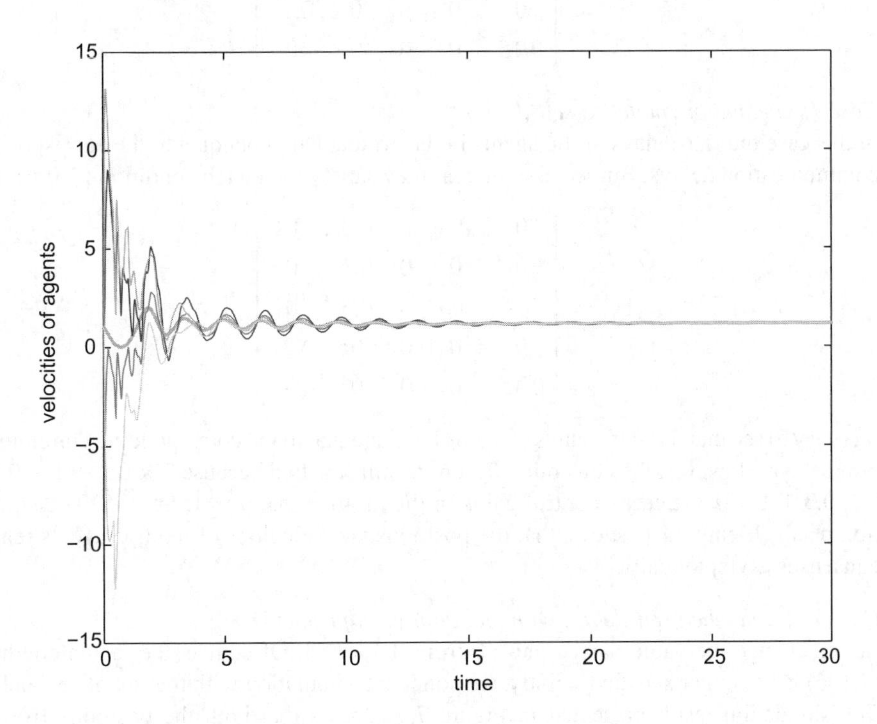

Figure 7.20 Velocities of the agents (with diverse delays).

Example 7.30 *Consensus with unknown communication delays.*

Now, we study the consensus protocols for unknown communication delays. To show the correctness of the theoretical results, we consider the case of identical double-integrator agents, which are extensively studied in the literature. Let

$$\frac{\mathrm{d}^2 x_i}{\mathrm{d}t^2} = \frac{\mathrm{d}u_i}{\mathrm{d}t} + u_i, \quad i = \overline{1,5}.$$

The interconnection topology graph is assumed to be the same as Figure 7.14. The adjacency matrix is given by

$$A = \begin{bmatrix} 0 & 2 & 0 & 1 & 0 \\ 0 & 0 & 0 & 3 & 0 \\ 1 & 1 & 0 & 1 & 0 \\ 0 & 0 & 0 & 0 & 1 \\ 1 & 0 & 0 & 0 & 0 \end{bmatrix}.$$

The diverse communication delays are set as

$$\{\tau_{ij}\} = \begin{bmatrix} 0 & 0.2 & 0 & 0.5 & 0 \\ 0 & 0 & 0 & 0.1 & 0 \\ 0.2 & 0.5 & 0 & 0.3 & 0 \\ 0 & 0 & 0 & 0 & 0.25 \\ 0.15 & 0 & 0 & 0 & 0 \end{bmatrix}.$$

Case 1: unequal but matched self-delays
In this case the self-delays of the agents in the protocol are not equal to the corresponding communication delays. But we assume that they satisfy the match condition (7.106). Let

$$\{\tau'_{ij}\} = \begin{bmatrix} 0 & 0.3 & 0 & 0.3 & 0 \\ 0 & 0 & 0 & 0.1 & 0 \\ 0.2 & 0.5 & 0 & 0.3 & 0 \\ 0 & 0 & 0 & 0 & 0.25 \\ 0.15 & 0 & 0 & 0 & 0 \end{bmatrix}.$$

It is easy to see that the self-delays of agent 1 are unequal to the corresponding communication delays. However, the constraint (7.106) is still satisfied because $2 \times 0.2 + 1 \times 0.5 = 2 \times 0.3 + 1 \times 0.3$. Let the control gains in the protocol be $\kappa_i = 1, i \in \overline{1,5}$. The simulation result (Figure 7.21) shows that the positions and velocities of the five agents tend to consensus asymptotically.

Case 2: Unmatched self-delays with an adaptive adjustment
Let us change the value of τ'_{12} and τ'_{14} from 0.3 to 0.8. Of course, the match condition (7.106) is no longer satisfied. Firstly, we conduct a simulation with this set of self-delays. The simulation result presented in Figure 7.22 shows that both the position error and

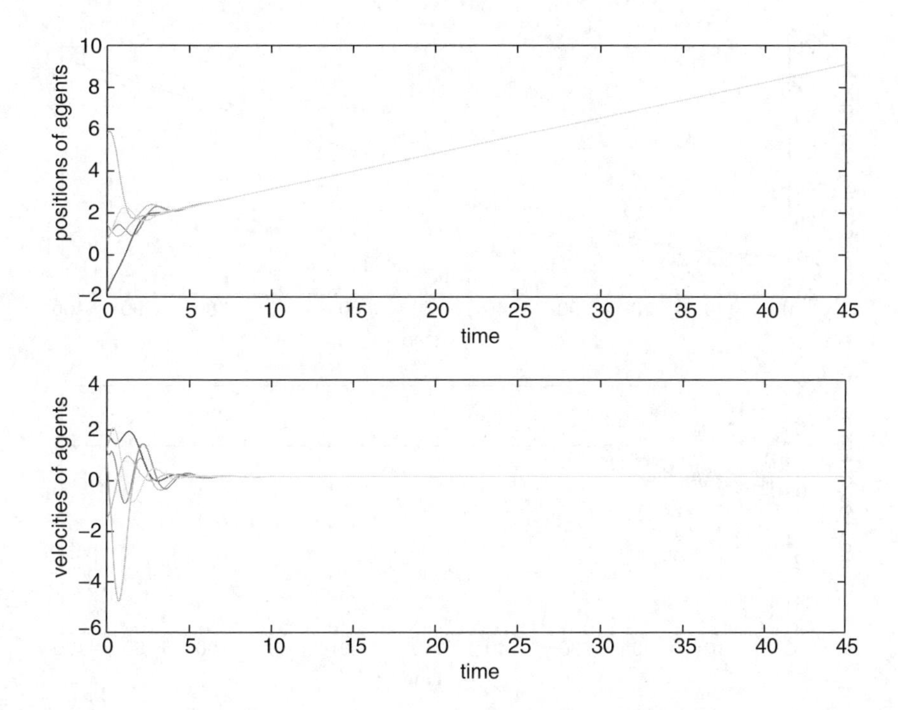

Figure 7.21 Positions and velocities of the agents (with unmatched but constrained delays).

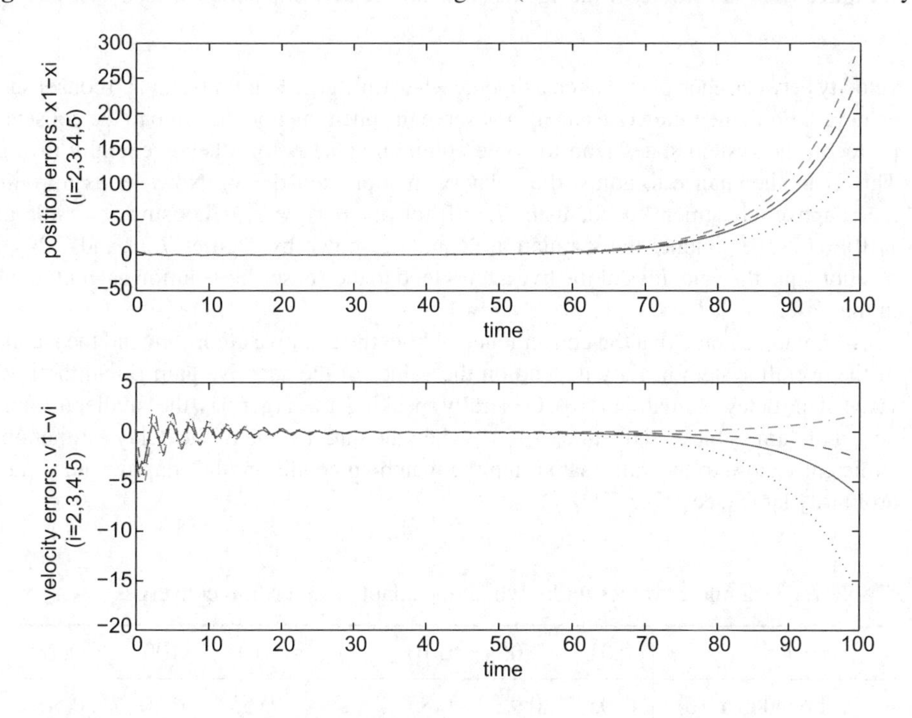

Figure 7.22 Position errors and velocity errors (with unconstrained delays).

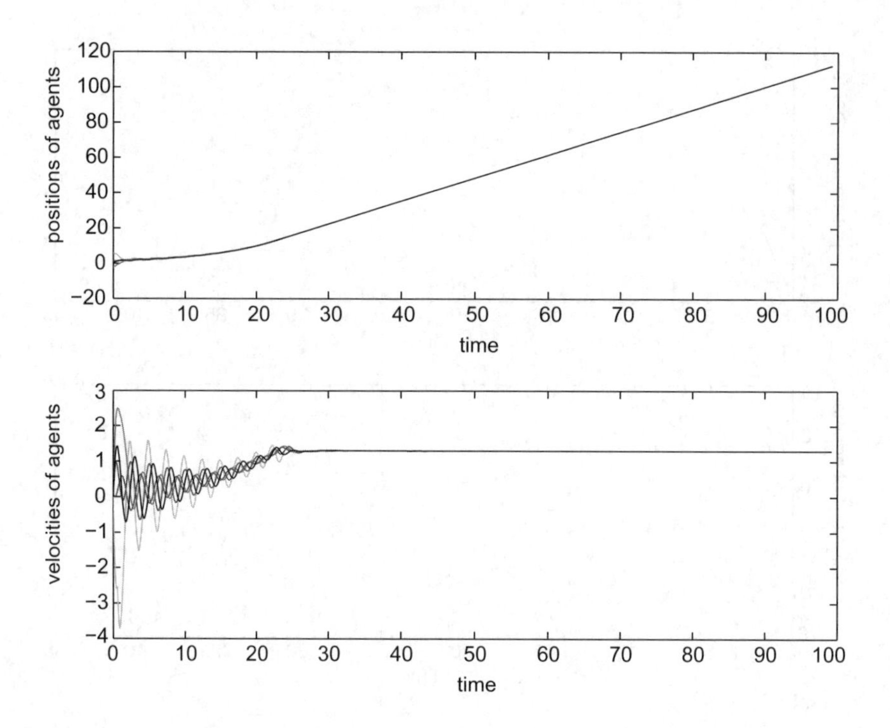

Figure 7.23 Positions of the agents (with an adaptive adjustment of self-delay).

velocity between each pair of agents diverge when time goes to infinity, i.e., no consensus is achieved. From the figure one can also observe the phenomenon that, due to the consensus protocol, the system states tend to some common value before they eventually diverge. This gives the chance to adjust the delay estimation (self-delay). Now, let us introduce the adaptive adjustment mechanism (7.110) for $\tau'_{12} = \tau'_{14} \triangleq \tau'_1$. Choosing $\beta = 0.05$ and $\tau'_1(0) = 0.8$ we conduct the simulation again. As shown by Figures 7.23 and 7.24, the positions and the velocities of the five agents tend to the consensus solution asymptotically in this case.

It should be noted that the convergence of both the adaptive algorithm and the stability of the overall system heavily depend on the values of the adaptive gain β and the initial value of the delay estimation $\tau'_1(0)$. Generally speaking, the larger β is, the smaller the upper bound of admissible initial value of τ'_1 is; when the initial value of the delay estimation is sufficiently close to the value satisfying the matching condition, the adaptive gain can be arbitrarily large (see Table 7.1).

Table 7.1 Parameter values under which the adaptive algorithm converges.

β	0.01	0.05	0.10	1	10	100	10000
upper bound of $\tau'_1(0)$	1.05	0.95	0.85	0.75	0.65	0.50	0.50

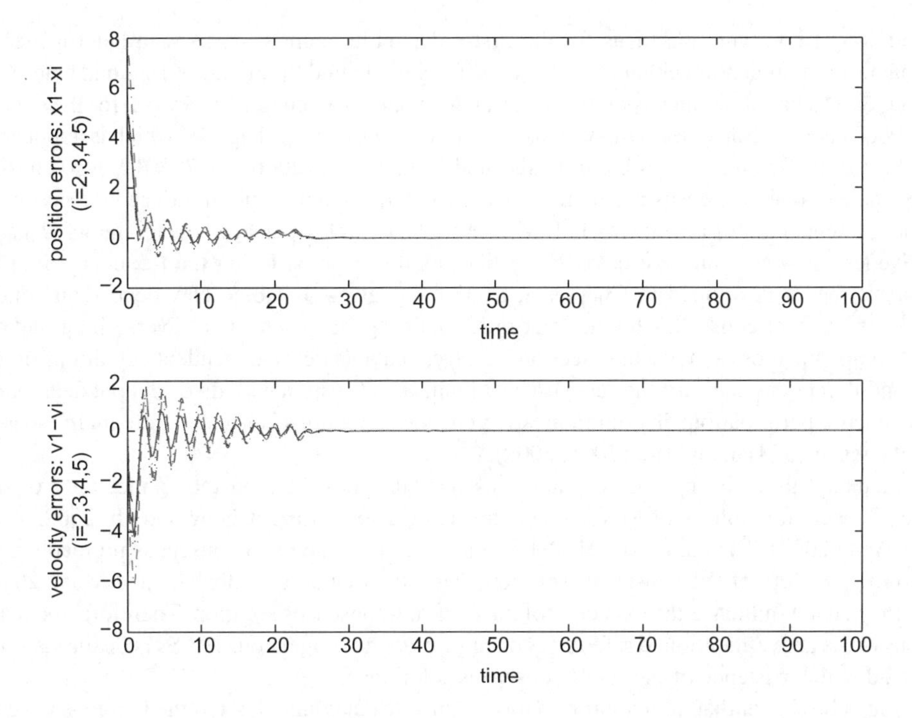

Figure 7.24 Position errors and velocity errors (with an adaptive adjustment of self-delay).

7.5 Notes and References

The study on consensus problems with communication delays can be traced in the research of distributed computation algorithms back to as early as the 1980s (Tsitsiklis, Bertsekas and Athans 1986). Similar problems are also studied in the context of synchronization of coupled oscillators (Yeung and Strogatz 1999). Recently, Olfati-Saber and Murray (2004) reviewed the consensus problem of the first-order multi-agent system and proposed a consensus protocol in which each agent delays its own measurement of state by the same value as the communication delay so that it could be matched by the delayed states of its neighbors. Olfati-Saber and Murray (2004) obtained a delay-dependent sufficient condition of constant consensus for the system with a uniform communication delay. Lin *et al.* (2007) extended the consensus protocol of Olfati-Saber and Murray (2004) to the second-order multi-agent system, and also obtained a delay-dependent sufficient condition of the second-order consensus. However, since each self-delay is equal to the corresponding communication delay in their protocol and only the unified delay bound is considered, the multi-agent systems studied in Olfati-Saber and Murray (2004) and Lin *et al.* (2007) are essentially homogeneous.

Lee and Spong (2006) considered a consensus protocol without self-delay and obtained a delay-independent sufficient condition of constant consensus for high-order heterogeneous systems with diverse communication delays by using Gershgorin's disc lemma and the small-gain theorem shown in Chapter 2. By using a similar technique, Wang and Elia (2008) also considered the consensus protocol without self-delay and gave a delay-independent sufficient condition of constant consensus for the first-order multi-agent systems with heterogeneous dynamic communication channels. Arcak (2007), Chopra and Spong (2006) studied heterogeneous MASs based on the passivity theory, and developed a general framework for the design of group coordination control of systems with nonlinear dynamical agents, which is applicable to the constant-consensus problem. Lestas and Vinnicombe (2006, 2007, 2010) also considered the constant-consensus problem for heterogeneous systems, and introduced the notion of S-hull, a relaxation of the convex hull of a set in the complex plane, to overcome the conservativeness of small-gain-like or passivity-like stability results. Using the frequency-domain analysis theory developed in Chapter 3, Section 7.1 gives scalable delay-dependent conditions of constant consensus for the first-order multi-agent system with diverse input delays and communication delays while Section 7.2 gives scalable delay-dependent conditions of the second-order consensus for the second-order multi-agent system with diverse input delays and asymmetric perturbations in communication channels. Sections 7.1 and 7.2 are mainly based on the results of Tian and Liu (2008, 2009).

Actually, the existence of constant consensus depends only on the connectivity of the interconnection topology of MASs (Ren and Beard 2005; Wang, Cheng and Hu 2008; Xiao and Wang 2007). The values of self-delays introduced by agents in consensus protocols may lead to instability of the consensus solution (Papachristodoulou, Jadbabaie and Münz 2010) but they do not influence the existence of the constant-consensus solution. Therefore, the main focus of the above-mentioned references on high-order heterogeneous MASs is on the stability instead of the existence of the set of consensus solutions.

It can be shown that an inappropriate value of self-delay may lead to the non-existence of high-order consensus solution. To guarantee the existence of high-order consensus solutions, currently existing consensus protocols introduce self-delays which are exactly equal to the corresponding communication delays (see, e.g., Hu and Hong 2007). In practice, however, communication delays can only be estimated approximately. Section 7.3 proposes a high-order consensus protocol based on the estimation of communication delays and investigates the existence of high-order consensus solution of high-order heterogeneous MASs under such a protocol. Section 7.4 presents a simple algorithm for on-line adjusting self-delays to guarantee the existence of high-order consensus solutions. These two sections are mainly taken from Tian and Zhang (2012) except for Section 7.3.4.

References

Arcak M (2007). Passivity as a design tool for group coordination. *IEEE Transactions on Automatic Control*, 52, 1380–1390.

Blondel VD, Hendrickx JM, Olshevsky A and Tsitsiklis JN (2005). Convergence in multiagent coordination, consensus, and flocking. *Proceeding of the Joint 44th IEEE Conference on Decision and Control and European Control Conference*, Seville, Spain, 2996–3000.

Cao M, Morse AS and Anderson BDO (2006). Reaching an agreement using delayed information. *IEEE Conference on Decision and Control*, San Diego, CA, USA, 3375–3380.

Chopra N and Spong MW (2006). Passivity-based control of multi-agent systems. *Advances in Robot Control: From Everyday Physics to Human-like Movements*, S. Kawamura and M. Svinin, Editors, 107–134, Spinger-Verlag, Berlin.

Hu J and Hong Y (2007). Leader-following coordination of multi-agent systems with coupling time delays. *Physica A*, 374, 853–863.

Jadbabaie A, Lin J and Morse AS (2003). Coordination of groups of mobile autonomous agents using nearest neighbor rules. *IEEE Transactions on Automatic Control*, 48, 988–1001.

Lee D and Spong MW (2006). Agreement with non-uniform information delays. *Proceedings of the American Control Conference*, Minneapolis, Minnesota, USA, 756–761.

Lestas I and Vinnicombe G (2006). Scalable decentralized robust stability certificates networks of interconnected heterogeneous dynamical systems. *IEEE Transactions on Automatic Control*, 51, 1613–1625.

Lestas I and Vinnicombe G (2007). The S-hull approach to consensus. *Proceedings of the 46th IEEE Conference on Decision and Control*, New Orleans, USA, 182–187.

Lestas I and Vinnicombe G (2010). Heterogeneity and scalability in group agreement protocols: Beyond small gain and passivity approaches. *Automatica*, 46, 1141–1151.

Lin P, Jia Y, Du J and Yuan S (2007). Distributed consensus control for second-order agents with fixed topology and time-delay. *Proceeding of the 26th Chinese Control Conference*, Zhangjiajie, Hunan, China, 577–581.

Moreau L (2005). Stability of multiagent systems with time-dependent communication links. *IEEE Transactions on Automatic Control*, 50, 169–182.

Olfati-Saber R and Murray RM (2004). Consensus problems in networks of agents with switching topology and time-delays. *IEEE Transactions on Automatic Control*, 49, 1520–1533.

Papachristodoulou A, Jadbabaie A and Münz U (2010). Effect of delay in multi-agent consensus and oscillator synchronization. *IEEE Transactions on Automatic Control*, 55, 1471–1477.

Ren W and Beard RW (2005). Consensus seeking in multiagent systems under dynamically changing interaction topologies. *IEEE Transactions on Automatic Control*, 50, 655–661.

Tian Y-P and Liu C-L (2008). Consensus of multi-agent systems with diverse input and communication delays. *IEEE Transactions on Automatic Control*, 53, 2122–2128.

Tian Y-P and Liu C-L (2009). Robust consensus of multi-agent systems with diverse input delays and asymmetric interconnection perturbations. *Automatica*, 45, 1347–1353.

Tian Y-P and Zhang Y (2012). High-order consensus of heterogeneous multi-agent systems with unknown communication delays. *Automatica*, 48, 1205–1212.

Tsitsiklis JN, Bertsekas DP and Athans M (1986). Distributed asynchronous deterministic and stochastic gradient optimisation algorithms. *IEEE Transactions on Automatic Control*, 31, 803–812.

Vicsek T, Czirok A, Ben Jacob E, *et al.*(1995). Novel type of phase transitions in a system of self-driven particles. *Physical Review Letters*, 75, 1226–1229.

Wang J, Cheng D and Hu X (2008). Consensus of multi-agent linear dynamic systems. *Asian Journal of Control*, 10, 144–155.

Wang J and Elia N (2008). Consensus over network with dynamic channels. *Proceeding of the American Control Conference*, Seattle, WA, 2637–2642.

Wang W and Slotine JJE (2006). Contraction analysis of time-delayed communications and group cooperation. *IEEE Transactions on Automatic Control*, 51, 712–717.

Xiao F and Wang L (2007). Consensus problems for high-dimensional multi-agent systems. *IET Control Theory and Applications*, 1, 830–837.

Yeung MKS and Strogatz SH (1999). Time delay in the Kuramoto model of coupled oscillators. *Physical Review Letter*, 82, 648–651.

Index

Frequency-Domain Analysis and Design of Distributed Control Systems, First Edition. Yu-Ping Tian.